ELEMENTARY LINEAR ALGEBRA

ELEMENTARY LINEAR ALGEBRA

SECOND EDITION

PAUL C. SHIELDS

DEPARTMENT OF MATHEMATICS

STANFORD UNIVERSITY

WORTH PUBLISHERS, INC.

444 PARK AVENUE SOUTH, NEW YORK, N.Y. 10016

LIBRARY OF CONGRESS CATALOG NUMBER 72-94204

ISBN: 0-87901-025-8

AMS 1968 SUBJECT CLASSIFICATIONS 1501, 1505, 1510, 1515, 1520,

1525, 1530, 1535, 1570, 3420, 4760

DESIGNED BY MALCOLM GREAR/GORDON BRETT

FOURTH PRINTING MAY 1977

Printed in the United States of America

PREFACE TO THE SECOND EDITION

The text is designed to introduce some of the basic concepts and techniques of linear algebra and to make them accessible to freshmen and sophomores, many of whom are not mathematics majors. With this audience in mind, I have not stressed sophisticated proof techniques, which can be left to later courses for mathematics majors. Instead, the emphasis is on interpretation and on the development of computational tools. Systems of equations and matrices are used as vehicles for the introduction and interpretation of vector spaces, subspaces, independence, and dimension. My experience indicates that this enables the student to move easily from familiar material into new ideas and, furthermore, provides him with techniques for problem solving while absorbing subsequent abstractions. This interplay is continued in the discussion of linear transformations, similarity, and linear differential equations, where students are again provided with computational tools as the general theory is presented. Discussions are generally confined to low-dimensional coordinate spaces, often omitting general proofs of the more technical results. Geometric interpretations are frequently given as a further aid to conceptual development.

The book is explicitly designed for classroom use on the basis of my own experience with beginning students. With only a few exceptions, the material has been divided into sections which can be discussed in one or two lectures. Each section consists of short theoretical discussions with numerous examples and a large number of exercises, most of which are extensions of the examples. *The examples and exercises are the most important part of the book.* My experience indicates that beginning students progress most easily via careful explication of examples combined with homework which further illuminates the concepts and techniques. Wherever possible answers to the exercises are given at the back of the book. This method of presenting material by examples and exercises allows for considerable flexibility of use, the instructor being able to select those parts of the discussion in each section which he finds most relevant while insuring the student's development of conceptual understanding and technique.

The primary difference between this and the first edition is that systematic use is now made of the concept of the rank of a matrix. The rank is defined as the number of nonzero rows in the reduced form, and the principal theorems about linear systems, independence, and invertibility are stated in terms of rank. I have also added a section (Section 4 of Chapter 2), which relates rank and the row and column space of a matrix. This section can be covered rather quickly, merely stating the theorems with a few examples, or one can discuss these ideas quite thoroughly. The appendix also contains a discussion of rank as the dimension of the range of a linear transformation.

The first four chapters of this edition correspond to the first two chapters of the first edition. This new division of the material was quite natural, and corresponds more closely to the way many schools have used the book. I have also added supplements on vector geometry to each of the first four chapters. These supplements contain the material which appeared in Appendix 1 of the first edition. I hope that the placing of this material within the main body of the text will encourage its wider use. I made no major changes elsewhere, although much of the material has been rewritten. This rewrite was guided by the desire to make the ideas more explicit, and to help the student understand the general goals of the course.

The first chapter develops the elimination method for solving linear systems, and then shows how such systems can be expressed as matrix-vector equations. The concepts of independence and dimension are described in Chapter 2. Linear transformations are introduced in Chapter 3, and the inverse is discussed in Chapter 4. The problem of finding a diagonal representation for a matrix is the subject of Chapter 5. Optional examples and exercises (marked with □) extend these ideas to spaces of functions and to complex spaces.

The sixth, and final, chapter is on the theory of constant-coefficient linear differential operators. This material, which uses only elementary-calculus results, is accessible to beginning students and provides a significant application of linear algebra. Such concepts as independence, null spaces, operator multiplication, and operator polynomials are further illuminated in this context.

An appendix on dimension theory has been included to introduce the basic techniques of advanced linear algebra. It can be used as a supplement for the better students or in a parallel honors course. The short annotated bibliography will direct interested students to further topics in linear algebra. For ease of reference, I have also included at the back of the book a list of the basic theorems used.

The arrangement of the book allows for a number of possible uses. These include:

> *A one-quarter course*, using the first five chapters and omitting the optional material

A one-semester course, using (1) the first six chapters less optional material, (2) the first five chapters with optional material, or (3) most of the first five chapters plus the appendix

A parallel honors course, using the material of the appendix where appropriate

A short course, using only the first four chapters

A supplement for other courses, using various parts of the book. Material from the first five chapters can be used to supplement a calculus program. Chapter 6 provides a useful supplement on linear differential equations.

By the time a book reaches a second edition, so many people have contributed to its development that a list of acknowledgments would be quite long but I would like to thank those who were particularly helpful: J. E. Schneider and Mary McCammon of Pennsylvania State University, Douglas Dickson and Forrest Richen of the University of Michigan, and Fred Gass and David Kullman of Miami University. I would like to express special thanks to my colleague and friend, Professor David Jonah, who read the manuscript of each edition at each stage of development, and whose criticisms and suggestions have resulted in major improvements in both content and style. I am also grateful to Ted Cary, Alan Smith and my wife Dorothy for assistance in proofreading. Thanks are also due to Elizabeth Surzyn, Isolde Field and Elizabeth Plowman for typewriting assistance.

This, my second book, is for Betsy, my second daughter.

Paul C. Shields
Palo Alto, California
October 1972

TABLE OF CONTENTS

ELEMENTARY LINEAR ALGEBRA

CHAPTER 1

MATRICES AND SYSTEMS OF EQUATIONS

This book is an introduction to some of the elementary concepts and results of linear algebra. Linear algebra provides a basic framework and language for the study of calculus of several variables, differential equations, and modern algebra. In addition, many problems in engineering and the physical and social sciences are now treated using linear concepts.

Linear algebra began as the study of systems of linear equations and this is where we begin. Our first goal is to learn a method for solving such systems. The basic method involves adding multiples of equations to other equations, chosen so as to eliminate as many variables as possible. This "elimination" method is described in Sections 1, 2, and 3.

Our second, and more important goal, is to express a system of equations involving numbers as a single equation involving arrays of numbers. This will enable us to simplify many expressions and to easily perform formal algebraic operations. This also provides a useful geometric framework, because certain arrays of numbers, such as pairs or triples, can be thought of as points in a plane or in space. These ideas are described in Sections 4, 5, and 6.

SECTION 1 The Elimination Method

We shall often find it necessary to solve systems of linear equations such as

1
$$\begin{aligned} x - y + 2z &= 0 \\ 3x + y - z &= 0 \\ 2x + 2y - 3z &= 0 \end{aligned}$$

Such a system, in which the constant terms are all zero, is called a **homogeneous** system, and several different methods can be used to solve it. We shall describe a procedure known as **elimination** or **reduction** for determining the solutions.

The trick of the elimination method is to add to one equation a suitable multiple of another, chosen so as to eliminate one of the variables. For example, we can add to the second equation of the above system -3 times the first equation:

$$\begin{aligned} -3(x - y + 2z = 0) = -3x + 3y - 6z &= 0 \\ 3x + y - z &= 0 \\ \hline 4y - 7z &= 0 \end{aligned}$$

If we then add to the third equation -2 times the first equation, we obtain

$$\begin{aligned} -2(x - y + 2z = 0) = -2x + 2y - 4z &= 0 \\ 2x + 2y - 3z &= 0 \\ \hline 4y - 7z &= 0 \end{aligned}$$

When we replace the original second and third equations with the new ones obtained in this way, our system of three linear equations becomes

2

$$\begin{aligned} x - y + 2z &= 0 \\ 4y - 7z &= 0 \\ 4y - 7z &= 0 \end{aligned}$$

Note that x has been eliminated from all except the first equation. System **2** has the same solutions as system **1**. A formal proof of this will be given in Chapter 3. We can now proceed to eliminate y from the first and third equations by adding multiples of the second equation to each.

$$\begin{aligned} \tfrac{1}{4}(4y - 7z = 0) = y - \tfrac{7}{4}z &= 0 \\ x - y + 2z &= 0 \\ \hline x + \tfrac{1}{4}z &= 0 \end{aligned}$$

$$\begin{aligned} -1(4y - 7z = 0) = -4y + 7z &= 0 \\ 4y - 7z &= 0 \\ \hline 0 &= 0 \end{aligned}$$

As the addition of -1 times the second equation to the third equation has reduced it to zero, we are left with the following two equations:

3

$$\begin{aligned} x + \tfrac{1}{4}z &= 0 \\ 4y - 7z &= 0 \end{aligned}$$

We have now proceeded as far as necessary to find solutions, for we can assign an arbitrary value to z and easily calculate x and y from system **3** to obtain a solution to system **1**. For example, if $z = 1$, equations **3** give

$$x = -\tfrac{1}{4} \qquad y = \tfrac{7}{4}$$

Thus $\qquad x = -\frac{1}{4} \qquad y = \frac{7}{4} \qquad z = 1$

is a solution to system **1**. We can express the dependence of the solution on z by rewriting system **3** as

4
$$\begin{aligned} x &= -\tfrac{1}{4}z \\ y &= \tfrac{7}{4}z \\ z &= z \end{aligned}$$

Note that there is little reason to continue writing down the variables x, y, and z. All we need to do is keep track of the coefficients that belong to each variable in each equation. The standard device for doing this is called a **matrix**, which is a rectangular array of numbers.

The **matrix of coefficients** for system **1**

$$\begin{aligned} x - y + 2z &= 0 \\ 3x + y - z &= 0 \\ 2x + 2y - 3z &= 0 \end{aligned} \qquad \text{is} \qquad \begin{bmatrix} 1 & -1 & 2 \\ 3 & 1 & -1 \\ 2 & 2 & -3 \end{bmatrix}$$

Note that the numbers in each row are the coefficients of x, y, and z (in that order) in the corresponding equation of system **1**. For example, the numbers in the *second* row are 3, 1, and -1, which are the coefficients of x, y, and z in the *second* equation.

The operations used to reduce system **1** to system **3** can now be described as operations on the rows of the matrix of coefficients. For example, if we add to the second row of the matrix -3 times the first row and then add to the third row -2 times the first row we obtain

$$\begin{bmatrix} 1 & -1 & 2 \\ 3 + (-3)(1) & 1 + (-3)(-1) & -1 + (-3)(2) \\ 2 + (-2)(1) & 2 + (-2)(-1) & -3 + (-2)(2) \end{bmatrix} = \begin{bmatrix} 1 & -1 & 2 \\ 0 & 4 & -7 \\ 0 & 4 & -7 \end{bmatrix}$$

which is the matrix of coefficients of system **2**.

Now add to the first row $\frac{1}{4}$ times the second row and then add to the third row -1 times the second row and obtain the matrix

$$\begin{bmatrix} 1 & 0 & \frac{1}{4} \\ 0 & 4 & -7 \\ 0 & 0 & 0 \end{bmatrix}$$

The bottom row of zeros shows that the third equation has been eliminated, while the zeros in the first and second columns indicate that x has been eliminated from all except the first equation and y from all except the second equation. The first two rows are just the coefficient matrix of

$$\begin{aligned} x \qquad + \tfrac{1}{4}z &= 0 \\ 4y - 7z &= 0 \end{aligned}$$

from which, as we have seen, the solutions are easy to determine. The following examples illustrate this procedure of matrix elimination.

EXAMPLE 1 This method of elimination with the use of matrices will be used to solve

5 $$\begin{aligned} 3x + y &= 0 \\ x - 2y &= 0 \end{aligned}$$

Form the matrix of coefficients

$$\begin{bmatrix} 3 & 1 \\ 1 & -2 \end{bmatrix}$$

Add to the first row of this matrix -3 times the second row; the resulting matrix will be

$$\begin{bmatrix} 0 & 7 \\ 1 & -2 \end{bmatrix}$$

Now add to the second row $\frac{2}{7}$ times the new first row, which gives

$$\begin{bmatrix} 0 & 7 \\ 1 & 0 \end{bmatrix}$$

This latter matrix is the coefficient matrix for

$$\begin{aligned} 0 \cdot x + 7 \cdot y &= 0 \\ 1 \cdot x + 0 \cdot y &= 0 \end{aligned} \qquad \text{that is} \qquad \begin{aligned} 7y &= 0 \\ x &= 0 \end{aligned}$$

Clearly the only solution is

$$x = y = 0$$

This is therefore the unique solution to system **5**.

EXAMPLE 2 The coefficient matrix of

6 $$\begin{aligned} 2x - y + z &= 0 \\ -7x + \tfrac{7}{2}y - \tfrac{7}{2}z &= 0 \\ 4x + y - 2z &= 0 \end{aligned} \qquad \text{is} \qquad \begin{bmatrix} 2 & -1 & 1 \\ -7 & \frac{7}{2} & -\frac{7}{2} \\ 4 & 1 & -2 \end{bmatrix}$$

Add to the second row $\frac{7}{2}$ times the first row; then add to the third row -2 times the first row. The result is the matrix

7 $$\begin{bmatrix} 2 & -1 & 1 \\ 0 & 0 & 0 \\ 0 & 3 & -4 \end{bmatrix}$$

Note the row of zeros, which indicates that the second equation has been entirely eliminated, as it is superfluous. Now add to the first row $\frac{1}{3}$ times the third row to obtain

8 $$\begin{bmatrix} 2 & 0 & -\frac{1}{3} \\ 0 & 0 & 0 \\ 0 & 3 & -4 \end{bmatrix}$$

which is the coefficient matrix (after dropping the second row) of

9 $$\begin{aligned} 2x \qquad - \tfrac{1}{3}z &= 0 \\ 3y - 4z &= 0 \end{aligned}$$

This system is easy to solve. For example, put $z = 2$ and calculate x and y from equations **9** to obtain one solution to the system **9** *and* to the original system **6**:

$$x = \tfrac{1}{3} \qquad y = \tfrac{8}{3} \qquad z = 2$$

The dependence of the solutions of the original system **6** upon z can be expressed by rewriting system **9** as

10 $$\begin{aligned} x &= \tfrac{1}{6}z \\ y &= \tfrac{4}{3}z \\ z &= z \end{aligned}$$

We could, of course, eliminate in a different fashion, obtaining other forms for the solutions. For example, first reduce matrix **6** to matrix **7**, then eliminate in the *third* column of matrix **7** by adding $\frac{1}{4}$ times the third row of matrix **7** to its first row, obtaining

$$\begin{bmatrix} 2 & -\frac{1}{4} & 0 \\ 0 & 0 & 0 \\ 0 & 3 & -4 \end{bmatrix} \quad \text{which corresponds to} \quad \begin{aligned} 2x - \tfrac{1}{4}y &= 0 \\ 3y - 4z &= 0 \end{aligned}$$

This system can be rewritten as

$$\begin{aligned} x &= \tfrac{1}{8}y \\ y &= \phantom{\tfrac{1}{8}}y \\ z &= \tfrac{3}{4}y \end{aligned}$$

which is quite different than equations **10** and expresses the dependency of the solutions of the system **6** upon the variable y, rather than z.

EXAMPLE 3 The coefficient matrix of

11
$$\begin{aligned} 2x - y + z &= 0 \\ -4x + 2y + z &= 0 \end{aligned} \quad \text{is} \quad \begin{bmatrix} 2 & -1 & 1 \\ -4 & 2 & 1 \end{bmatrix}$$

Adding 2 times the first row to the second row gives

$$\begin{bmatrix} 2 & -1 & 1 \\ 0 & 0 & 3 \end{bmatrix}$$

Thus, in the process of eliminating x from the second equation we have also eliminated y. Now add to the first row $-\frac{1}{3}$ times the second row to obtain

$$\begin{bmatrix} 2 & -1 & 0 \\ 0 & 0 & 3 \end{bmatrix}$$

which is the matrix of coefficients of

$$\begin{aligned} 2x - y &= 0 \\ 3z &= 0 \end{aligned}$$

We see that z must be zero, while y can be assigned arbitrary values and x calculated in terms of y. Rewriting this to express the dependence on y, we have

12
$$\begin{aligned} x &= \tfrac{1}{2}y \\ y &= y \\ z &= 0y \end{aligned}$$

as a general form of the solutions to system **11**.

DISCUSSION The elimination method is not always the easiest or fastest way to solve a system, but it does have the advantage of being a systematic procedure. We shall also later find that this process of reducing a matrix is useful in other situations. The student may want to work some of the exercises at this stage to make certain that he understands the procedures given in the above examples.

As shown in Example **2** there may be several ways to describe the form of the solutions, depending upon the way we eliminate in the matrix. For convenience in stating results it is helpful to obtain a uniform form of the solutions. The kind of matrix that will give this form will be called a reduced matrix.

A **reduced matrix** is a matrix such that

a The first nonzero entry in each row is 1; all other entries in that column are zeros.

13 **b** Each row that consists entirely of zeros is below each row which contains a nonzero entry.

c The first nonzero entry in each row is to the right of the first nonzero entry of each preceding row.

Here are three examples of reduced matrices:

$$\begin{bmatrix} 0 & 0 & 1 \\ 0 & 0 & 0 \\ 0 & 0 & 0 \end{bmatrix} \quad \begin{bmatrix} 1 & 0 & 0 \\ 0 & 1 & 0 \\ 0 & 0 & 1 \end{bmatrix} \quad \begin{bmatrix} 1 & 2 & 0 \\ 0 & 0 & 1 \\ 0 & 0 & 0 \end{bmatrix}$$

The matrix **8** of Example **2**,

$$\begin{bmatrix} 2 & 0 & -\frac{1}{3} \\ 0 & 0 & 0 \\ 0 & 3 & -4 \end{bmatrix}$$

is *not* in reduced form, for it violates both **a** and **b** of the definition **13**. It can, however, be put in reduced form by multiplying row one by $\frac{1}{2}$, row three by $\frac{1}{3}$, and then interchanging rows two and three to obtain

$$\begin{bmatrix} 1 & 0 & -\frac{1}{6} \\ 0 & 1 & -\frac{4}{3} \\ 0 & 0 & 0 \end{bmatrix}$$

This matrix *is* reduced. It is the coefficient matrix of (omitting the superfluous row)

$$\begin{aligned} x \qquad - \tfrac{1}{6}z &= 0 \\ y - \tfrac{4}{3}z &= 0 \end{aligned}$$

from which the solutions to the original system **6** are easy to describe.

After one has worked a few examples it becomes clear that any given matrix A can be converted to a reduced matrix R by systematically applying the three row operations of adding multiples of rows to other rows, interchanging rows, and multiplying rows by nonzero numbers. (See Exercises **12**, **13**, and **14**.) It is a much deeper result that the final reduced form R is unique; that is, if two persons independently reduce A, using the row operations in any way they desire until they reach a reduced matrix, they will both get the *same* reduced matrix (see the Bibliography for a reference).

The remaining examples of this section illustrate the ease with which solutions to a system can be ascertained from the reduced form.

EXAMPLE 4 The matrix of coefficients of

14
$$\begin{aligned} 4x - 2y + 2z + w &= 0 \\ 2x - y - 3z - w &= 0 \end{aligned} \qquad \text{is} \qquad \begin{bmatrix} 4 & -2 & 2 & 1 \\ 2 & -1 & -3 & -1 \end{bmatrix}$$

We proceed to reduce this as follows: first multiply row one by $\frac{1}{4}$; then add to row two -2 times row one. This gives first

$$\begin{bmatrix} 1 & -\frac{1}{2} & \frac{1}{2} & \frac{1}{4} \\ 2 & -1 & -3 & -1 \end{bmatrix} \quad \text{then} \quad \begin{bmatrix} 1 & -\frac{1}{2} & \frac{1}{2} & \frac{1}{4} \\ 0 & 0 & -4 & -\frac{3}{2} \end{bmatrix}$$

Now multiply row two by $-\frac{1}{4}$; then add to row one $-\frac{1}{2}$ times row two. This gives first

$$\begin{bmatrix} 1 & -\frac{1}{2} & \frac{1}{2} & \frac{1}{4} \\ 0 & 0 & 1 & \frac{3}{8} \end{bmatrix} \quad \text{then} \quad \begin{bmatrix} 1 & -\frac{1}{2} & 0 & \frac{1}{16} \\ 0 & 0 & 1 & \frac{3}{8} \end{bmatrix}$$

The latter matrix is reduced, for the first nonzero entry in each row is a 1, all other entries in that column being zeros; the zero rows are below the nonzero rows (for there are no zero rows); and the leading nonzero entry in the second row is to the right of the leading nonzero entry in the first row. The reduced matrix is the coefficient matrix of

$$\begin{aligned} x - \tfrac{1}{2}y \qquad + \tfrac{1}{16}w &= 0 \\ z + \tfrac{3}{8}w &= 0 \end{aligned}$$

Now transfer all except the leading variable in each row to the right side to obtain

$$\begin{aligned} x &= \tfrac{1}{2}y - \tfrac{1}{16}w \\ z &= 0 \cdot y - \tfrac{3}{8}w \end{aligned}$$

which clearly shows the dependence of the solutions to the original system **14** upon y and w.

The method of reduction can be used for systems with more equations than unknowns such as

$$\begin{aligned} x - y + z &= 0 \\ x + y - z &= 0 \\ -x \qquad + z &= 0 \\ x + 2y + z &= 0 \end{aligned}$$

EXAMPLE 5 The coefficient matrix is

$$\begin{bmatrix} 1 & -1 & 1 \\ 1 & 1 & -1 \\ -1 & 0 & 1 \\ 1 & 2 & 1 \end{bmatrix}$$

To make the first column obey rule **a** for a reduced matrix, subtract the first row from the second and fourth rows and add it to the third row to yield

$$\begin{bmatrix} 1 & -1 & 1 \\ 0 & 2 & -2 \\ 0 & -1 & 2 \\ 0 & 3 & 0 \end{bmatrix}$$

Now in order to conform to rules **b** and **c**, multiply the fourth row by $\frac{1}{3}$; add the fourth row to the first and third rows; and add -2 times the fourth row to the second row. After interchanging the fourth and second rows we have

$$\begin{bmatrix} 1 & 0 & 1 \\ 0 & 1 & 0 \\ 0 & 0 & 2 \\ 0 & 0 & -2 \end{bmatrix}$$

Now multiply the third row by $\frac{1}{2}$; subtract the third row from the first; and add twice the third row to the fourth row. We now have

$$\begin{bmatrix} 1 & 0 & 0 \\ 0 & 1 & 0 \\ 0 & 0 & 1 \\ 0 & 0 & 0 \end{bmatrix}$$

This matrix is reduced and gives the system

$$\begin{aligned} x \qquad &= 0 \\ y \quad &= 0 \\ z &= 0 \end{aligned}$$

The unique solution to our original system, then, is $x = y = z = 0$.

EXAMPLE 6 The method we have illustrated works even if there is a large number of equations with a large number of unknowns. Of course, the process of elimination will take much longer. Consider the system (subscripts, rather than separate letters, have been used on the unknowns)

15

$$\begin{aligned} x_1 + x_2 - x_3 + x_4 \qquad\quad &= 0 \\ x_3 + x_4 + x_5 &= 0 \\ 2x_1 + 2x_2 - x_3 \qquad + x_5 &= 0 \\ x_1 + x_2 - 2x_3 \qquad - x_5 &= 0 \\ 2x_3 - 4x_4 + 2x_5 &= 0 \\ -x_1 - x_2 + 2x_3 - 3x_4 + x_5 &= 0 \end{aligned}$$

Form the matrix of coefficients:

$$\begin{bmatrix} 1 & 1 & -1 & 1 & 0 \\ 0 & 0 & 1 & 1 & 1 \\ 2 & 2 & -1 & 0 & 1 \\ 1 & 1 & -2 & 0 & -1 \\ 0 & 0 & 2 & -4 & 2 \\ -1 & -1 & 2 & -3 & 1 \end{bmatrix}$$

One possible sequence of steps for reducing this matrix is

1 Add to the third row -2 times the first row.
2 Add to the fourth row -1 times the first row.
3 Add to the sixth row 1 times the first row.

This finishes the elimination in the first column. Notice that in this case we have also eliminated as much as we can in the second column, for we now have

$$\begin{bmatrix} 1 & 1 & -1 & 1 & 0 \\ 0 & 0 & 1 & 1 & 1 \\ 0 & 0 & 1 & -2 & 1 \\ 0 & 0 & -1 & -1 & -1 \\ 0 & 0 & 2 & -4 & 2 \\ 0 & 0 & 1 & -2 & 1 \end{bmatrix}$$

Proceed to eliminate in the third column:

4 Add to the first row 1 times the second row.
5 Add to the third row -1 times the second row.
6 Add to the fourth row 1 times the second row.
7 Add to the fifth row -2 times the second row.
8 Add to the sixth row -1 times the second row.

This sequence gives

$$\begin{bmatrix} 1 & 1 & 0 & 2 & 1 \\ 0 & 0 & 1 & 1 & 1 \\ 0 & 0 & 0 & -3 & 0 \\ 0 & 0 & 0 & 0 & 0 \\ 0 & 0 & 0 & -6 & 0 \\ 0 & 0 & 0 & -3 & 0 \end{bmatrix}$$

Proceed to eliminate in the fourth column by adding multiples of the third row to the fifth and sixth rows.

9 Add to the fifth row -2 times the third row.
10 Add to the sixth row -1 times the third row.

For ease of further elimination,

11 Multiply the third row by $-\frac{1}{3}$.

We now have

$$\begin{bmatrix} 1 & 1 & 0 & 2 & 1 \\ 0 & 0 & 1 & 1 & 1 \\ 0 & 0 & 0 & 1 & 0 \\ 0 & 0 & 0 & 0 & 0 \\ 0 & 0 & 0 & 0 & 0 \\ 0 & 0 & 0 & 0 & 0 \end{bmatrix}$$

The reduction to reduced form is completed as follows:

12 Add to the first row -2 times the third row.
13 Add to the second row -1 times the third row.

The student should check definition **13**, page 7, to see whether the matrix we now have is a reduced matrix.

$$\begin{bmatrix} 1 & 1 & 0 & 0 & 1 \\ 0 & 0 & 1 & 0 & 1 \\ 0 & 0 & 0 & 1 & 0 \\ 0 & 0 & 0 & 0 & 0 \\ 0 & 0 & 0 & 0 & 0 \\ 0 & 0 & 0 & 0 & 0 \end{bmatrix}$$

This is the matrix of coefficients of

16
$$\begin{aligned} x_1 + x_2 \qquad\qquad + x_5 &= 0 \\ x_3 \qquad + x_5 &= 0 \\ x_4 \qquad &= 0 \\ 0 &= 0 \\ 0 &= 0 \\ 0 &= 0 \end{aligned}$$

We can now assign arbitrary values to the unknowns x_2 and x_5 and compute x_1, x_3, and x_4 from these equations to obtain a solution to system **15**.

For example, put $x_2 = 0$ and $x_5 = 0$, and obtain the solution

$$x_1 = x_2 = x_3 = x_4 = x_5 = 0$$

If we put $x_2 = -1$ and $x_5 = 2$, we obtain the solution

$$x_1 = -1 \qquad x_2 = -1 \qquad x_3 = -2 \qquad x_4 = 0 \qquad x_5 = 2$$

Just as was done in Example **4**, the dependence of the solutions upon x_2 and x_5 can be described merely by transferring all but the leading variable in each row of the reduced system **16** to the right side:

$$\begin{aligned} x_1 &= -x_2 - x_5 \\ x_3 &= -x_5 \\ x_4 &= 0 \end{aligned}$$

EXERCISES

1 Each of the matrices on the right is the coefficient matrix for one or more of the systems on the left. Match them.

a $\begin{aligned} 3x + 2y - z &= 0 \\ 2x - y + 2z &= 0 \end{aligned}$

b $\begin{aligned} 2x - y &= 0 \\ x + 4y &= 0 \\ 0x - 0y &= 0 \\ 3x - 3y &= 0 \end{aligned}$

c $\begin{aligned} 3x_1 + 2x_2 - x_3 &= 0 \\ 2x_1 - x_2 + 2x_3 &= 0 \end{aligned}$

d $\begin{aligned} 2x - y &= 0 \\ x + 4y &= 0 \\ 3x - 3y &= 0 \end{aligned}$

e $x + 2y + 3z + w = 0$

f $\begin{aligned} 4x_1 + x_2 + 2x_3 \qquad\qquad + 6x_5 &= 0 \\ x_1 \qquad - x_3 + x_4 + 2x_5 &= 0 \\ \tfrac{1}{2}x_1 \qquad\qquad + \tfrac{1}{3}x_4 + x_5 &= 0 \\ 2x_1 + 2x_2 + 2x_3 + x_4 \qquad &= 0 \\ 2x_1 \qquad - 2x_3 + 2x_4 + 4x_5 &= 0 \end{aligned}$

g $\begin{aligned} 2x - y &= 0 \\ 3x - 3y &= 0 \\ x + 4y &= 0 \end{aligned}$

A $\begin{bmatrix} 2 & -1 \\ 1 & 4 \\ 3 & -3 \end{bmatrix}$

B $\begin{bmatrix} 4 & 1 & 2 & 0 & 6 \\ 1 & 0 & -1 & 1 & 2 \\ \tfrac{1}{2} & 0 & 0 & \tfrac{1}{3} & 1 \\ 2 & 2 & 2 & 1 & 0 \\ 2 & 0 & -2 & 2 & 4 \end{bmatrix}$

C $[1 \quad 2 \quad 3 \quad 1]$

D $\begin{bmatrix} 2 & -1 \\ 3 & -3 \\ 1 & 4 \end{bmatrix}$

E $\begin{bmatrix} 2 & -1 \\ 1 & 4 \\ 0 & 0 \\ 3 & -3 \end{bmatrix}$

F $\begin{bmatrix} 3 & 2 & -1 \\ 2 & -1 & 2 \end{bmatrix}$

2 Form the matrix of coefficients and use the elimination method to find one description of the solutions to each of the following (as in Examples **1**, **2**, and **3**).

a $$\begin{aligned} x - 3y &= 0 \\ 2x + y &= 0 \end{aligned}$$

b $$\begin{aligned} 7x - 2y &= 0 \\ 2x + y &= 0 \end{aligned}$$

c $$\begin{aligned} 2x - \tfrac{1}{2}y &= 0 \\ 4x - y &= 0 \end{aligned}$$

d $$\begin{aligned} 3x - 2y + 7z &= 0 \\ 3x + 2y - 7z &= 0 \end{aligned}$$

e $$\begin{aligned} 6x + y - z &= 0 \\ x + \tfrac{1}{6}y + 2z &= 0 \end{aligned}$$

f $$\begin{aligned} 2x - y &= 0 \\ x + 2y &= 0 \\ 3x - 2y &= 0 \end{aligned}$$

g $$\begin{aligned} 6x - y + 2z &= 0 \\ -x + 3y - z &= 0 \\ 2x + 7y &= 0 \end{aligned}$$

h $$\begin{aligned} 2x + y + 2z &= 0 \\ -x - \tfrac{1}{2}y - z &= 0 \\ 2x + y &= 0 \end{aligned}$$

i $$\begin{aligned} 10x + 12y - z &= 0 \\ 3x - 7y + 6z &= 0 \end{aligned}$$

j $$\begin{aligned} 6x - 3y + 4z &= 0 \\ x + 6y - 7z &= 0 \\ 5x - 5y &= 0 \end{aligned}$$

k $3x - 2y + z = 0$

3 For the system

$$\begin{aligned} x - 2y + z &= 0 \\ 2x - y - 3z &= 0 \end{aligned}$$

form the matrix of coefficients. Reduce to express the solutions in a form showing the dependence upon z. Reduce in a different manner to show the dependence upon y. Then reduce in a third way to show the dependence upon x. (See Example **2**.)

4 Find the form of the solutions to each of the systems of Exercise **1** by reducing the coefficient matrix to a *reduced* matrix.

5 Find the form of the solutions to each of the systems of Exercise **2** by reducing the coefficient matrix to a *reduced* matrix.

6 Proceed as in Exercise **5** for each of the systems

a $$\begin{aligned} x + y + z + w &= 0 \\ x - y + z + w &= 0 \\ x - y - z + w &= 0 \\ x - y - z - w &= 0 \end{aligned}$$

b $$\begin{aligned} 2x_1 + x_2 + 3x_3 + x_5 &= 0 \\ 2x_1 - 2x_2 - 6x_4 + x_6 &= 0 \\ x_1 - x_2 - 3x_4 - x_5 &= 0 \\ -2x_1 + 4x_2 - x_3 + 6x_4 + x_5 + x_6 &= 0 \end{aligned}$$

c $$\begin{aligned} 3x_1 + 2x_2 - 7x_3 + 2x_4 + x_5 &= 0 \\ -x_1 + x_2 - 7x_4 + x_5 + 3x_6 &= 0 \end{aligned}$$

7 Does a homogeneous system always have at least one solution? If so, what is this solution?

8 Which of the following matrices are reduced matrices?

a $\begin{bmatrix} 0 & 0 \\ 0 & 0 \end{bmatrix}$ **b** $[1 \quad 2 \quad 1]$

c $\begin{bmatrix} 0 & 1 & 1 \\ 0 & 0 & 1 \end{bmatrix}$ **d** $\begin{bmatrix} 0 & 1 & 1 \\ 1 & 0 & 0 \end{bmatrix}$

e $\begin{bmatrix} 1 & 0 & 1 \\ 0 & 1 & 0 \\ 0 & 0 & 1 \end{bmatrix}$ **f** $\begin{bmatrix} 1 & 2 & 0 & 0 \\ 0 & 0 & 0 & 0 \\ 0 & 0 & 1 & 0 \\ 0 & 0 & 0 & 1 \end{bmatrix}$

9 List all possible reduced matrices with two rows and two columns; with two rows and three columns; with three rows and three columns. For example, every reduced matrix with two rows and two columns which has one zero row must be of the form

$$\begin{bmatrix} 0 & 1 \\ 0 & 0 \end{bmatrix} \quad \text{or} \quad \begin{bmatrix} 1 & a \\ 0 & 0 \end{bmatrix}$$

10 Suppose $A = \begin{bmatrix} a & b \\ c & d \end{bmatrix}$ and that A can be reduced to $\begin{bmatrix} 1 & 0 \\ 0 & 1 \end{bmatrix}$. Find the solutions to

$$\begin{aligned} ax + by &= 0 \\ cx + dy &= 0 \end{aligned}$$

11 Show that $A = \begin{bmatrix} a & b \\ c & d \end{bmatrix}$ reduces to $\begin{bmatrix} 1 & 0 \\ 0 & 1 \end{bmatrix}$ if $ad - bc \neq 0$. Show that if $ad - bc = 0$, then A reduces to a matrix with one row consisting of zeros.

12 Let $A = \begin{bmatrix} a & b \\ c & d \end{bmatrix}$. Show that the following procedure converts A to a reduced matrix.

Step 1 If $a = b = c = d = 0$, stop. Otherwise go to step 2.
Step 2 If $a = c = d = 0$, $b \neq 0$, multiply row 1 by $1/b$ and stop. Otherwise go to step 3.
Step 3 If $a = b = c = 0$, $d \neq 0$, interchange rows 1 and 2, multiply the new row 1 by $1/d$, and stop. Otherwise go to step 4.

Step 4 If $a = c = 0$, $b \neq 0$, $d \neq 0$, add $-d/b$ times row 1 to row 2, then multiply row 1 by $1/b$ and stop. Otherwise go to step 5.

Step 5 If $a = 0$ and $c \neq 0$, interchange rows 1 and 2, then relabel so the first entry in the new row 1 is labeled a and the first entry in the new row 2 is labeled c. Go to step 6.

Step 6 Add $-c/a$ times row 1 to row 2, then multiply row 1 by $1/a$. Relabel so the second entry in row 1 is b and the second entry in row 2 is d. Go to step 7.

Step 7 If $d = 0$ stop. Otherwise go to step 8.

Step 8 Add $-b/d$ times row 2 to row 1, then multiply row 2 by $1/d$ and stop.

13 Describe a procedure for reducing matrices with three rows and columns to reduced form. (See Exercise **12**.)

☐ **14** Describe a procedure for reducing a matrix with m rows and n columns to reduced form. (See Exercise **12**.)

15 Show that if the second row of A is a sum of multiples of the third, fifth, and sixth rows of A, then we can reduce A to a matrix with a row of zeros.

16 Suppose we interchange two *columns* of A to obtain B. How is the system whose matrix of coefficients is B related to the system whose matrix is A?

17 Show that if by a sequence of row operations A is reduced to B, then there is a sequence of row operations that reduces B to A.

18 Suppose R_1 and R_2 are each reduced matrices with two rows and two columns such that the corresponding systems have the same solutions. Show that $R_1 = R_2$.

19 Show that if

$$c_0 + c_1x + c_2x^2 + c_3x^3 = 0$$

for all x then $c_0 = c_1 = c_2 = c_3 = 0$. (Hint: Put $x = 0, 1, 2, 3$ to obtain a system of four equations in the unknowns c_0, c_1, c_2, c_3. By reduction show that this system has only one solution.)

20 Show that if $c_1 \sin x + c_2 \sin 2x + c_3 \sin 3x = 0$ for all x, then $c_1 = c_2 = c_3 = 0$.

☐ **21** The method of elimination given also works for systems with complex coefficients. Find the matrix of coefficients and use this to find the form of solutions to

a $$\begin{aligned} ix + (1 + i)y + 2z &= 0 \\ 2x - (1 - i)y + iz &= 0 \end{aligned}$$

b $$\begin{aligned} (1 + i)x - iy + 2iz &= 0 \\ (1 - 2i)x + y \qquad &= 0 \end{aligned}$$

[Hint: Remember that $1/(a + ib) = (a - ib)/(a^2 + b^2)$.]

SECTION 2 Nonhomogeneous Systems

Matrix elimination methods can also be used to solve systems in which the constants are not all zero such as

1
$$\begin{aligned} 3x + y &= 4 \\ x - y &= 2 \end{aligned}$$

Such a system is called a **nonhomogeneous linear system.** We now need to carry the constant terms along in the computations, using some device that will keep track of the constants that go with each equation. The usual way to do this is to form the so-called **augmented matrix** of the system.

2
$$\left[\begin{array}{cc:c} 3 & 1 & 4 \\ 1 & -1 & 2 \end{array}\right]$$

The first two columns of this matrix consist of the coefficient matrix of system **1**, while the *final* column consists of the corresponding constants. Now we need only apply the row operations of Section 1 to this matrix until we obtain the augmented matrix of a simpler system. We therefore proceed to reduce matrix **2**. One possible sequence of operations is to add to the first row -3 times the second row. This gives

$$\left[\begin{array}{cc:c} 0 & 4 & -2 \\ 1 & -1 & 2 \end{array}\right]$$

Multiply the first row by $\frac{1}{4}$; then interchange the first and second rows. We now have

$$\left[\begin{array}{cc:c} 1 & -1 & 2 \\ 0 & 1 & -\frac{1}{2} \end{array}\right]$$

Add to the first row 1 times the second row. We now have obtained the reduced matrix

$$\left[\begin{array}{cc:c} 1 & 0 & \frac{3}{2} \\ 0 & 1 & -\frac{1}{2} \end{array}\right]$$

which is the augmented matrix of

$$\begin{aligned} x &= \tfrac{3}{2} \\ y &= -\tfrac{1}{2} \end{aligned}$$

We conclude that this is the only solution to system **1**.

A homogeneous system always has at least one solution (namely, the one in which all the variables are equal to zero). It is possible, however, that a *nonhomogeneous system may have no solution.* This situation will be made evident during the process of reduction of the augmented matrix by the occurrence of a row in which all entries, *except* the last entry are zeros. The following example illustrates this.

EXAMPLE 1 The augmented matrix of

3
$$\begin{aligned} 3x - y + z &= 1 \\ 7x + y - z &= 6 \\ 2x + y - z &= 2 \end{aligned} \qquad \text{is} \qquad \left[\begin{array}{ccc:c} 3 & -1 & 1 & 1 \\ 7 & 1 & -1 & 6 \\ 2 & 1 & -1 & 2 \end{array}\right]$$

which can be reduced by using the following sequence of row operations: multiply row one by $\frac{1}{3}$; then add to row two -7 times row one; and then add to row three -2 times row one. We now have

$$\left[\begin{array}{ccc:c} 1 & -\frac{1}{3} & \frac{1}{3} & \frac{1}{3} \\ 0 & \frac{10}{3} & -\frac{10}{3} & \frac{11}{3} \\ 0 & \frac{5}{3} & -\frac{5}{3} & \frac{4}{3} \end{array}\right]$$

and can use the second row to eliminate. To continue the row operations, multiply row two by $\frac{3}{10}$; then add to row three $-\frac{5}{3}$ times row two. The result is

$$\left[\begin{array}{ccc:c} 1 & -\frac{1}{3} & \frac{1}{3} & \frac{1}{3} \\ 0 & 1 & -1 & \frac{11}{10} \\ 0 & 0 & 0 & -\frac{1}{2} \end{array}\right]$$

which is the augmented matrix of

4
$$\begin{aligned} x - \tfrac{1}{3}y + \tfrac{1}{3}z &= \tfrac{1}{3} \\ y - z &= \tfrac{11}{10} \\ 0 &= -\tfrac{1}{2} \end{aligned}$$

System **4** clearly has no solution, for the last equation cannot hold for any choice of x, y, and z.

DISCUSSION The procedure for solving nonhomogeneous systems is thus the following: form the augmented matrix of the system and proceed to reduce this matrix to a reduced matrix. Suppose during this reduction we obtain a matrix in which

5 There is a row in which the first nonzero entry appears in the last column.

If this occurs, *stop*; for such a matrix is the augmented matrix of a system with *no solutions*.

If property **5** does *not* occur, we finally obtain a reduced matrix with the property that

6 No row has its first nonzero entry in the last column.

In this case the system has *one or more* solutions, and we can easily determine the solution or solutions from the reduced form.

Of course, it is possible for a nonhomogeneous system to have more than one solution, as the next example illustrates.

EXAMPLE 2 The augmented matrix of

7 $$\begin{aligned} x - y + 2z &= 1 \\ 3x + y - z &= 2 \end{aligned} \qquad \text{is} \qquad \left[\begin{array}{ccc:c} 1 & -1 & 2 & 1 \\ 3 & 1 & -1 & 2 \end{array}\right]$$

This matrix can be reduced to (the student should work this out)

$$\left[\begin{array}{ccc:c} 1 & 0 & \frac{1}{4} & \frac{3}{4} \\ 0 & 1 & -\frac{7}{4} & -\frac{1}{4} \end{array}\right]$$

which is the augmented matrix of

$$\begin{aligned} x + \tfrac{1}{4}z &= \tfrac{3}{4} \\ y - \tfrac{7}{4}z &= -\tfrac{1}{4} \end{aligned}$$

We can assign to z any arbitrary value, then compute x and y from this to obtain a solution to system **7**. For example, if we put $z = 0$, we obtain the solution $x = \frac{3}{4}$, $y = -\frac{1}{4}$, $z = 0$; and if we put $z = 1$, we have the solution $x = \frac{1}{2}$, $y = \frac{3}{2}$, $z = 1$.

We can express the dependence of these solutions upon z by rewriting to obtain

8 $$\begin{aligned} x &= -\tfrac{1}{4}z + \tfrac{3}{4} \\ y &= \tfrac{7}{4}z - \tfrac{1}{4} \end{aligned}$$

DISCUSSION We see that three cases can occur for nonhomogeneous systems: no solutions, exactly one solution, and more than one solution. This first case *cannot* occur for a homogeneous system because such a system always has at least one solution.

Further discussion of the relation between homogeneous systems and nonhomogeneous systems will be given in later sections. The next example indicates one of the possibilities.

EXAMPLE 3 Consider the systems

9
$$\begin{aligned} 2x - y + z &= 1 \\ 3x + 2z &= -1 \\ 4x + y + 2z &= 2 \end{aligned} \qquad \text{and} \qquad \begin{aligned} 2x - y + z &= 0 \\ 3x + 2z &= 0 \\ 4x + y + 2z &= 0 \end{aligned}$$

Form the matrix of coefficients (which is the same for both systems):

10
$$\begin{bmatrix} 2 & -1 & 1 \\ 3 & 0 & 2 \\ 4 & 1 & 2 \end{bmatrix}$$

and reduce it with the following sequence of operations:

1 Multiply row two by $\frac{1}{3}$.
2 Add to row one -2 times row two; then add to row three -4 times row two.
3 Interchange rows one and two.
4 Add to row two 1 times row three; then interchange rows two and three.
5 Add to row one $\frac{2}{3}$ times row three; then add to row two $-\frac{2}{3}$ times row three.
6 Multiply row three by -1.

This procedure results in the matrix (the student should check this)

11
$$\begin{bmatrix} 1 & 0 & 0 \\ 0 & 1 & 0 \\ 0 & 0 & 1 \end{bmatrix}$$

so that the second system of the systems **9** has the unique solution

$$x = y = z = 0$$

Now apply the same operations, in the same order, to the augmented matrix of the first system of the systems **9**.

12
$$\left[\begin{array}{ccc|c} 2 & -1 & 1 & 1 \\ 3 & 0 & 2 & -1 \\ 4 & 1 & 2 & 2 \end{array}\right]$$

The resulting matrix (which should be checked by the student) is

$$\left[\begin{array}{ccc:c} 1 & 0 & 0 & 3 \\ 0 & 1 & 0 & 0 \\ 0 & 0 & 1 & -5 \end{array}\right]$$

so that the unique solution to the first system of the systems **9** is $x = 3$, $y = 0$, and $z = -5$.

Note what has happened here. If the reduction of the coefficient matrix results in a matrix *without* any row consisting of zeros (such as matrix **11**), then the same sequence of operations applied to any corresponding augmented matrix (such as matrix **12**) must give a reduced matrix in which no row can have its first nonzero entry in the last column. In this case, the fact that the homogeneous system has a unique solution (which follows from the form of matrix **11**) necessarily means that the nonhomogeneous system also has a unique solution. This relation between homogeneous and nonhomogeneous systems will be explored further in subsequent sections.

EXERCISES

1 For each of the following systems find the augmented matrix; then, by reducing, determine whether the system has a solution. If the system has a solution, reduce the augmented matrix to a reduced matrix and determine the form of the solutions.

a
$$\begin{aligned} x - 3y &= 1 \\ 2x - y &= 4 \end{aligned}$$

b
$$\begin{aligned} 3x + y + 6z &= 6 \\ x + y + 2z &= 2 \\ 2x + y + 4z &= 3 \end{aligned}$$

c
$$\begin{aligned} 2x_1 + x_2 \phantom{{}+3x_3} + x_4 &= 2 \\ 3x_1 + 3x_2 + 3x_3 + 5x_4 &= 4 \\ 3x_1 \phantom{{}+3x_2} - 3x_3 - 2x_4 &= 3 \end{aligned}$$

d
$$\begin{aligned} 2x + y \phantom{{}+3z} + w &= 2 \\ 3x + 3y + 3z + 5w &= 3 \\ 3x \phantom{{}+3y} - 3z - 2w &= 3 \end{aligned}$$

e
$$\begin{aligned} 2x + y + z &= 1 \\ 4x + 2y + 3z &= 1 \\ -2x - y + z &= 2 \end{aligned}$$

f
$$\begin{aligned} 2x_1 + x_2 + 2x_3 + 3x_4 + x_5 + 4x_6 &= 1 \\ x_2 + x_3 + 2x_4 + 2x_5 &= -1 \\ 2x_1 + x_2 + 3x_3 + x_4 + 5x_6 &= 1 \\ 4x_1 + 3x_2 + 6x_3 + 6x_4 + 3x_5 + 10x_6 &= -1 \\ 3x_4 + x_5 + x_6 &= 1 \\ x_2 + x_3 + 2x_4 + 4x_5 &= 0 \end{aligned}$$

g
$$\begin{aligned} 2x_1 + x_2 + 2x_3 + 3x_4 + x_5 + 4x_6 &= 1 \\ x_2 + x_3 + 2x_4 + 2x_5 &= -1 \\ 2x_1 + x_2 + 3x_3 + x_4 + 5x_6 &= 1 \end{aligned}$$

2 Form the matrix of coefficients of

$$\begin{aligned} 3x + y - z &= 0 \\ 6x + 2y - z &= 0 \end{aligned}$$

and reduce to determine the solutions. Then apply the same row operations to the augmented matrix of

$$\begin{aligned} 3x + y - z &= 2 \\ 6x + 2y - z &= -1 \end{aligned}$$

and determine whether a solution exists and is unique. Compare the first three columns of the reduced form of each of these matrices.

3 Show that if

$$\begin{aligned} ax + by &= 0 \\ cx + dy &= 0 \end{aligned}$$

has a unique solution, then for any choice of c_1 and c_2, the system

$$\begin{aligned} ax + by &= c_1 \\ cx + dy &= c_2 \end{aligned}$$

has a unique solution. (Hint: Show that the coefficient matrix must reduce to

$$\begin{bmatrix} 1 & 0 \\ 0 & 1 \end{bmatrix}$$

and proceed as in Example **3**.)

4 Assume that each of the following matrices is the matrix of the coefficients of a homogeneous system and decide whether the system has a solution and, if so, whether the solution is unique.

a $\begin{bmatrix} 1 & 0 & 0 \\ 0 & 1 & 1 \\ 0 & 0 & 1 \end{bmatrix}$ **b** $\begin{bmatrix} 1 & 2 & 1 \\ 0 & 1 & 1 \\ 0 & 0 & 0 \end{bmatrix}$

c $\begin{bmatrix} 1 & 2 & 0 & 0 \\ 0 & 0 & 1 & 1 \\ 0 & 0 & 0 & 1 \end{bmatrix}$ **d** $\begin{bmatrix} 1 & 3 & 2 \\ 0 & 1 & 1 \\ 0 & 0 & 2 \\ 0 & 0 & 0 \end{bmatrix}$

5 Assume that each of the matrices of Exercise **4** is the augmented matrix of a nonhomogeneous system and decide whether the system has a solution and, if so, whether the solution is unique.

6 Assume that each of the matrices of Exercise **4** is the matrix of coefficients of a nonhomogeneous system and decide whether the system can always be solved (no matter what the constant terms may be). If the system can be solved (sometimes or always), must the solutions be unique?

7 Certain algebraic questions can be answered by converting them to questions about the solutions to systems of equations. For example, suppose we wish to know whether there are numbers a, b, and c such that the equation

$$x^3 = a + bx + cx^2$$

holds for *all* x. This question can be converted to a question about a linear system of equations as follows: Choose four values of x, say $x = 0, 1, 2,$ and 3, and substitute these in the above to obtain the system

$$\begin{aligned} 0^3 &= a + b \cdot 0 + c \cdot 0^2 \\ 1^3 &= a + b \cdot 1 + c \cdot 1^2 \\ 2^3 &= a + b \cdot 2 + c \cdot 2^2 \\ 3^3 &= a + b \cdot 3 + c \cdot 3^2 \end{aligned} \qquad \text{that is} \qquad \begin{aligned} 0 &= a \\ 1 &= a + b + c \\ 8 &= a + 2b + 4c \\ 27 &= a + 3b + 9c. \end{aligned}$$

The methods of this section can then be used to show that this system has *no* solution. In particular, therefore, there can be no numbers a, b, and c such that $x^3 = a + bx + cx^2$ is true for all x.

a Verify that the above system indeed has no solution.

b Show that there are no numbers a and b such that

$$\sqrt{x} = a + bx \qquad \text{holds for all } x$$

c Show that there are no numbers a, b, c such that

$$x^4 = a + bx + cx^3 \qquad \text{holds for all } x$$

d Show that there are no numbers a and b such that

$$e^{3x} = ae^x + be^{2x} \qquad \text{holds for all } x$$

e Show that there are no numbers a and b such that

$$\sin 3x = a \sin x + b \sin 2x \qquad \text{holds for all } x$$

8 **a** The questions discussed in the previous exercise can be answered even more quickly using the techniques of calculus. For example, $x^3 = a + bx + cx^2$ cannot hold for all x, since we could then differentiate three times to obtain $6 = 0$. This method can be applied to Exercises **7b** and **7c** to give immediate answers. To answer **7d**, one can differentiate the given equation twice, obtaining the three equations

$$\begin{aligned} e^{3x} &= ae^x + be^{2x} \\ 3e^{3x} &= ae^x + 2be^{2x} \\ 9e^{3x} &= ae^x + 4be^{2x} \end{aligned}$$

Let $x = 0$ in each of these and show that the resulting system has no solution.

b Do Exercise **7e** using the above differentiation method.

c Show that there are no numbers a, b, and c such that

$$e^x = a + bx + cx^2 \qquad \text{holds for all } x$$

□ 9 The technique of forming the augmented matrix and reducing also works for systems with complex coefficients and constants. For each of the following systems find the augmented matrix and determine whether a solution exists. If a solution exists, reduce to a reduced matrix and determine the solutions.

a
$$\begin{aligned} ix + 2y &= 1 + i \\ (1 + 3i)x - iy &= 0 \end{aligned}$$

b
$$\begin{aligned} (1 + i)x + iy &= 0 \\ 2x + (1 + i)y &= 1 \end{aligned}$$

c
$$\begin{aligned} ix + (1 + i)y + 2iz &= 3 \\ (1 - 2i)x - y &= 1 \\ (2 - i)x + y + (2 + 2i)z &= 4 - 3i \end{aligned}$$

SECTION 3 Existence and Uniqueness Theorems for Homogeneous Systems

A homogeneous system of linear equations in the unknowns $x_1, x_2, \ldots, x_n$ always has at least one solution, namely

$$x_1 = 0, x_2 = 0, \ldots, x_n = 0$$

For such systems the primary question is whether there are any other solutions, that is, whether there are solutions in which at least one of the x_i is *not* zero. The answer to this question is determined by the relationship between the number of variables and the number of equations in the reduced form, the latter being called the rank of the system.

The **rank** of a matrix is the number of rows with nonzero entries in its reduced form. The rank of a homogeneous linear system is the rank of its coefficient matrix.

Thus to determine the rank of a matrix, first use the elimination method to obtain a reduced matrix, then count the number of rows with nonzero entries. This number does not depend upon the particular sequence of operations used to obtain a reduced matrix, as will be shown in Section 4, Chapter 2. Our basic result is the following theorem.

THEOREM 1

a If the rank of a homogeneous linear system is the *same* as the number of variables then the system has a *unique* solution.

b If the rank of a homogeneous linear system is *less than* the number of variables then the system has *more than one* solution.

This result can be reformulated in a number of ways. A homogeneous linear system in n variables always has one solution, that in which each variable is 0. We can therefore restate the theorem in the following form:

A homogeneous linear system in n variables $x_1, x_2, \ldots, x_n$:

1

a Has the unique solution $x_1 = x_2 = \cdots = x_n = 0$ if its reduced matrix of coefficients has exactly n rows with nonzero entries.

b Has a solution in which at least one variable is *not* zero if its reduced matrix of coefficients has no more than $n - 1$ rows with nonzero entries.

The exercises include other ways of stating these results. The proof of Theorem **1b** is not too difficult for it depends upon the fact that if the rank is smaller than the number of unknowns then at least one variable is not the lead variable of some equation of the reduced form, and hence the leading variables can be expressed in terms of it. The following example will clarify this. Example **2** will indicate what happens when the rank and the number of variables are the same. A proof of Theorem **1** will be outlined after the discussion of Example **2**. Analogous results for nonhomogeneous systems will be discussed in Exercise **11**, below.

EXAMPLE 1 Suppose one is given the system

2

$$\begin{aligned} x - y + 2z &= 0 \\ 3x + y - z &= 0 \\ 2x + 2y - 3z &= 0 \end{aligned}$$

and one wishes to know if $x = 0$, $y = 0$, $z = 0$ is the only solution or whether there are also solutions with at least one of the variables x, y, or z *not* zero. The matrix of coefficients

$$\begin{bmatrix} 1 & -1 & 2 \\ 3 & 1 & -1 \\ 2 & 2 & -3 \end{bmatrix} \quad \text{reduces to} \quad R = \begin{bmatrix} 1 & 0 & \frac{1}{4} \\ 0 & 1 & -\frac{7}{4} \\ 0 & 0 & 0 \end{bmatrix}$$

The reduced matrix has *two* nonzero rows (that is, rows with at least one nonzero entry), so the system has rank 2. Since the number of variables is three, the theorem tells us that the original system **2** has more than one solution. In particular there must be at least one solution in which one of x, y, or z is not zero. Let us look at the reduced system and see why this is so.

The reduced matrix R corresponds to the reduced system

3 $$\begin{aligned} x + \tfrac{1}{4}z &= 0 \\ y - \tfrac{7}{4}z &= 0 \end{aligned}$$

after discarding the superfluous equation corresponding to the bottom row of zeros in R. This can be solved directly for x and y in terms of z:

4 $$\begin{aligned} x &= -\tfrac{1}{4}z \\ y &= \tfrac{7}{4}z \end{aligned}$$

so that for any value of z a solution to the original system **1** is obtained by calculating x and y from the formulas **4**.

The key observation here is that the variable z is *not the leading variable* in either equation of the reduced system **3**; hence this system can be rewritten in the form **4**, which expresses the leading variables x and y in terms of the remaining variable z. Since z can then be given arbitrary values, the original system must have infinitely many solutions.

EXAMPLE 2 Let us see what the reduced system looks like when the system has only one solution. For example, the coefficient matrix of the system

5 $$\begin{aligned} 2x - 7y &= 0 \\ 3x + 5y &= 0 \end{aligned} \quad \text{is} \quad \begin{bmatrix} 2 & -7 \\ 3 & 5 \end{bmatrix}$$

which reduces to

6 $$\begin{bmatrix} 1 & 0 \\ 0 & 1 \end{bmatrix} \quad \text{the coefficient matrix of} \quad \begin{aligned} x \quad &= 0 \\ y &= 0 \end{aligned}$$

This reduced matrix has two rows with nonzero entries; that is, the system has rank 2. The system has two variables, so Theorem **1** tells us that $x = y = 0$ is the only solution to the original system **5**, a fact clearly shown in the reduced system **6**. In this case, *each* variable is the leading variable of some equation of the reduced form and has been eliminated from all other equations of the reduced system.

DISCUSSION The key to the proof of Theorem **1** is indicated in the above examples. For a given homogeneous linear system form the matrix of coefficients and reduce to a reduced matrix R. The matrix R is the matrix of coefficients of a reduce system which has the same solutions as the original system. *After* the redundant equations which correspond to the zero rows of R are discarded, the reduced system obviously has the following property:

7 The number of equations of the reduced system is the same as the rank of the original system.

Furthermore,

8 Each equation of the reduced system has a leading variable whose coefficient is 1 and that leading variable appears in no other equation of the system.

Property **8** is merely a restatement of the fact that the leading nonzero entry of each row of R is a 1 and all other entries in the column in which that 1 appears are zeros (see definition **13**, page 7).

Theorem **1**, part **a**, follows immediately from properties **7** and **8**, for if the rank of the system is equal to the number of variables of the system, then these properties imply that *each variable is the leading variable of some equation of the reduced system and appears in no other equation of the reduced system.*

Thus if the system has n variables $x_1, x_2, \ldots, x_n$ and has rank n, then each equation of the reduced system contains exactly one of the variables and each variable appears in exactly one equation of the reduced system, where it has coefficient 1. The reduced system must therefore be

$$\begin{aligned} x_1 \qquad\qquad\qquad &= 0 \\ x_2 \qquad\qquad &= 0 \\ x_3 \qquad &= 0 \\ \cdot \quad &\;\; \cdot \\ \cdot \quad &\;\; \cdot \\ \cdot \quad &\;\; \cdot \\ x_n &= 0 \end{aligned}$$

which obviously has the unique solution

$$x_1 = 0, x_2 = 0, \ldots, x_n = 0$$

This proves part **a** of Theorem **1**.

The proof of part **b** is quite similar to the discussion in Example **1**. If the rank of the system is less than the number of variables, then properties **7** and **8** imply that some variables are *not* leading variables of equations of the reduced system. These nonleading variables can be transferred to the right-hand side (as we did in equations **4**). Since each leading variable appears in exactly one equation of the reduced system (property **8**), each leading variable is now expressed in terms of the nonleading variables. The nonleading variables can be assigned arbitrary values, and a solution to the original system is then obtained by calculating the values of the lead variables from the reduced system. This proves the following stronger version of Theorem **1a**.

9 If the number of variables of a homogeneous linear system exceeds its rank, then the system has infinitely many solutions.

This completes the proof of Theorem **1**.

It is not always necessary to determine the reduced matrix in order to apply Theorem **1**. In fact, there is one case of frequent occurrence where it is known in advance that a homogeneous system has more than one solution.

10 A homogeneous linear system with *more unknowns than equations* always has at least one solution in which at least one variable is not zero.

This is a simple consequence of Theorem **1b**, for if the system has more unknowns than equations, so will the reduced system (since "elimination" is designed to eliminate redundant equations), so that the rank of such a system is always less than the number of unknowns.

EXAMPLE 3 The result **10** tells us immediately that the system

$$\begin{aligned} 2x - y + z &= 0 \\ x + 4y - z &= 0 \end{aligned}$$

must have more than one solution. In particular, there must be a solution in which one of the variables is not zero. Such a solution can be found by reducing

$$\begin{bmatrix} 2 & -1 & 1 \\ 1 & 4 & -1 \end{bmatrix} \quad \text{to} \quad \begin{bmatrix} 1 & 0 & \frac{1}{3} \\ 0 & 1 & -\frac{1}{3} \end{bmatrix}$$

the coefficient matrix of the system

$$\begin{aligned} x + \tfrac{1}{3}z &= 0 \\ y - \tfrac{1}{3}z &= 0 \end{aligned}$$

Put $z = 1$ in this to obtain the solution

$$x = -\tfrac{1}{3} \qquad y = \tfrac{1}{3} \qquad z = 1$$

in which at least one of the variables is *not* zero. In this case it also happens that if x, y, z is any solution other than $x = y = z = 0$, then $x \neq 0$,

$y \neq 0$, and $z \neq 0$. There are systems for which this does not happen. (See Exercise **10** below.)

EXAMPLE 4 As we noted above it is usually not necessary to reduce completely to determine the rank of a matrix. For example, subtraction of suitable multiples of the first row from the remaining rows reduces

$$A = \begin{bmatrix} 1 & 1 & 1 \\ 2 & -1 & 1 \\ 4 & 1 & 3 \\ 7 & 1 & 5 \end{bmatrix} \quad \text{to} \quad \begin{bmatrix} 1 & 1 & 1 \\ 0 & -3 & -1 \\ 0 & -3 & -1 \\ 0 & -6 & -2 \end{bmatrix}$$

Subtracting suitable multiples of the second row from the third and fourth gives

$$\begin{bmatrix} 1 & 1 & 1 \\ 0 & -3 & -1 \\ 0 & 0 & 0 \\ 0 & 0 & 0 \end{bmatrix}$$

from which it is clear that the reduced form has two nonzero rows, that is, rank $A = 2$.

EXERCISES

1 Determine the rank of each of the following matrices

a $\begin{bmatrix} 1 & -3 \\ 2 & 1 \end{bmatrix}$ **b** $\begin{bmatrix} 3 & -2 & 7 \\ -3 & 2 & -7 \end{bmatrix}$

c $\begin{bmatrix} 2 & 1 & 2 \\ -1 & -\frac{1}{2} & -1 \\ 2 & 1 & 0 \end{bmatrix}$ **d** $\begin{bmatrix} 2 & 5 \\ 6 & -1 \\ 0 & 3 \end{bmatrix}$

e $\begin{bmatrix} 1 & 1 & 1 & 1 \\ 1 & -1 & 1 & 1 \\ 1 & -1 & -1 & 1 \\ 1 & -1 & -1 & -1 \end{bmatrix}$

2 By determining the rank of each of the following systems show that the system has infinitely many solutions.

a $2x - y = 0$

b $\begin{aligned} 2x - y &= 0 \\ 4x - 2y &= 0 \\ -x + \tfrac{1}{2}y &= 0 \end{aligned}$

c
$$\begin{aligned} 3x - 2y + z + w &= 0 \\ 4x + 2y + z - w &= 0 \\ y - 3z + 2w &= 0 \end{aligned}$$

d
$$\begin{aligned} x + y + z &= 0 \\ x + 2y + z &= 0 \\ x + 4y + z &= 0 \\ -x + y - z &= 0 \end{aligned}$$

e
$$\begin{aligned} x_1 + x_2 - x_3 + x_4 &= 0 \\ x_3 + x_4 + x_5 &= 0 \\ 2x_1 + 2x_2 - x_3 + x_5 &= 0 \\ x_1 + x_2 - 2x_3 - x_5 &= 0 \\ 2x_3 - 4x_4 + 2x_5 &= 0 \\ x_1 - x_2 + 2x_3 - 3x_4 + x_5 &= 0 \end{aligned}$$

3 For each of the systems of Exercise **2**, find at least one solution in which not all of the variables are zero.

4 Use Theorem **1** to show that each of the following systems has a unique solution. What is this solution? (It is necessary to reduce only far enough to see that the final reduced form has the right number of nonzero rows, as in Example **4**.)

a
$$\begin{aligned} 2x - y &= 0 \\ x + 3y &= 0 \end{aligned}$$

b
$$\begin{aligned} 3x - 2y + z &= 0 \\ 2x - 4y + 2z &= 0 \\ -x - y + 3z &= 0 \end{aligned}$$

c
$$\begin{aligned} 4x + 3y &= 0 \\ 7x - y &= 0 \\ x + 2y &= 0 \end{aligned}$$

d
$$\begin{aligned} 2x - y + z - w &= 0 \\ 3x + y + w &= 0 \\ 2x + 2z + w &= 0 \\ 4x - 2y - z - w &= 0 \\ 5x - y - 2z &= 0 \end{aligned}$$

e
$$\begin{aligned} x_1 + x_2 + x_3 + x_4 + x_5 &= 0 \\ x_1 - x_2 + x_3 + x_4 - x_5 &= 0 \\ -x_1 + x_2 - x_3 + x_4 - x_5 &= 0 \\ x_1 + x_2 - x_3 + x_4 + x_5 &= 0 \\ x_1 + x_2 + x_3 - x_4 + x_5 &= 0 \end{aligned}$$

5 Without reducing determine the form of the reduced matrix and use Theorem **1** to decide which of the following systems have a unique solution.

a
$$\begin{aligned} x + 2y &= 0 \\ 3y &= 0 \end{aligned}$$

b
$$\begin{aligned} x_1 + 2x_2 + x_3 + x_4 &= 0 \\ x_3 - x_4 &= 0 \\ x_3 + x_4 &= 0 \\ 2x_4 &= 0 \end{aligned}$$

c
$$\begin{aligned} x_1 + x_2 + x_3 + x_4 &= 0 \\ 2x_2 - x_3 + x_4 &= 0 \\ 3x_3 - x_4 &= 0 \\ 4x_4 &= 0 \end{aligned}$$

6 **a** If a homogeneous system has rank 5 and at least two solutions, what is the minimum number of variables the system could have? State reasons.

b If a homogeneous system has five variables and exactly one solution, what are the minimum and maximum possible values for its rank?

7 If a homogeneous system has two solutions, must it have infinitely many solutions? Give reasons.

8 **a** Can the first nonzero entry of row five of a reduced matrix R appear in column six? Column four? (Either give an example or state why.)

b If a matrix A has five columns, what is the smallest its rank can be? What is the largest its rank can be? Give your reasons.

9 We often use equivalent formulations of the conclusion of Theorem **1b**. Which of the following statements are the same as the conclusion of Theorem **1b**?

a There is a solution $x_1, x_2, \ldots, x_n$ such that for each i, $x_i \neq 0$.
b There is a solution $x_1, x_2, \ldots, x_n$ such that for some i, $x_i \neq 0$.
c The system has a unique solution.
d The system has more than one solution.
e Any two solutions of the system are identical.
f There is a solution $x_1, x_2, \ldots, x_n$ such that $x_1^2 + x_2^2 + \cdots + x_n^2 > 0$.
g There is a solution $x_1, x_2, \ldots, x_n$ such that the product $x_1x_2 \ldots x_n \neq 0$.
h There is a solution $x_1, x_2, \ldots, x_n$ such that $1/x_i$ exists for some i.

10 Can you find a solution to

$$\begin{aligned} x + 2y + z &= 0 \\ -x - 2y + 3z &= 0 \end{aligned}$$

in which $z \neq 0$? Why does this not contradict statement **10**, page 28?

Theorem 1 has an extension to nonhomogeneous systems, stated here as follows.

THEOREM 1A A nonhomogeneous linear system has a solution if and only if the rank of the augmented matrix equals the rank of the coefficient matrix. If these ranks are equal, then the system has a unique solution if and only if this common rank is equal to the number of unknowns.

11 **a** Give an argument like that given on pages 26–28 to establish Theorem **1A**. (It may be helpful to review the discussion in Section 2, especially Example 3, page 20.)

b Show that, if the coefficient matrix of a nonhomogeneous system with n unknowns and n equations has rank n, then the system has a unique solution. (Hint: Show that the augmented matrix has rank n.)

12 A linear equation in two unknowns is the equation of a line in the two-dimensional plane. A linear equation in three unknowns is the equation of a plane in space.

a What can you say about the line of a homogeneous equation in two unknowns? The plane of a homogeneous equation in three unknowns? What can be said about the intersection of the lines (or planes) of a homogeneous system if the solution is unique? If there is more than one solution?

b Show by geometric arguments that a homogeneous system of two equations in three unknowns has infinitely many solutions.

c What can be said about the intersection of the lines (or planes) of a nonhomogeneous system if the system has no solution? Exactly one solution? More than one solution?

d If you change the constant term in the equation of a line (or plane), how does the line (or plane) move? Deduce from this and the above by geometric arguments that if a homogeneous system of two equations in two unknowns (or three equations in three unknowns) has a unique solution, then so does a nonhomogeneous system with the same coefficient matrix.

13 Show that

$$\begin{aligned} ax + by &= 0 \\ cx + dy &= 0 \end{aligned}$$

has more than one solution if and only if $ad - bc = 0$.

14 Show that

$$\begin{aligned} a_1x + a_2y &= 0 \\ b_1x + b_2y &= 0 \\ c_1x + c_2y &= 0 \end{aligned}$$

has more than one solution if and only if

$$a_1b_2 - a_2b_1 = a_1c_2 - a_2c_1 = b_1c_2 - b_2c_1 = 0$$

☐ **15** The theorems of this section can be extended to the case of complex systems.

a Does the following system have a unique solution?

$$\begin{aligned} ix + (1 + i)y &= 0 \\ 2x - (3i + 1)y &= 0 \end{aligned}$$

If so, what must this solution be?

b Does the following system have a unique solution?

$$\begin{aligned} x - \qquad iy + 2iz &= 0 \\ 3x + (1 - i)y - \quad iz &= 0 \end{aligned}$$

SECTION 4 Vector Spaces

Homogeneous systems of linear equations have the property that sums and multiples of solutions are again solutions. For example, if

1 $$x = a \qquad y = b \qquad z = c$$

and

2 $$x = a' \qquad y = b' \qquad z = c'$$

are each solutions to

3 $$\begin{aligned} x - 7y + 2z &= 0 \\ 3x + \tfrac{1}{4}y - \quad z &= 0 \end{aligned}$$

then so is their sum

4 $$x = a + a' \qquad y = b + b' \qquad z = c + c'$$

and any multiple of either solution, such as

$$x = 9a \qquad y = 9b \qquad z = 9c$$

or

$$x = \tfrac{3}{2}a' \qquad y = \tfrac{3}{2}b' \qquad z = \tfrac{3}{2}c'$$

This observation amounts to little more than noting that the associative, commutative, and distributive laws of ordinary arithmetic hold. Thus, for example, the assumption that equations **1** and **2** each give a solution to system **3** tells us that

$$\begin{aligned} a - 7b + 2c &= 0 \\ 3a + \tfrac{1}{4}b - c &= 0 \end{aligned} \quad \text{and} \quad \begin{aligned} a' - 7b' + 2c' &= 0 \\ 3a' + \tfrac{1}{4}b' - c' &= 0 \end{aligned}$$

from which it certainly follows that

$$\begin{aligned} (a + a') - 7(b + b') + 2(c + c') &= 0 \\ 3(a + a') + \tfrac{1}{4}(b + b') - (c + c') &= 0 \end{aligned}$$

so that the sum **4** is indeed a solution to the system **3**.

This suggests the possibility that the study of "addition" of triples and "multiplication" of triples by numbers would be fruitful in studying systems of equations in three unknowns. In order not to confine our discussion to three unknowns we first introduce a suitable generalization of triples.

An ***n*-tuple** of real numbers is an ordered set $(x_1,x_2, \ldots, x_n)$ where each x_i is a real number; in other words, an n-tuple is a matrix with one row and n columns.

A 2-tuple (x_1,x_2) is thus just an ordered pair, while a 3-tuple is an ordered triple (x_1,x_2,x_3). The entries of an n-tuple $(x_1,x_2, \ldots, x_n)$ are called *coordinates*; for example, 2 is the third coordinate of the 4-tuple (3,1,2,0).

Two n-tuples $(x_1,x_2, \ldots, x_n)$ and $(y_1,y_2, \ldots, y_n)$ are **equal** if all corresponding coordinates are equal, that is,

$$x_1 = y_1,\ x_2 = y_2, \ldots, x_n = y_n$$

Thus, for example,

$$(2,-1,0,3) = (4 - 2,\ 0 - 1,\ 2 - 2,\ 1 + 2)$$

The **sum** of two n-tuples $(x_1,x_2,\ldots,x_n)$ and $(y_1,y_2,\ldots,y_n)$ is defined by

$$(x_1,x_2,\ldots,x_n) + (y_1,y_2,\ldots,y_n) = (x_1 + y_1, x_2 + y_2, \ldots, x_n + y_n)$$

The **multiple** of $(x_1,x_2,\ldots,x_n)$ by the real number t is defined by

$$t(x_1,x_2,\ldots,x_n) = (tx_1,tx_2,\ldots,tx_n)$$

This is usually called the **scalar** multiple, for it is common practice in physics to refer to numbers as scalars.

The **zero** n-tuple is $(0,0,\ldots,0)$.

Subtraction is defined by

$$(x_1,x_2,\ldots,x_n) - (y_1,y_2,\ldots,y_n) = (x_1 - y_1, x_2 - y_2, \ldots, x_n - y_n)$$

We shall generally find it useful to replace the lengthy expression for an n-tuple $(x_1,x_2,\ldots,x_n)$ by a single letter with an overbar, such as $\bar{x}$. (Other authors use $\vec{x}$ or $\mathbf{x}$.) We would then write $\bar{x} + \bar{y}$, $\bar{x} - \bar{y}$, or $t\bar{x}$ for the sum, difference, or multiple and $\bar{0}$ for the zero n-tuple.

EXAMPLE 1 Some examples of addition, scalar multiplication, and subtraction are:

$$(3,2,-1,1) + (-8,5,0,7) = (-5,7,-1,8)$$
$$\tfrac{3}{2}(-4,1,6) = (-6,\tfrac{3}{2},9)$$
$$(\tfrac{5}{3},\tfrac{1}{2},0,-7,\tfrac{1}{4}) - (\tfrac{3}{8},4,\tfrac{6}{5},-8,1) = (\tfrac{31}{24},-\tfrac{7}{2},-\tfrac{6}{5},1,-\tfrac{3}{4})$$

Note that $(2,1,0) + (4,2)$ is *not* defined because addition (and subtraction) have been defined only for ordered sets with the *same* number of entries.

EXAMPLE 2 It is a simple matter to do algebraic manipulations with n-tuples just as though they were numbers. For example, if $\bar{u}$, $\bar{x}$, and $\bar{c}$ are n-tuples and

$$\bar{u} + 5\bar{x} = \bar{c}$$

then

$$\bar{x} = \tfrac{1}{5}(\bar{c} - \bar{u})$$

Thus if

$$\bar{x} = (x_1,x_2,x_3) \qquad \bar{u} = (2,1,-1) \qquad \bar{c} = (0,3,-1)$$

and $\bar{u} + 5\bar{x} = \bar{c}$, then

$$\begin{aligned}(x_1,x_2,x_3) &= \tfrac{1}{5}[(0,3,-1) - (2,1,-1)]\\ &= (-\tfrac{2}{5},\tfrac{2}{5},0)\end{aligned}$$

so that

$$x_1 = -\tfrac{2}{5}, \qquad x_2 = \tfrac{2}{5}, \qquad x_3 = 0$$

DISCUSSION The set of solutions to a given homogeneous linear system in n variables is a collection of n-tuples. For example, the set of solutions to

$$\begin{aligned} 2x - y &= 0 \\ x + 3y &= 0 \end{aligned}$$

consists of the single pair $\bar{0} = (0,0)$ (since the system has rank 2), while the solutions to

$$\begin{aligned} 4x - 2y + 2z + w &= 0 \\ 2x - y - 3z - w &= 0 \end{aligned}$$

consists of all 4-tuples (x,y,z,w) such that

$$x = \tfrac{1}{2}y - \tfrac{1}{16}w \qquad \text{and} \qquad z = 0y - \tfrac{3}{8}w$$

(See page 9.)

It is *not true* that an arbitrary set of n-tuples can be the set of solutions to a homogeneous linear system, for we have already noted that sums and multiples of solutions must also be solutions. Collections which have this property are commonly known as vector spaces. While there is a very general definition of this concept which does not refer to n-tuples, we do not require such generality at this point and will use the following definition.

A **vector space** is a collection V of n-tuples such that:

a The zero n-tuple $\bar{0} = (0,0,\ldots,0)$ belongs to V.
b If $\bar{x}$ and $\bar{y}$ belong to V so does their sum $\bar{x} + \bar{y}$; that is, V is *closed* under addition.
5 c If $\bar{x}$ belongs to V and t is any real number, then the scalar multiple $t\bar{x}$ also belongs to V; that is, V is *closed* under scalar multiplication.

The members of V are often called **vectors** or **points**.

REMARK The above names may be confusing, for they appear to have little to do with the physical concept of vector as a quantity with both magnitude and direction, the representation of such vectors by arrows, or the geometric concepts of "point" and "space." For pairs and triples such geometric and physical representations can be made, however, and hence the terminology has been adopted for n-tuples to serve as a guide for our intuition.

One final piece of terminology is useful. A vector space V of n-tuples is a **subspace** of a vector space V' of m-tuples if $m = n$ and each vector in V also belongs to V'.

For example, we could now say that the set of solutions to a homogeneous linear system in n variables is a subspace of the set of *all* n-tuples. This tells us that the solutions consist of n-tuples, that the zero n-tuple is a solution, and that sums and multiples of solutions are solutions.

The following examples indicate some of the additional notation and terminology we will use, and describe some of the possible geometric representations of vectors.

EXAMPLE 3 *The Space R^2.*

We denote the collection of *all* ordered pairs of real numbers by R^2. This collection is certainly a vector space for it contains (0,0), and the sum and multiple of any pairs. This is the largest vector space we can construct using pairs, in the sense that any vector space of pairs of real numbers is a subspace of R^2.

Figure 1a *The parallelogram law for the sum in R^2.*

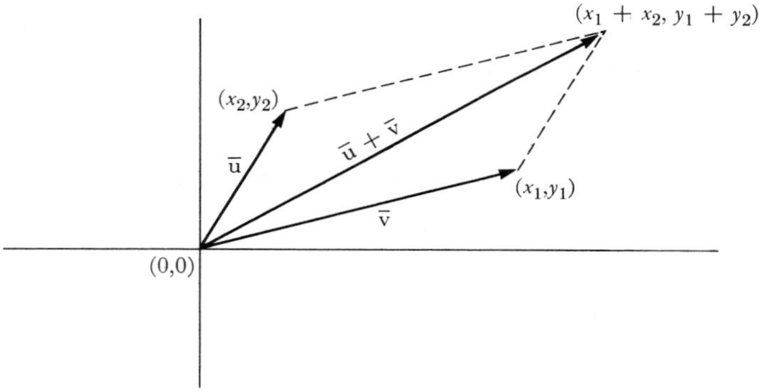

Figure 1b *A multiple of a vector.*

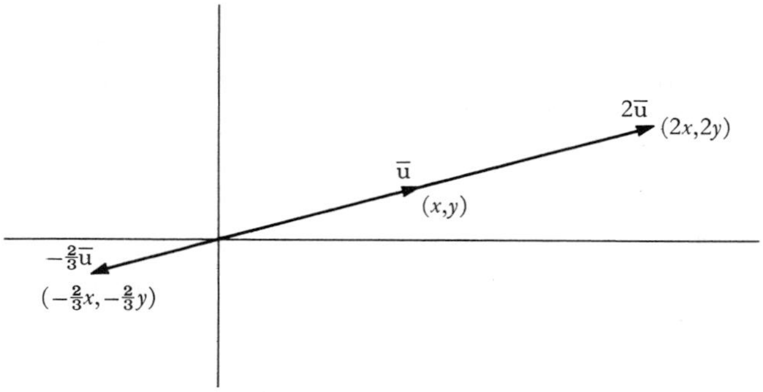

The space R^2 does have a convenient geometric representation. For the usual selection of coordinate axes in a plane, a pair (x,y) can be represented in the usual manner as the point whose coordinates are (x,y). In this representation, the sum and multiple of a pair have simple geometric interpretations, as shown in Figures 1a and 1b.

In physics, a vector is a physical quantity that possesses both magnitude and direction. If such quantities act in a plane, they can be represented by arrows in a plane, the length of the arrow being the magnitude of the quantity and the direction of the arrow being the direction in which the quantity acts.

We usually say that two such arrows represent the same quantity if they have the same length and the same direction. Hence, each such quantity can be uniquely represented by an arrow issuing from (0,0). In this case, such an arrow is completely described by giving the coordinates of its terminal point. Therefore, we can also represent pairs (x,y) in R^2 as arrows; namely, the pair (x,y) corresponds to the arrow from (0,0) to (x,y). This is the origin of the word "vector" in our setting.

If (x_1,y_1) is represented as the arrow from (0,0) to (x_1,y_1), and (x_2,y_2) is represented as the arrow from (0,0) to (x_2,y_2), then $(x_1 + x_2, y_1 + y_2)$ is represented as the arrow from (0,0) to $(x_1 + x_2, y_1 + y_2)$. (See Figure 1a.)

A quite useful fact about this representation is the following:

6 The arrow from (0,0) to $(x_2 - x_1, y_2 - y_1)$ points in the same direction as and has the same length as the arrow from (x_1,y_1) to (x_2,y_2).

This is shown in Figure 2. Thus we often find it useful to represent $(x_2 - x_1, y_2 - y_1)$ as the arrow from (x_1,y_1) to (x_2,y_2).

Figure 2

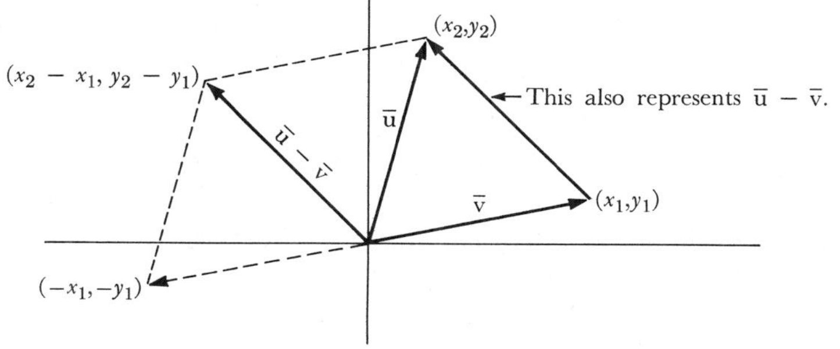

One kind of subspace of R^2 has a convenient geometric interpretation: if M consists of all multiples of a single point (x_0,y_0), M is then a subspace of R^2. Certainly M will contain (0,0) because $(0,0) = 0(x_0,y_0)$, and M is closed under addition and multiplication by numbers.

The set of all multiples of (x_0,y_0) is just the line through (0,0) and (x_0,y_0), as Figure 1b indicates. Thus, the subspace M can be represented as a line through the origin. In summary:

7 If M consists of all multiples of (x_0,y_0), M is a subspace of R^2 which can be represented as the line through (0,0) and (x_0,y_0).

It is also common practice to denote the vector from (0,0) to (1,0) as $\bar{\text{i}}$ and the vector from (0,0) to (0,1) as $\bar{\text{j}}$. The vector from (0,0) to (x,y) is thus denoted by $x\bar{\text{i}} + y\bar{\text{j}}$.

EXAMPLE 4 *The Space R^3.*

Any vector space of triples of real numbers is a subspace of the set of *all* triples of real numbers, which we shall denote by R^3. This space also has a convenient representation as points in three-dimensional space. (This is the origin of the word "space" in the phrase "vector space.") In fact, by selecting suitable coordinate axes in space, we can represent a triple (x,y,z) in the usual manner as the point whose coordinates are (x,y,z).

In a manner similar to that given for R^2 we can also represent (x,y,z) as the arrow from (0,0,0) to (x,y,z) or as any arrow with the same direction and length. In this way R^3 can be used as a representation of physical vectors which act in space.

It is common practice to denote the vector from (0,0,0) to (1,0,0) by $\bar{\text{i}}$, the vector from (0,0,0) to (0,1,0) by $\bar{\text{j}}$, and the vector from (0,0,0) to (0,0,1) by $\bar{\text{k}}$, so that $x\bar{\text{i}} + y\bar{\text{j}} + z\bar{\text{k}}$ represents the vector from (0,0,0) to (x,y,z).

In this situation the subtraction rule **6** can be given as

$$(x_2 - x_1)\bar{\text{i}} + (y_2 - y_1)\bar{\text{j}} + (z_2 - z_1)\bar{\text{k}}$$

which represents the arrow from (x_1,y_1,z_1) to (x_2,y_2,z_2).

If M is the subset of R^3 consisting of all multiples of a single vector (x_0,y_0,z_0) then, as in result 7, M is a subspace of R^3 and can be represented as the line through (0,0,0) and (x_0,y_0,z_0).

If the points (0,0,0), (x_1,y_1,z_1), and (x_2,y_2,z_2) are not collinear, and if M is the set of all vectors of the form

$$a(x_1,y_1,z_1) + b(x_2,y_2,z_2)$$

M is a subspace of R^3. This subspace can be represented as the plane determined by (x_1,y_1,z_1), (x_2,y_2,z_2), and $(0,0,0)$.

A number of applications to geometry that use these interpretations of R^2 and R^3 as collections of arrows are given at the end of this chapter.

EXAMPLE 5 *The Space R^n.*
The collection of all ordered n-tuples of real numbers will be denoted by R^n. It is a vector space (for it includes the zero n-tuple and sums and multiples of all n-tuples) and any vector space of n-tuples is a subspace of R^n. In subsequent sections we shall make a detailed study of various subspaces of R^n. This study will generally use algebraic properties or geometric concepts which have been given algebraic form so they can apply to R^n when $n > 3$.

For example let $\bar{x} = (x_1,x_2, \ldots,x_n)$ be a fixed n-tuple and let M be the set of all multiples of $\bar{x}$. Then M is a vector space for the equation $\bar{0} = 0\bar{x}$ expresses the zero n-tuple as a multiple of $\bar{x}$. Furthermore, the sum

$$t\bar{x} + s\bar{x} = (t + s)\bar{x}$$

of multiples of $\bar{x}$ is again a multiple of $\bar{x}$ and any multiple, say s, of a multiple, say t, of $\bar{x}$ is again a multiple of $\bar{x}$, for

$$s(t\bar{x}) = (st)\bar{x}$$

The subspace M is usually called the *line* through $\bar{0}$ and $\bar{x}$, by analogy with the situation in R^2 or R^3.

EXAMPLE 6 ☐ *Vector Function Spaces.*
In Chapter 6, we shall discuss solutions to linear differential equations, such as

8 $$y'' - y = 0$$

The ordinary laws of differentiation show that sums and multiples of solutions to this equation are again solutions. For example

$$y_1 = e^x \qquad \text{and} \qquad y_2 = e^{-x}$$

are both solutions to equation **8** and so are all linear combinations

$$c_1y_1 + c_2y_2 = c_1e^x + c_2e^{-x}$$

It appears from this that the properties of the collections of solutions to such differential equations may be quite similar to the properties of collections of n-tuples. It is therefore convenient to extend our definitions to include collections of functions.

A *vector space of functions* (or merely *function space*) is a collection V of functions, all with a common domain, such that:

a The zero function is in V.
9 **b** If f and g belong to V so does $f + g$.
c If f belongs to V and t is any real number then tf belongs to V.

For example, let $C[0,1]$ denote the collection of all functions f that are defined and continuous for $0 \leq x \leq 1$. The sum and multiple are then given by the following (see Figure 3):

$f + g$ is the function whose value at x is $f(x) + g(x)$.
tf is the function whose value at x is $tf(x)$.

Figure 3

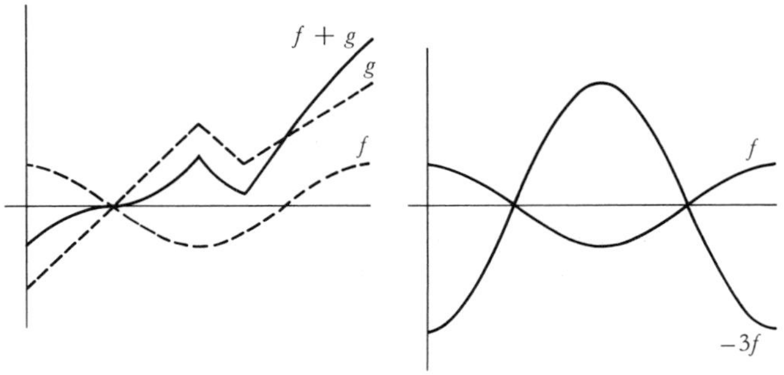

In calculus it is shown that if f and g are continuous at x, then $f + g$ and af are also continuous at x. In particular, if f and g belong to $C[0,1]$, then $f + g$ and af also belong to $C[0,1]$. Therefore $C[0,1]$ is a vector space. The realization that collections of continuous functions have an *algebraic structure* similar to R^2 and R^3 has made it possible to study many questions about continuous functions by methods originally geometric in form.

Among the subspaces of $C[0,1]$ that are of interest are the following two: the set M of all functions f that belong to $C[0,1]$ and are such that the derivative f' also belongs to $C[0,1]$ and the set M_1 of all polynomial functions that belong to $C[0,1]$. Certainly the zero function belongs to each of these. Furthermore, it is shown in calculus that

$$(f + g)' = f' + g' \qquad \text{and} \qquad (af)' = af'$$

Therefore M is closed under addition and scalar multiplication. Since sums and multiples of polynomials are again polynomials we see that M_1 is also a subspace of $C[0,1]$.

EXERCISES

1 Let $\bar{u} = (1,1,0)$, $\bar{v} = (-\frac{1}{2},0,\frac{2}{3})$, $\bar{w} = (0,\frac{1}{4},2)$.
  a Find $\bar{u} + (2\bar{v} + \bar{w})$ and $(\bar{u} + 2\bar{v}) + \bar{w}$.
  b Find $3(\bar{u} + \bar{v}) - 3(\bar{u} - \bar{w})$.
  c Find $\bar{x}$ so that $\bar{u} + \bar{x} = \bar{w}$.

2 Let $\bar{u} = (1,-1,2,1,0)$, $\bar{v} = (3,1,2,0,0)$, $\bar{w} = (0,0,1,2,-1)$.
  a Find $\bar{u} - (2\bar{v} + \bar{w})$.
  b Find $3(\bar{u} + 2\bar{v}) + 3(\bar{w} - 2\bar{v})$.
  c Find $\bar{x}$ so that $2\bar{u} + \bar{x} = 2\bar{w}$.

3 Suppose $\bar{u} = (1,2)$ and $\bar{v} = (3,-1)$. Represent $\bar{u}$, $\bar{v}$, $\bar{u} - \bar{v}$, $\bar{u} + \bar{v}$, $2\bar{u}$, and $-3\bar{v}$ as points in a plane relative to the usual choice of coordinate axes.

4 Represent each of the points of Exercise **3** as sums of multiples of $\bar{i}$ and $\bar{j}$ and draw the arrows that represent these vectors.

5 a Express $(x - y,\ 2x + y,\ x - y,\ 2y)$ as a sum of multiples of $(1,2,1,0)$ and $(-1,1,-1,2)$.
  b Express $(-\frac{5}{3}x_3 + 0 \cdot x_4 + \frac{1}{5}x_5,\ -\frac{1}{5}x_3 + 0 \cdot x_4 - \frac{3}{5}x_5,\ x_3,x_4,x_5)$ as a sum of multiples of $(-\frac{5}{3},-\frac{1}{5},1,0,0)$, $(0,0,0,1,0)$, and $(\frac{1}{5},-\frac{3}{5},0,0,1)$.

6 a Express the line whose equation is $x + 2y = 0$ as the subspace of R^2 which consists of multiples of a single vector.
  b Suppose M consists of all multiples of $(3,1)$. Find an equation of the line which represents M.
  c Express the plane whose equation is $x - 2y + z = 0$ as the subspace of R^3 which consists of sums of multiples of two vectors.
  d Can the line $x + 2y = 1$ be represented as a subspace of R^2?

7 Express $2(x_1,x_2,x_3,x_4) - 3(y_1,y_2,y_3,y_4) + (z_1,0,1,z_4)$ as a single 4-tuple.

8 Show that the set V of all triples which can be expressed in the form $a(1,0,1) + b(1,-1,0)$ is a subspace of R^3. To do this, one must show that four properties hold: (*i*) the vectors in V are all 3-tuples; (*ii*) the vector $(0,0,0)$ belongs to V; (*iii*) if $\bar{x}$ and $\bar{y}$ are in V so is $\bar{x} + \bar{y}$; and (*iv*) if $\bar{x}$ is in V and t is any real number then $t\bar{x}$ is in V. For example, to verify (*iv*) it must be shown that if $\bar{x} = a(1,0,1) + b(1,-1,0)$ for some choice of a and b, then $t\bar{x}$ is *also* a sum of multiples of $(1,0,1)$ and $(1,-1,0)$.

9 Show that the set of all vectors of the form $a(1,1,-1,1) + b(1,2,0,1) + c(1,2,1,1)$ is a subspace of R^4. (Hint: Proceed as in Exercise **8**.)

10 The set V of all triples which satisfy the equation $x + y - z = 1$ is *not* a subspace of R^3, for it violates all three of the conditions of the definition **5**.

Show that each of the following collections is not a subspace of R^3 by showing that at least one of the conditions **5** is not true for the given collection.

a The set of all (x,y,z) such that $x^2 + y^2 - z^2 = 0$

b The set of all (x,y,z) such that $x \geq 0$

c The set of all (x,y,z) such that $x^2 + y^2 + z^2 > 0$

☐ **11** Let $\bar{u} = 1 - 2x + x^3$, $\bar{v} = 3x - x^2$, $\bar{w} = 2x - 1$ be vectors in $C[0,1]$. (See Example **6**.)

a Find $\bar{u} - 2\bar{v} + 3\bar{w}$.

b Find $3(\bar{u} + \bar{v}) - 3(\bar{u} - \bar{v})$.

c Find $\bar{z}$ so that $\bar{u} + \bar{z} = \bar{w} + \bar{v}$.

d Graph $\bar{v}$, $\bar{w}$, and $2\bar{v} - \bar{w}$ on the same axes.

☐ **12** Show that $(0,0,0)$ is a solution to $2x - y + z = 0$ and that sums and multiples of solutions to this equation are also solutions to this equation and hence form a subspace of R^3.

☐ **13** Show that the set of all functions f such that $2f'' - 3f' + f = 0$ is a subspace of $C[0,1]$. (Hint: Proceed as in Exercise **12**.)

☐ **14** The set C^n of all n-tuples of complex numbers is a **complex vector space**. For $\bar{u} = (i, 2 + i, 0)$, $\bar{v} = (-1,i,2i)$, $\bar{w} = (i,0,1)$, find $\bar{u} - 2i\bar{v} + (1 + i)\bar{w}$. Also find an $\bar{x}$ in C^3 such that $\bar{u} + (1 + i)\bar{x} = \bar{w}$.

SECTION 5 Matrix Equations

We now introduce a formal way to interpret a system of equations as a matrix equation involving vectors in R^n. This will result in a considerable simplification of notation, which often enables us to give simple proofs of fairly complicated results.

For convenience vectors in R^n will be written as column matrices. For example, the form

$$\begin{bmatrix} x \\ y \end{bmatrix}$$

will be used, rather than (x,y). Of course, then the sum and multiple become

$$\begin{bmatrix} x_1 \\ y_1 \end{bmatrix} + \begin{bmatrix} x_2 \\ y_2 \end{bmatrix} = \begin{bmatrix} x_1 + x_2 \\ y_1 + y_2 \end{bmatrix} \quad \text{and} \quad a\begin{bmatrix} x \\ y \end{bmatrix} = \begin{bmatrix} ax \\ ay \end{bmatrix}$$

We can express the system

1
$$\begin{aligned} ax + by &= 0 \\ cx + dy &= 0 \end{aligned}$$

as a matrix equation by suitably defining a matrix product. If

$$A = \begin{bmatrix} a & b \\ c & d \end{bmatrix} \quad \text{and} \quad \bar{u} = \begin{bmatrix} x \\ y \end{bmatrix}$$

the **product** $A\bar{u}$ is defined by

2 $$A\bar{u} = \begin{bmatrix} ax + by \\ cx + dy \end{bmatrix}$$

The system 1 can then be expressed as the vector equation

3 $$A\bar{u} = \bar{0}$$

In other words, once we agree that $A\bar{u}$ is defined by equation 2, we can rewrite system 1 in the shorter form, 3. This shorter vector form is particularly useful for treating large systems. The general definition of the product $A\bar{u}$ is as follows:

If $$A = \begin{bmatrix} A_{11} & A_{12} & \cdots & A_{1n} \\ A_{21} & A_{22} & \cdots & A_{2n} \\ \vdots & & & \\ A_{m1} & A_{m2} & \cdots & A_{mn} \end{bmatrix} \quad \text{and} \quad \bar{u} = \begin{bmatrix} x_1 \\ x_2 \\ \vdots \\ x_n \end{bmatrix}$$

the **product** $A\bar{u}$ is defined by

4 $$A\bar{u} = \begin{bmatrix} A_{11}x_1 + A_{12}x_2 + \cdots + A_{1n}x_n \\ A_{21}x_1 + A_{22}x_2 + \cdots + A_{2n}x_n \\ \vdots \\ A_{m1}x_1 + A_{m2}x_2 + \cdots + A_{mn}x_n \end{bmatrix}$$

Notice that the right-hand side has m entries (the number of rows of A) so that $A\bar{u}$ is a vector in R^m. The ith coordinate of $A\bar{u}$ is obtained by adding together the products of the ith row of A with the corresponding entries of $\bar{u}$. We observe that the product is defined only when the number of entries of $\bar{u}$ *equals* the number of columns of A. You should learn to calculate products quickly and accurately.

This product may seem strange and arbitrary. At this point our only justification for the definition is that it provides a simple way to express a system of equations as a single equation. A more general form of this matrix product will be discussed in Chapter 3.

EXAMPLE 1 Formula **4** gives

$$\begin{bmatrix} 3 & 1 & 2 \\ 0 & 1 & 1 \end{bmatrix} \begin{bmatrix} x_1 \\ x_2 \\ x_3 \end{bmatrix} = \begin{bmatrix} 3x_1 + x_2 + 2x_3 \\ x_2 + x_3 \end{bmatrix}$$

We can therefore express

5
$$\begin{aligned} 3x_1 + x_2 + 2x_3 &= 6 \\ x_2 + x_3 &= 8 \end{aligned}$$

as the equation

6
$$A\bar{u} = \bar{c} \qquad \text{where } A = \begin{bmatrix} 3 & 1 & 2 \\ 0 & 1 & 1 \end{bmatrix}, \qquad \bar{u} = \begin{bmatrix} x_1 \\ x_2 \\ x_3 \end{bmatrix}, \qquad \bar{c} = \begin{bmatrix} 6 \\ 8 \end{bmatrix}$$

The same terminology is used for vector equations as for systems. For example, A is the *matrix of coefficients* of equation **6**, and this equation is *nonhomogeneous* because $\bar{c} \neq \bar{0}$.

The vector

$$\bar{u} = \begin{bmatrix} -1 \\ 7 \\ 1 \end{bmatrix}$$

is a *solution* to equation **6** since

$$\begin{bmatrix} 3 & 1 & 2 \\ 0 & 1 & 1 \end{bmatrix} \begin{bmatrix} -1 \\ 7 \\ 1 \end{bmatrix} = \begin{bmatrix} 3(-1) + 1 \cdot 7 + 2 \cdot 1 \\ 0(-1) + 1 \cdot 7 + 1 \cdot 1 \end{bmatrix} = \begin{bmatrix} 6 \\ 8 \end{bmatrix}$$

The vector

$$\bar{u} = \begin{bmatrix} \frac{1}{2} \\ -\frac{3}{2} \\ 0 \end{bmatrix}$$

is *not* a solution to equation **6**, for we have

$$\begin{bmatrix} 3 & 1 & 2 \\ 0 & 1 & 1 \end{bmatrix} \begin{bmatrix} \frac{1}{2} \\ -\frac{3}{2} \\ 0 \end{bmatrix} = \begin{bmatrix} 3(\frac{1}{2}) + 1(-\frac{3}{2}) + 2 \cdot 0 \\ 0(\frac{1}{2}) + 1(-\frac{3}{2}) + 1 \cdot 0 \end{bmatrix} = \begin{bmatrix} 0 \\ -\frac{3}{2} \end{bmatrix}$$

and this last vector is not the same as $\bar{c}$.

EXAMPLE 2 Further examples of products:

a $\begin{bmatrix} -1 & 0 \\ 1 & 2 \\ 6 & 1 \end{bmatrix} \begin{bmatrix} 0 \\ -3 \end{bmatrix} = \begin{bmatrix} 0 \\ -6 \\ -3 \end{bmatrix}$ **b** $[2 \quad 1 \quad -3 \quad 5] \begin{bmatrix} 1 \\ 2 \\ -1 \\ 0 \end{bmatrix} = 7$

c $\begin{bmatrix} -1 & \frac{1}{2} \\ \frac{1}{3} & -4 \end{bmatrix} \begin{bmatrix} \frac{2}{3} \\ \frac{1}{4} \end{bmatrix} = \begin{bmatrix} -\frac{13}{24} \\ -\frac{7}{9} \end{bmatrix}$ **d** $\begin{bmatrix} -1 & 0 \\ 1 & 2 \end{bmatrix} \begin{bmatrix} 1 \\ 3 \\ 1 \end{bmatrix}$ is not defined

As part **d** shows, to know that $A\bar{u}$ is defined, you need to know that the number of rows of $\bar{u}$ is the same as the number of columns of A.

EXAMPLE 3 If the matrices have many rows and columns it is easy to make errors, so be certain to calculate products carefully.

$$\begin{bmatrix} 7 & 0 & 1 & 1 & 2 & 1 \\ 3 & 1 & 0 & 1 & -1 & 0 \\ 4 & 0 & 1 & 1 & 6 & 0 \\ 1 & 1 & 2 & 1 & 0 & 3 \\ 0 & 0 & 0 & 1 & 0 & 0 \\ 0 & 0 & 0 & 0 & 0 & 0 \\ 1 & 0 & 0 & 0 & 0 & 0 \end{bmatrix} \begin{bmatrix} x_1 \\ x_2 \\ x_3 \\ x_4 \\ x_5 \\ x_6 \end{bmatrix} = \begin{bmatrix} 7x_1 + x_3 + x_4 + 2x_5 + x_6 \\ 3x_1 + x_2 + x_4 - x_5 \\ 4x_1 + x_3 + x_4 + 6x_5 \\ x_1 + x_2 + 2x_3 + x_4 + 3x_6 \\ x_4 \\ 0 \\ x_1 \end{bmatrix}$$

Note that the sixth row of the coefficient matrix consists of zeros, resulting in the zero in the sixth row of the right-hand matrix. Using an example with numbers rather than letters we have

$$\begin{bmatrix} 7 & 0 & 1 & 1 & 2 & 1 \\ 3 & 1 & 0 & 1 & -1 & 0 \\ 4 & 0 & 1 & 1 & 6 & 0 \\ 1 & 1 & 2 & 1 & 0 & 3 \\ 0 & 0 & 0 & 1 & 0 & 0 \\ 0 & 0 & 0 & 0 & 0 & 0 \\ 1 & 0 & 0 & 0 & 0 & 0 \end{bmatrix} \begin{bmatrix} -1 \\ 1 \\ 2 \\ 0 \\ 1 \\ 0 \end{bmatrix} = \begin{bmatrix} -3 \\ -3 \\ 4 \\ 4 \\ 0 \\ 0 \\ -1 \end{bmatrix}$$

DISCUSSION The following theorem summarizes two useful properties of this multiplication. Note the similarity between these results and the distributive and associative laws of ordinary multiplication.

THEOREM 2 If A is a matrix with m rows and n columns, $\bar{u}$ and $\bar{v}$ are in R^n, and a is a number, then

7 $$A(\bar{u} + \bar{v}) = A\bar{u} + A\bar{v}$$

8 $$A(a\bar{u}) = a(A\bar{u})$$

We give the proof only for the case when A has 2 rows and 2 columns. First add $\bar{u}$ and $\bar{v}$, then multiply by A:

$$\begin{bmatrix} A_{11} & A_{12} \\ A_{21} & A_{22} \end{bmatrix} \left\{ \begin{bmatrix} u_1 \\ u_2 \end{bmatrix} + \begin{bmatrix} v_1 \\ v_2 \end{bmatrix} \right\} = \begin{bmatrix} A_{11} & A_{12} \\ A_{21} & A_{22} \end{bmatrix} \begin{bmatrix} u_1 + v_1 \\ u_2 + v_2 \end{bmatrix}$$
$$= \begin{bmatrix} A_{11}(u_1 + v_1) + A_{12}(u_2 + v_2) \\ A_{21}(u_1 + v_1) + A_{22}(u_2 + v_2) \end{bmatrix}$$

Now multiply out each term and regroup to obtain:

$$\begin{bmatrix} (A_{11}u_1 + A_{12}u_2) + (A_{11}v_1 + A_{12}v_2) \\ (A_{21}u_1 + A_{22}u_2) + (A_{21}v_1 + A_{22}v_2) \end{bmatrix}$$

This is just the sum:

$$\begin{bmatrix} A_{11}u_1 + A_{12}u_2 \\ A_{21}u_1 + A_{22}u_2 \end{bmatrix} + \begin{bmatrix} A_{11}v_1 + A_{12}v_2 \\ A_{21}v_1 + A_{22}v_2 \end{bmatrix}$$

Each term can now be written as a product

$$\begin{bmatrix} A_{11} & A_{12} \\ A_{21} & A_{22} \end{bmatrix} \begin{bmatrix} u_1 \\ u_2 \end{bmatrix} + \begin{bmatrix} A_{11} & A_{12} \\ A_{21} & A_{22} \end{bmatrix} \begin{bmatrix} v_1 \\ v_2 \end{bmatrix}$$

so that, indeed,

$$A(\bar{u} + \bar{v}) = A\bar{u} + A\bar{v}$$

This proves part 7 for this case. The proof is just a consequence of the definition and the usual laws of arithmetic. The proof of part **8** is even easier and will be omitted.

In the next section we shall show how the rules of Theorem 2 allow us to deduce a number of results about matrix equations which would otherwise involve quite complicated notation. In Chapter 3 we give a general discussion of operations on vectors that satisfy rules such as those of Theorem 2.

EXAMPLE 4 The rules of Theorem 2 enable us to establish connections between the solutions to the two equations

$$A\bar{u} = \bar{c} \qquad \text{and} \qquad A\bar{u} = \bar{0}$$

For example, suppose

$$A = \begin{bmatrix} 1 & -1 & 2 \\ 3 & 1 & -1 \end{bmatrix} \qquad \text{and} \qquad \bar{c} = \begin{bmatrix} 4 \\ 8 \end{bmatrix}$$

If

$$\bar{u}_1 = \begin{bmatrix} 3 \\ -1 \\ 0 \end{bmatrix} \qquad \text{and} \qquad \bar{v} = \begin{bmatrix} -1 \\ 7 \\ 4 \end{bmatrix}$$

then

$$A\bar{u}_1 = \begin{bmatrix} 1 & -1 & 2 \\ 3 & 1 & -1 \end{bmatrix} \begin{bmatrix} 3 \\ -1 \\ 0 \end{bmatrix} = \begin{bmatrix} 4 \\ 8 \end{bmatrix} = \bar{c}$$

and

$$A\bar{v} = \begin{bmatrix} 1 & -1 & 2 \\ 3 & 1 & -1 \end{bmatrix} \begin{bmatrix} -1 \\ 7 \\ 4 \end{bmatrix} = \begin{bmatrix} 0 \\ 0 \end{bmatrix} = \bar{0}$$

In other words, $\bar{u}_1$ is a solution to $A\bar{u} = \bar{c}$ and $\bar{v}$ is a solution to $A\bar{u} = \bar{0}$.

By applying row reduction to the coefficient matrix or augmented matrix, we can determine completely the solutions to each of these equations. Instead let us see how much information we can obtain using Theorem 2.

Put $\bar{u}_2 = \bar{u}_1 + \bar{v}$, and then apply Theorem 2 to obtain

$$A\bar{u}_2 = A(\bar{u}_1 + \bar{v}) = A\bar{u}_1 + A\bar{v}$$

The latter is just $\bar{c}$, for $A\bar{u}_1 = \bar{c}$, while $A\bar{v} = \bar{0}$. In other words, $A\bar{u}_2 = \bar{c}$; that is

> The sum of a solution to $A\bar{u} = \bar{c}$ and a solution to $A\bar{v} = \bar{0}$ is a solution to $A\bar{u} = \bar{c}$.

This consequence of Theorem 2 can also be verified by direct calculation, for

$$\bar{u}_2 = \bar{u}_1 + \bar{v} = \begin{bmatrix} 3 \\ -1 \\ 0 \end{bmatrix} + \begin{bmatrix} -1 \\ 7 \\ 4 \end{bmatrix} = \begin{bmatrix} 2 \\ 6 \\ 4 \end{bmatrix}$$

and, indeed,

$$A\bar{u}_2 = \begin{bmatrix} 1 & -1 & 2 \\ 3 & 1 & -1 \end{bmatrix} \begin{bmatrix} 2 \\ 6 \\ 4 \end{bmatrix} = \begin{bmatrix} 4 \\ 8 \end{bmatrix} = \bar{c}$$

The equation $A\bar{u} = \bar{c}$ has many other solutions. For example, $2\bar{v}$ is a solution to $A\bar{u} = \bar{0}$, since

$$A(2\bar{v}) = 2A\bar{v} = 2 \cdot \bar{0} = \bar{0}$$

It follows that

$$A(\bar{u}_1 + 2\bar{v}) = A\bar{u}_1 + A(2\bar{v}) = \bar{c} + \bar{0} = \bar{c}$$

so that $\bar{u}_1 + 2\bar{v}$ is also a solution to the equation $A\bar{u} = \bar{c}$.

Similar calculations show that each of the vectors

$$\bar{u}_1 + 3\bar{v} \qquad \bar{u}_1 + 7\bar{v} \qquad \bar{u}_1 - \tfrac{3}{4}\bar{v}$$

is also a solution to $A\bar{u} = \bar{c}$. In fact, if t is any real number, then

$$A(\bar{u}_1 + t\bar{v}) = A\bar{u}_1 + A(t\bar{v}) = A\bar{u}_1 + tA\bar{v} = \bar{c} + t\bar{0} = \bar{c}$$

so that $\bar{u}_1 + t\bar{v}$ is a solution to $A\bar{u} = \bar{c}$.

Subsequent sections will make frequent use of the laws of Theorem 2 to establish further properties of solutions to systems of equations.

EXERCISES

1 Suppose $\bar{u} = \begin{bmatrix} 1 \\ 0 \\ 1 \end{bmatrix}$, $\bar{v} = \begin{bmatrix} 3 \\ 2 \\ -1 \end{bmatrix}$, and $\bar{w} = \begin{bmatrix} 0 \\ 1 \\ 1 \end{bmatrix}$. Find:

a $\bar{u} - 2\bar{v} + \bar{w}$
b $3\bar{v} - 2\bar{w}$
c A vector $\bar{x}$ such that $\bar{u} + \bar{x} = 2\bar{w}$

2 For $\bar{u} = (1,0,1)$, $\bar{v} = (3,2,-1)$, and $\bar{w} = (0,1,1)$ find
a $\bar{u} - 2\bar{v} + \bar{w}$
b $3\bar{v} - 2\bar{w}$
c A vector $\bar{x}$ such that $\bar{u} + \bar{x} = 2\bar{w}$
How do your answers compare with those of Exercise **1**?

3 Find each of the following products.

a $\begin{bmatrix} 3 & 1 & 2 \\ -1 & 1 & 1 \end{bmatrix} \begin{bmatrix} 1 \\ 2 \\ 1 \end{bmatrix}$ **b** $\begin{bmatrix} 2 & -1 & 0 & 1 \\ 1 & 0 & 1 & 1 \end{bmatrix} \begin{bmatrix} 1 \\ 2 \\ 1 \\ 4 \end{bmatrix}$

c $\begin{bmatrix} 2 & 1 & 0 \\ 1 & 1 & 1 \\ 0 & 1 & 1 \end{bmatrix} \begin{bmatrix} 1 \\ 0 \\ 1 \end{bmatrix}$ **d** $\begin{bmatrix} 0 & 0 & 0 \\ 0 & 0 & 0 \\ 0 & 0 & 0 \end{bmatrix} \begin{bmatrix} 1 \\ 2 \\ 1 \end{bmatrix}$

e $\begin{bmatrix} 1 & 0 & 0 \\ 0 & 1 & 0 \\ 0 & 0 & 1 \end{bmatrix} \begin{bmatrix} 1 \\ 3 \\ 1 \end{bmatrix}$ **f** $\begin{bmatrix} 0 & 0 & 0 & 0 & 0 \\ 1 & 1 & 1 & 1 & 1 \\ 2 & 2 & 2 & 2 & 2 \end{bmatrix} \left\{ \begin{bmatrix} 1 \\ 2 \\ 3 \\ 4 \\ 5 \end{bmatrix} + \begin{bmatrix} -1 \\ 0 \\ 1 \\ 1 \\ 1 \end{bmatrix} \right\}$

g $\begin{bmatrix} 2 & 1 & 0 & 1 & 1 \\ 1 & 0 & 1 & 0 & 2 \\ 0 & 0 & 3 & 0 & 0 \\ 4 & 1 & 2 & 1 & 0 \\ 0 & 0 & 0 & 0 & 0 \end{bmatrix} \left\{ \begin{bmatrix} 1 \\ 1 \\ 1 \\ -1 \\ 2 \end{bmatrix} + 3 \begin{bmatrix} 4 \\ 1 \\ 0 \\ 1 \\ 1 \end{bmatrix} \right\}$

4 Express each of the following systems as a matrix equation.

a $\begin{aligned} 2x - y + z &= 1 \\ x - 3y + z &= 2 \\ x + y - z &= 0 \\ 4x - y + 2z &= 1 \end{aligned}$ **b** $x - y = 0$

c $\begin{aligned} 0 \cdot x + 0 \cdot y + 0 \cdot z &= 0 \\ 0 \cdot x + 0 \cdot y + 0 \cdot z &= 0 \\ 0 \cdot x + 0 \cdot y + 0 \cdot z &= 0 \end{aligned}$

5 Suppose $A = \begin{bmatrix} 2 & 1 & 0 \\ 1 & 1 & 1 \end{bmatrix}$ and $\bar{c} = \begin{bmatrix} 3 \\ 1 \end{bmatrix}$. Which of the following vectors are solutions to the equation $A\bar{x} = \bar{c}$?

a $\begin{bmatrix} 3 \\ 1 \\ 2 \end{bmatrix}$ **b** $\begin{bmatrix} 2 \\ -1 \\ 0 \end{bmatrix} + 4 \begin{bmatrix} 1 \\ -2 \\ 1 \end{bmatrix}$ **c** $\begin{bmatrix} 2 \\ -1 \\ 0 \end{bmatrix}$ **d** $3 \begin{bmatrix} 2 \\ -1 \\ 0 \end{bmatrix}$

6 Suppose $\bar{x} = \begin{bmatrix} x_1 \\ x_2 \end{bmatrix}$ is *not* a solution to $\begin{bmatrix} 1 & 3 \\ 1 & 2 \end{bmatrix} \bar{x} = \begin{bmatrix} 0 \\ 0 \end{bmatrix}$. Is $2\bar{x}$ a solution to this equation?

7 Show that if $\bar{x}$ is the zero vector in R^3 and A has two rows and three columns, then $A\bar{x}$ is the zero vector in R^2. Generalize this result.

8 For $A = \begin{bmatrix} 2 & 1 \\ 1 & 3 \end{bmatrix}$, $\bar{u} = \begin{bmatrix} 1 \\ -1 \end{bmatrix}$, and $\bar{v} = \begin{bmatrix} 2 \\ 1 \end{bmatrix}$, calculate $A(\bar{u} + \bar{v})$ and $A(3\bar{u})$ in two ways, using Theorem **2**.

9 Suppose $A\bar{u} = \bar{0}$ and $A\bar{v} = \bar{0}$. Show that $A(\bar{u} + \bar{v}) = \bar{0}$ and that $A(a\bar{u}) = \bar{0}$ for any scalar a.

10 Suppose A has three rows and four columns. Does the equation $A\bar{u} = \bar{0}$ have a solution $\bar{u} \neq \bar{0}$?

11 Suppose A can be reduced to $\begin{bmatrix} 1 & 0 & 0 \\ 0 & 1 & 0 \\ 0 & 0 & 1 \end{bmatrix}$. Does the equation $A\bar{u} = \bar{0}$ have a solution $\bar{u} \neq \bar{0}$?

12 Express the equation $c_1(1,2) + c_2(-1,3) = (4,2)$ as a matrix equation. (Hint: First write the vectors as column matrices.)

13 Suppose the second row of A consists of zeros. What can you say about the second coordinate of $A\bar{u}$?

14 Suppose $A = \begin{bmatrix} 1 & 1 & 2 \\ 1 & -1 & 1 \end{bmatrix}$, $\bar{u}_1 = \begin{bmatrix} \frac{1}{2} \\ \frac{1}{2} \\ -3 \end{bmatrix}$, $\bar{v} = \begin{bmatrix} 3 \\ 1 \\ -2 \end{bmatrix}$, and $\bar{c} = \begin{bmatrix} -5 \\ -3 \end{bmatrix}$.

Show that $A\bar{u}_1 = \bar{c}$ and $A\bar{v} = \bar{0}$. Without further calculations with matrix entries show that each of the vectors

$$\bar{u}_1 + 2\bar{v} \qquad \bar{u}_1 - 3\bar{v} \qquad \bar{u}_1 + 4(\bar{v} - 5\bar{v})$$

is also a solution to $A\bar{u}_1 = \bar{c}$. (See Example **4**.)

15 Suppose $A = \begin{bmatrix} 1 & 4 & 1 & 2 \\ 2 & 8 & -1 & 1 \end{bmatrix}$, $\bar{u}_1 = \begin{bmatrix} -4 \\ 1 \\ 0 \\ 0 \end{bmatrix}$, and $\bar{u}_2 = \begin{bmatrix} -1 \\ 0 \\ -1 \\ 1 \end{bmatrix}$.

Show that $\bar{u}_1$ and $\bar{u}_2$ are both solutions to $A\bar{x} = \bar{0}$. Are $\bar{u}_1 + \bar{u}_2$ and $3\bar{u}_1 - 7\bar{u}_2$ also solutions to $A\bar{x} = \bar{0}$?

□ 16 The definitions and results of this section extend immediately to the complex case. Find

a $\begin{bmatrix} i & 1 - i \\ 2 & 3 + i \end{bmatrix} \begin{bmatrix} 1 + i \\ 0 \end{bmatrix}$ **b** $\begin{bmatrix} 2 - i & 0 \\ 1 & i \end{bmatrix} \left\{ i \begin{bmatrix} 1 \\ -i \end{bmatrix} + (1 + i) \begin{bmatrix} i \\ 0 \end{bmatrix} \right\}$

c The matrix form of the system

$$\begin{aligned} ix - 2iy &= i \\ (3 - i)x + (i - 1)y &= i/2 \end{aligned}$$

SECTION 6 The Theory of Matrix Equations

Let us summarize the results of the previous sections. A homogeneous system of linear equations in the n unknowns $x_1, x_2, \ldots, x_n$

1

$$\begin{aligned} A_{11}x_1 + A_{12}x_2 + \cdots + A_{1n}x_n &= 0 \\ A_{21}x_1 + A_{22}x_2 + \cdots + A_{2n}x_n &= 0 \\ \cdots\cdots\cdots\cdots\cdots\cdots\cdots\cdots\cdots \\ A_{m1}x_1 + A_{m2}x_2 + \cdots + A_{mn}x_n &= 0 \end{aligned}$$

can be expressed as a single matrix equation $A\bar{x} = \bar{0}$, where $\bar{0}$ is the column matrix with m entries, all zeros, and

$$A = \begin{bmatrix} A_{11} & A_{12} & \ldots & A_{1n} \\ A_{21} & A_{22} & \ldots & A_{2n} \\ \cdots & \cdots & \cdots & \cdots \\ A_{m1} & A_{m2} & \ldots & A_{mn} \end{bmatrix} \qquad \bar{x} = \begin{bmatrix} x_1 \\ x_2 \\ \vdots \\ x_n \end{bmatrix}$$

Note that

2
- a The number of *rows* of A is the same as the number of equations of the system. This number is also the same as the number of entries in the product $A\bar{x}$.
- b The number of *columns* of A is the same as the number of unknowns of the system. This number is also the same as the number of entries in the column matrix $\bar{x}$.

As was observed in Section 4, the set of solutions to the system **1** is a subspace of R^n. This is just a compact way of saying that

3
- a Each solution of system **1** is an n-tuple.
- b The zero n-tuple is a solution to system **1**.
- c Sums and multiples of solutions to system **1** are solutions to system **1**.

If n-tuples are expressed as column matrices, then the set of solutions to system **1** is just the set of solutions to the corresponding matrix equation $A\bar{x} = \bar{0}$. In this case the set of solutions is called the null space of A.

The **null space** of the matrix A is the set of solutions to the equation $A\bar{x} = \bar{0}$.

The null space of A is certainly a subspace of R^n where n is the number of columns of A, for it is just the set of solutions, written as column matrices, of a homogeneous linear system in n unknowns. It is worth noting at this point that a simple proof that the null space is a subspace of R^n can be given using only the fundamental distributive laws of Theorem 2 (page 46). The following paragraphs give such a proof.

Certainly if $A\bar{x} = \bar{0}$ then $\bar{x}$ is a column matrix with n entries, for otherwise the product $A\bar{x}$ would not be defined. Hence each solution is an n-tuple. Furthermore, if $\bar{0}$ is the zero matrix with n entries then the product $A\bar{0}$ must be a zero matrix, so that $\bar{0}$ belongs to the null space of A.

To complete the proof that the null space of A is a subspace, it must now be shown that sums and multiples of vectors in the null space of A *also belong to the null space of A*. To show this, suppose $\bar{u}$ and $\bar{v}$ are in the null space of A, so that

$$A\bar{u} = \bar{0} \qquad \text{and} \qquad A\bar{v} = \bar{0}$$

Using the laws of Theorem 2 (page 46) we have

$$A(\bar{u} + \bar{v}) = A\bar{u} + A\bar{v} = \bar{0} + \bar{0} = \bar{0}$$

so that the sum $\bar{u} + \bar{v}$ must also belong to the null space of A. Likewise, any multiple of a vector in the null space of A must belong to the null space of A, for if $A\bar{u} = \bar{0}$ and a is any real number then

$$A(a\bar{u}) = aA\bar{u} = a\bar{0} = \bar{0}$$

This completes the formal proof that the null space is a subspace.

The reduced form of a matrix enables us to describe the vectors in its null space. In particular the relation between the rank of A, that is, the number of nonzero rows in the reduced form of A, and the number of columns of A tells us whether the null space contains one or infinitely many vectors. This is summarized in the following matrix version of Theorem 1, which is obtained by merely translating from system to matrix language.

THEOREM 1

a If the rank of A is equal to the number of columns of A then the null space of A contains only the zero vector.

b If the rank of A is less than the number of columns of A, then the null space of A contains at least one nonzero vector (that is, a vector with at least one nonzero coordinate).

Let us turn now to an examination of the case when the null space of A contains at least two and hence infinitely many vectors. A closer look at the reduced form in this case will enable us to express the null space as the set of all sums and multiples of a *finite* number of solutions. The following terminology is useful in carrying out this task.

A vector $\bar{u}$ is a **linear combination** of the vectors $\bar{u}_1, \bar{u}_2, \ldots, \bar{u}_k$ if there are scalars $a_1, a_2, \ldots, a_k$ such that

$$\bar{u} = a_1\bar{u}_1 + a_2\bar{u}_2 + \cdots + a_k\bar{u}_k$$

The following example shows how to express the null space of a matrix as a set of all linear combinations of a finite number of vectors.

EXAMPLE 1 The system

4
$$\begin{aligned} x_1 + 2x_2 + x_3 + x_4 + x_5 &= 0 \\ 2x_1 + 4x_2 - x_3 \qquad + x_5 &= 0 \end{aligned}$$

can be rewritten as $A\bar{x} = \bar{0}$, where

$$A = \begin{bmatrix} 1 & 2 & 1 & 1 & 1 \\ 2 & 4 & -1 & 0 & 1 \end{bmatrix} \qquad \bar{x} = \begin{bmatrix} x_1 \\ x_2 \\ x_3 \\ x_4 \\ x_5 \end{bmatrix} \qquad \bar{0} = \begin{bmatrix} 0 \\ 0 \end{bmatrix}$$

The matrix A reduces to

$$R = \begin{bmatrix} 1 & 2 & 0 & \frac{1}{3} & \frac{2}{3} \\ 0 & 0 & 1 & \frac{2}{3} & \frac{1}{3} \end{bmatrix}$$

so that the null space of A consists of the set of solutions to the reduced system

5
$$\begin{aligned} x_1 + 2x_2 \qquad + \tfrac{1}{3}x_4 + \tfrac{2}{3}x_5 &= 0 \\ x_3 + \tfrac{2}{3}x_4 + \tfrac{1}{3}x_5 &= 0 \end{aligned}$$

Transfer all but the leading variables x_1 and x_3 to the right-hand side to obtain

$$\begin{aligned} x_1 &= -2x_2 - \tfrac{1}{3}x_4 - \tfrac{2}{3}x_5 \\ x_3 &= \qquad\quad -\tfrac{2}{3}x_4 - \tfrac{1}{3}x_5 \end{aligned}$$

To obtain a vector expression of this merely insert the redundant equations $x_2 = x_2$, $x_4 = x_4$, and $x_5 = x_5$, then put in matrix brackets to obtain

$$\begin{bmatrix} x_1 \\ x_2 \\ x_3 \\ x_4 \\ x_5 \end{bmatrix} = \begin{bmatrix} -2x_2 & -\frac{1}{3}x_4 & -\frac{2}{3}x_5 \\ x_2 & & \\ & -\frac{2}{3}x_4 & -\frac{1}{3}x_5 \\ & x_4 & \\ & & x_5 \end{bmatrix}$$

The right-hand side can now be expressed in the form

6 $$x_2\begin{bmatrix}-2\\1\\0\\0\\0\end{bmatrix}+x_4\begin{bmatrix}-\frac{1}{3}\\0\\-\frac{2}{3}\\1\\0\end{bmatrix}+x_5\begin{bmatrix}-\frac{2}{3}\\0\\-\frac{1}{3}\\0\\1\end{bmatrix}$$

Thus every solution to the equation $A\bar{x} = \bar{0}$ can be written in the form **6**. In other words if we put

7 $$\bar{u}_1=\begin{bmatrix}-2\\1\\0\\0\\0\end{bmatrix}\qquad \bar{u}_2=\begin{bmatrix}-\frac{1}{3}\\0\\-\frac{2}{3}\\1\\0\end{bmatrix}\qquad \bar{u}_3=\begin{bmatrix}-\frac{2}{3}\\0\\-\frac{1}{3}\\0\\1\end{bmatrix}$$

and it is known that $A\bar{x} = \bar{0}$, then $\bar{x}$, $\bar{u}_1$, $\bar{u}_2$, and $\bar{u}_3$ are related by the equation

$$\bar{x} = x_2\bar{u}_1 + x_4\bar{u}_2 + x_5\bar{u}_3$$

This shows that *every solution to equation* $A\bar{x} = \bar{0}$ *is a linear combination of the three vectors* $\bar{u}_1$, $\bar{u}_2$, and $\bar{u}_3$.

It is easy to verify that each of the vectors $\bar{u}_1$, $\bar{u}_2$, and $\bar{u}_3$ is a solution to $A\bar{x} = \bar{0}$ (merely calculate each of the products $A\bar{u}_1$, $A\bar{u}_2$, $A\bar{u}_3$ and see that each is $\bar{0}$). From this fact and the laws of Theorem **2** (page 46) it follows that any vector of the form

$$a_1\bar{u}_1 + a_2\bar{u}_2 + a_3\bar{u}_3$$

is a solution to $A\bar{x} = \bar{0}$, for

$$\begin{aligned} A(a_1\bar{u}_1 + a_2\bar{u}_2 + a_3\bar{u}_3) &= A(a_1\bar{u}_1) + A(a_2\bar{u}_2) + A(a_3\bar{u}_3)\\ &= a_1A\bar{u}_1 + a_2A\bar{u}_2 + a_3A\bar{u}_3\\ &= a_1\ \bar{0} + a_2\cdot\bar{0} + a_3\cdot\bar{0}\\ &= \bar{0}\end{aligned}$$

In summary we have shown that

8 The null space of A consists of all linear combinations of the three vectors $\bar{u}_1$, $\bar{u}_2$, and $\bar{u}_3$, (given by the formulas 7).

Note that A has five columns, its reduced form has two nonzero rows (so that rank $A = 2$), and the null space of A is the collection of all linear combinations of three vectors $(3 = 5 - 2)$. The vectors $\bar{u}_1$, $\bar{u}_2$, and $\bar{u}_3$ correspond to the three variables x_2, x_4, and x_5 which are *not* leading variables in the reduced form 5. This is shown more clearly by the following alternative description of $\bar{u}_1$, $\bar{u}_2$, and $\bar{u}_3$.

Set the first remaining variable x_2 equal to 1, and the other two, x_4 and x_5, equal to 0. Calculate x_1 and x_3 from equations 5. This gives

$$x_1 = -2 \qquad x_2 = 1 \qquad x_3 = 0 \qquad x_4 = 0 \qquad x_5 = 0$$

which is the solution $\bar{u}_1$. To obtain $\bar{u}_2$, set the second remaining variable x_4 equal to 1, and the others, x_2 and x_5, equal to 0. To obtain $\bar{u}_3$, set the third remaining variable x_5 equal to 1, and the other two, x_2 and x_4, equal to 0.

We will explore this type of description of the null space more fully in the next chapter.

EXAMPLE 2 Some simple observations will enable us to adopt the above ideas to obtain a description of the solutions to a nonhomogeneous system. For example, the system

9
$$\begin{aligned} x_1 + 2x_2 + x_3 + x_4 + x_5 &= -7 \\ 2x_1 + 4x_2 - x_3 + x_5 &= 3 \end{aligned}$$

can be written in the form $A\bar{x} = \bar{c}$, where

$$A = \begin{bmatrix} 1 & 2 & 1 & 1 & 1 \\ 2 & 4 & -1 & 0 & 1 \end{bmatrix} \qquad \text{and} \qquad \bar{c} = \begin{bmatrix} -7 \\ 3 \end{bmatrix}$$

The matrix A is the same as the one discussed in the previous example, but now the system is not homogeneous, since $\bar{c} \neq \bar{0}$. Following the pattern of Example 1, we can reduce the *augmented matrix* to reduced form and transfer the nonleading variables x_2, x_4, x_5 to the right, obtaining thereby the system

$$\begin{aligned} x_1 &= -\tfrac{4}{3} - 2x_2 - \tfrac{1}{3}x_4 - \tfrac{2}{3}x_5 \\ x_3 &= -\tfrac{17}{3} - \tfrac{2}{3}x_4 - \tfrac{1}{3}x_5 \end{aligned}$$

Now introduce the redundant equations $x_2 = x_2$, $x_4 = x_4$, $x_5 = x_5$, and rewrite in vector form

$$\begin{bmatrix} x_1 \\ x_2 \\ x_3 \\ x_4 \\ x_5 \end{bmatrix} = \bar{v} + \begin{bmatrix} -2x_2 & -\frac{1}{3}x_4 & -\frac{2}{3}x_5 \\ x_2 & & \\ & -\frac{2}{3}x_4 & -\frac{1}{3}x_5 \\ & x_4 & \\ & & x_5 \end{bmatrix}, \qquad \text{where} \qquad \bar{v} = \begin{bmatrix} -\frac{4}{3} \\ 0 \\ -\frac{17}{3} \\ 0 \\ 0 \end{bmatrix};$$

that is, if $A\bar{x} = \bar{c}$, then $\bar{x}$ and the vectors $\bar{u}_1$, $\bar{u}_2$, $\bar{u}_3$ of expressions **7** are related by the equation

10 $$\bar{x} = \bar{v} + x_2\bar{u}_1 + x_4\bar{u}_2 + x_5\bar{u}_3$$

Note the form of this expression: $\bar{x}$ is the sum of two vectors $\bar{v}$ and $\bar{w} = x_2\bar{u}_1 + x_4\bar{u}_2 + x_5\bar{u}_3$. The vector $\bar{v}$ satisfies the equation $A\bar{v} = \bar{c}$, and the vector $\bar{w}$ is in the null space of A. In other words, the general form of a solution to the equation $A\bar{x} = \bar{c}$ is the sum of a particular solution, $\bar{v}$, and some arbitrary solution, $\bar{w}$, to the homogeneous equation $A\bar{x} = \bar{0}$. The following discussion describes in more detail this important connection between homogeneous and nonhomogeneous systems.

DISCUSSION Let us use the distributive laws of Theorem **2** to describe the relation between homogeneous and nonhomogeneous equations. This discussion will be given without reference to the entries of the matrix and should provide some good practice in formal manipulations with matrix equations. Throughout this discussion A will denote a fixed matrix with m rows and n columns, and $\bar{c}$ will be a fixed nonzero column matrix with m rows.

First we prove:

11 If $A\bar{v}_1 = A\bar{v}_2 = \bar{c}$, then $\bar{v}_1 - \bar{v}_2$ is in the null space of A.

By hypothesis $A\bar{v}_1 = \bar{c}$ and $A\bar{v}_2 = \bar{c}$, so that

$$A(\bar{v}_1 - \bar{v}_2) = A\bar{v}_1 - A\bar{v}_2 = \bar{c} - \bar{c} = \bar{0}$$

and, indeed, $\bar{v}_1 - \bar{v}_2$ is in the null space of A. Hence statement **11** is proved.

On pages 47ff, it was shown that

12 If $A\bar{v} = \bar{c}$ and $\bar{u}$ is in the null space of A; then $A(\bar{v} + \bar{u}) = \bar{c}$.

The proof is so easy that we shall repeat it here. By assumption, $A\bar{v} = \bar{c}$ and $A\bar{u} = \bar{0}$. Hence

$$A(\bar{v} + \bar{u}) = A\bar{v} + A\bar{u} = \bar{c} + \bar{0} = \bar{c}$$

so that statement **12** is proved.

These combine to give the following basic result:

13 If $A\bar{x} = \bar{c}$ has a unique solution, then $A\bar{x} = \bar{0}$ has a unique solution.

To prove this suppose

$$A\bar{v} = \bar{c} \qquad \text{and} \qquad A\bar{u} = \bar{0}$$

By hypothesis, $\bar{v}$ is the *only* vector such that $A\bar{v} = \bar{c}$. According to statement **12**, we also know that

$$A(\bar{v} + \bar{u}) = \bar{c}$$

so that our hypothesis implies that $\bar{v} = \bar{v} + \bar{u}$. This means that $\bar{u} = \bar{0}$. We have therefore shown that the equation $A\bar{x} = \bar{0}$ has only the zero solution, which completes the proof of statement **13**.

The converse to statement **13** would assert that $A\bar{x} = \bar{c}$ has a unique solution whenever $A\bar{x} = \bar{0}$ has a unique solution. This is not quite true, for $A\bar{x} = \bar{c}$ may have *no* solution. The following, however, is true:

14 If $A\bar{x} = \bar{0}$ has a unique solution, then $A\bar{x} = \bar{c}$ *cannot* have two solutions.

The proof of this is left to the exercises. In Chapter 4 we shall show that if A is square (that is, $m = n$) and $A\bar{x} = \bar{0}$ has a unique solution, then for any $\bar{c}$, $A\bar{x} = \bar{c}$ has a unique solution.

EXERCISES

1 Which of the following vectors are in the null space of $A = \begin{bmatrix} 1 & 1 & 1 \\ 1 & -1 & 1 \end{bmatrix}$?

a $\begin{bmatrix} 1 \\ 1 \\ -2 \end{bmatrix}$ **b** $\begin{bmatrix} 1 \\ 0 \\ 1 \end{bmatrix}$ **c** $100\begin{bmatrix} 1 \\ 0 \\ -1 \end{bmatrix}$ **d** $\begin{bmatrix} 1 \\ 1 \\ -2 \end{bmatrix} + \begin{bmatrix} 1 \\ 0 \\ -1 \end{bmatrix}$ **e** $\begin{bmatrix} 0 \\ 0 \\ 0 \end{bmatrix}$

2 Which of the following vectors are solutions to

$$\begin{bmatrix} 1 & 1 & 1 \\ 1 & -1 & 1 \end{bmatrix} \begin{bmatrix} x \\ y \\ z \end{bmatrix} = \begin{bmatrix} 0 \\ -2 \end{bmatrix}$$

a $\begin{bmatrix} 1 \\ 1 \\ -2 \end{bmatrix}$ **b** $\begin{bmatrix} 1 \\ 0 \\ -1 \end{bmatrix}$ **c** $\begin{bmatrix} 1 \\ 1 \\ -2 \end{bmatrix} + 100 \begin{bmatrix} 1 \\ 0 \\ -1 \end{bmatrix}$

d $\begin{bmatrix} 1 \\ 1 \\ -2 \end{bmatrix} - \begin{bmatrix} 1 \\ 0 \\ -1 \end{bmatrix}$ **e** $\begin{bmatrix} 0 \\ 0 \\ 0 \end{bmatrix}$

3 Describe the null space of each of the following matrices as in Example **1**.

a $A = \begin{bmatrix} 1 & 0 & 3 \\ 2 & 1 & 1 \end{bmatrix}$ **b** $A = \begin{bmatrix} 3 & 1 & 1 & 0 \\ 6 & 2 & 2 & 1 \end{bmatrix}$

c $A = \begin{bmatrix} 1 & 1 \\ 1 & -1 \end{bmatrix}$ **d** $A = \begin{bmatrix} 2 & 1 & 0 & 6 & 1 & 1 & 2 \\ 4 & 1 & 0 & 6 & 2 & 2 & 1 \end{bmatrix}$

4 For each of the matrices A of Exercise **3** describe the general solution to $A\bar{x} = \begin{bmatrix} 1 \\ 1 \end{bmatrix}$ as in Example **2**.

5 Prove statement **14**. (Hint: Assume that $A\bar{x} = \bar{c}$ has two solutions $\bar{v}_1$ and $\bar{v}_2$ and apply statement **11**.)

6 Show that if $a_1 + a_2 = 1$ and $A\bar{v}_1 = A\bar{v}_2 = \bar{c}$ then $A(a_1\bar{v}_1 + a_2\bar{v}_2) = \bar{c}$; that is, a linear combination of solutions to a nonhomogeneous equation will be a solution if the sum of the coefficients is 1.

7 Show that if $A = \begin{bmatrix} 1 & 0 \\ 0 & 1 \\ 0 & 0 \end{bmatrix}$ then $A\bar{x} = \bar{0}$ has a unique solution. Find a vector $\bar{c}$ such that $A\bar{x} = \bar{c}$ has *no* solution. (This shows that the converse of statement **13** is not quite true.)

8 Show that if each solution to $A\bar{x} = \bar{0}$ is a linear combination of $\bar{u}_1$ and $\bar{u}_2$, and $A\bar{v} = \bar{c}$, then each solution to $A\bar{x} = \bar{c}$ is of the form $\bar{v} + \bar{w}$, where $\bar{w}$ is a linear combination of $\bar{u}_1$ and $\bar{u}_2$. (Hint: Adopt the arguments used in proving statements **13** and **14**.)

9 Without reducing decide whether the equation

$$\begin{bmatrix} 2 & 1 & 3 \\ 1 & 0 & 2 \end{bmatrix} \begin{bmatrix} x \\ y \\ z \end{bmatrix} = \begin{bmatrix} 1 \\ 1 \end{bmatrix}$$

has a unique solution.

10 Suppose A has one row and three columns, and $A \neq 0$.

a Describe geometrically the null space of A.

b Describe geometrically the solutions to $A\bar{u} = \bar{c}$.

11 Suppose A has two rows and three columns, and $A \neq 0$.

a Describe geometrically the null space of A.

b What does it mean geometrically when $A\bar{u} = \bar{c}$ has no solutions? Describe the possibilities when $A\bar{u} = \bar{c}$ has at least one solution.

12 Suppose A has three rows and three columns.

a Describe geometrically the null space of A.

b What does it mean geometrically when $A\bar{u} = \bar{c}$ has no solution?

c Suppose $A\bar{u} = \bar{c}$ has a unique solution. Describe geometrically the null space of A.

d Suppose the general solution to $A\bar{u} = \bar{c}$ is $\bar{u}_1 + a\bar{u}_2$, where $A\bar{u}_1 = \bar{c}$, $\bar{u}_2 \neq 0$, and $A\bar{u}_2 = \bar{0}$. Describe geometrically the set of solutions to $A\bar{u} = \bar{c}$.

13 Show that $2(2,1,0) - 3(1,-1,2) + (1,2,1) = (2,7,-5)$

a How many solutions does $c_1(2,1,0) + c_2(1,-1,2) + c_3(1,2,1) = (0,0,0)$ have? (Hint: Express as a homogeneous system in c_1, c_2, c_3.)

b How many solutions does $c_1(2,1,0) + c_2(1,-1,2) + c_3(1,2,1) = (2,7,-5)$ have?

☐ **14** The results of this section also extend to the complex case. Describe the null space of A as in Example **1**, where

$$A = \begin{bmatrix} 1+i & i & 0 \\ 1-i & 1 & i \end{bmatrix}$$

Describe the solutions to $A\bar{x} = \begin{bmatrix} 1 \\ i \end{bmatrix}$ as in Example **2**.

☐ **15** The results of this section can be applied to other situations in which properties analogous to Theorem **2** hold, in particular to linear differential operators. Consider the operator L, defined for twice differentiable functions by

$$Lf = f'' - f' - 2f$$

a Show that $L(f + g) = Lf + Lg$.

b Show that $L(af) = aLf$.

c Show that the solutions to $Lf = 0$ are a subspace of $C[0,1]$. (See page 39.)

d Show that e^{-x} and e^{2x} are solutions to $Lf = 0$. Must $3e^{-x} + 4e^{2x}$ also be a solution?

e Show that if f_1 is a solution to $Lf = g$, and f_2 is a solution to $Lf = 0$, then $f_1 + f_2$ is a solution to $Lf = g$.

f Show that if f_1 and f_2 are solutions to $Lf = g$, then $f_1 - f_2$ is a solution to $Lf = 0$.

g Show that if f_1 is a solution to $Lf = g$, then all solutions to this equation are of the form $f_1 + f_2$, where f_2 is a solution to $Lf = 0$.

h In Chapter 6 it will be shown that the solutions to $Lf = 0$ are precisely those functions of the form $ae^{-x} + be^{2x}$. Show that $Lx = -1 - 2x$ and deduce that the solutions to $Lf = -1 - 2x$ are the functions of the form $x + ae^{-x} + be^{2x}$.

Supplement on Vector Geometry

Vector methods provide a powerful descriptive tool in analytic geometry. In this section we show how to describe lines, line segments, and parallelograms.

EQUATIONS OF LINES

Suppose L is a line which passes through (x_0,y_0) and is parallel to the vector

$$\bar{v} = a\bar{i} + b\bar{j} \qquad \text{with } \bar{v} \neq \bar{0}$$

Let (x,y) denote an arbitrary point on L. Then the vector

1 $$\bar{u} = (x - x_0)\bar{i} + (y - y_0)\bar{j}$$

is parallel to L and hence parallel to $\bar{v}$. Therefore $\bar{u}$ is a multiple of $\bar{v}$, so there is a number t such that

2 $$\bar{u} = t\bar{v}$$

Conversely if $\bar{u}$ is of the form **1** and satisfies equation **2** then (x,y) must lie on L; thus these relations completely determine L. (See Figure 4.)

Figure 4

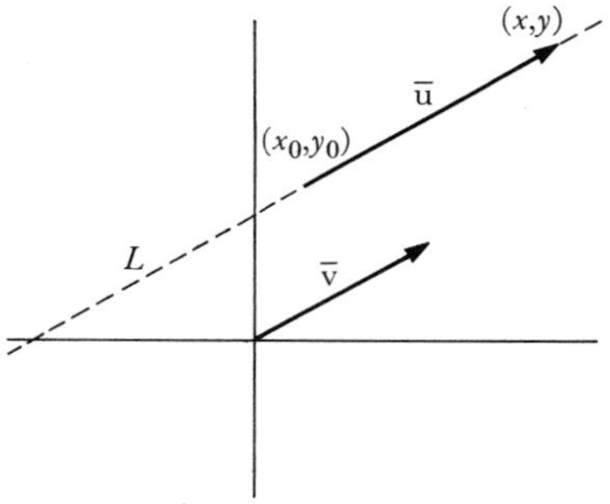

Put $\bar{u}_0 = x_0\bar{i} + y_0\bar{j}$ and $\bar{w} = x\bar{i} + y\bar{j}$ so that $\bar{u} = \bar{w} - \bar{u}_0$. We can then rewrite equation **2** as

3 $$\bar{w} = t\bar{v} + \bar{u}_0$$

Therefore the condition that the terminal point of $\bar{w}$ lies on L is the condition that $\bar{w}$ be the sum of $\bar{u}_0$ and a multiple of $\bar{v}$, as shown in Figure 5.

Figure 5

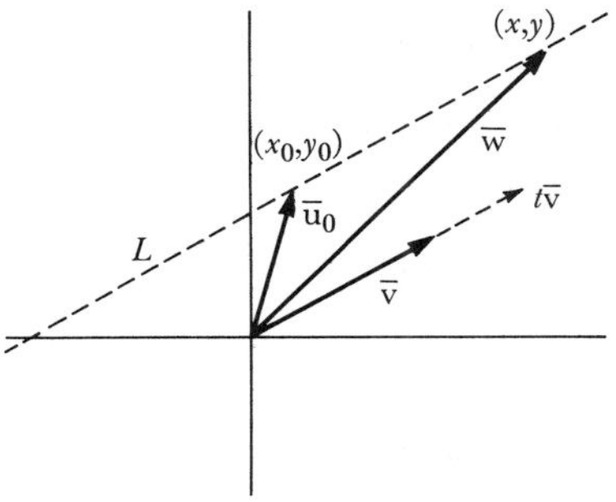

In other words, for a given $\bar{u}_0$ and $\bar{v}$, the *line through the terminal point of* $\bar{u}_0$ *which is parallel to* $\bar{v}$ *consists of the terminal points of the vectors* $t\bar{v} + \bar{u}_0$. We usually summarize this statement by saying that equation **3** is the *vector equation* of this line.

In terms of coordinates we can rewrite equation **3** as the two equations:

4
$$x = at + x_0$$
$$y = bt + y_0$$

Eliminating t we have, if $a \neq 0$ and $b \neq 0$,

5
$$\frac{x - x_0}{a} = \frac{y - y_0}{b}$$

We call the set of equations **4** a *parametric description* of L with *parameter* t, meaning that as t varies, the points (x,y) given by these equations are precisely the points of L. Equation **5** gives the familiar *Cartesian equation* of this line.

The vector description of a line in R^3 is similar to equation **3**. Suppose, for example,

$$\bar{u}_0 = x_0\bar{i} + y_0\bar{j} + z_0\bar{k}$$
$$\bar{v} = a\bar{i} + b\bar{j} + c\bar{k}$$
$$\bar{w} = x\bar{i} + y\bar{j} + z\bar{k}$$

Then the condition

6
$$\bar{w} = t\bar{v} + \bar{u}_0$$

is the same as the condition that *the terminal point of $\bar{w}$ lies on the line through the terminal point of $\bar{u}_0$ which is parallel to $\bar{v}$*. Thus equation **6** is a *vector equation* of this line. In coordinate form this becomes the parametric description with parameter t:

7
$$\begin{aligned} x &= at + x_0 \\ y &= bt + y_0 \\ z &= ct + z_0 \end{aligned}$$

Eliminating t and assuming that $a \neq 0$, $b \neq 0$, and $c \neq 0$ we have the Cartesian equations of this line:

8
$$\frac{x - x_0}{a} = \frac{y - y_0}{b} = \frac{z - z_0}{c}$$

We usually write these equations even if one or more of a, b, and c is zero, adopting the convention that this is to mean that the corresponding numerator is zero. For example,

$$\frac{x - 1}{2} = \frac{y + 3}{-4} = \frac{z - 2}{0}$$

means the line given in parametric form by

$$\begin{aligned} x &= 2t + 1 \\ y &= -4t - 3 \\ z &= 2 \end{aligned}$$

or, in other words, the line through $(1,-3,2)$ parallel to $2\bar{i} - 4\bar{j}$.

The line through the terminal points of $\bar{u}$ and $\bar{v}$ is just the line through the terminal point of $\bar{u}$ which is parallel to $\bar{v} - \bar{u}$. The vector equation of this line is

9
$$\bar{w} = t(\bar{v} - \bar{u}) + \bar{u} = t\bar{v} + (1 - t)\bar{u}$$

Thus, for example, the line through $(2,1,3)$ and $(-1,0,1)$ has vector form

$$\bar{w} = t(2\bar{i} + \bar{j} + 3\bar{k}) + (1 - t)(-\bar{i} + \bar{k})$$

Suppose $\bar{v} \neq \bar{0}$. Note that $\bar{w} = t\bar{v}$ is the equation of the line through the origin parallel to $\bar{v}$. This is just the subspace consisting of all multiples of $\bar{v}$. Thus equation **3** (or **6**) shows that a line *not* through the origin is just a translation of a line through the origin, or in other words, a translation of a one-dimensional subspace.

LINE SEGMENTS

The form **9** enables us to give a useful description of a line segment. The line through the terminal points of $\bar{u}$ and $\bar{v}$ has the vector equation

$$\bar{w} = t\bar{v} + (1 - t)\bar{u}$$

Suppose we let t vary with the restriction $0 \leq t \leq 1$. Then, as Figure 6 shows, the terminal point of $\bar{w}$ varies over the line segment from $\bar{u}$ to $\bar{v}$.

For example, if $t = 0$, then $\bar{w} = \bar{u}$; if $t = 1$, then $\bar{w} = \bar{v}$; while if $t = \frac{1}{2}$, then

$$\bar{w} = \tfrac{1}{2}\bar{v} + \tfrac{1}{2}\bar{u}$$

the terminal point of which *bisects* the line segment from $\bar{u}$ to $\bar{v}$. Also, the terminal points of $\frac{1}{3}\bar{v} + \frac{2}{3}\bar{u}$ and $\frac{2}{3}\bar{v} + \frac{1}{3}\bar{u}$ *trisect* this line segment.

In summary, the vector description of the *line segment* connecting the terminal points of $\bar{u}$ and $\bar{v}$ is

$$\bar{w} = t\bar{v} + (1 - t)\bar{u} \qquad 0 \leq t \leq 1$$

The rays R_1 and R_2 of Figure 7 have the respective descriptions

$$t\bar{v} + (1 - t)\bar{u} \qquad t \leq 0$$
$$t\bar{v} + (1 - t)\bar{u} \qquad t \geq 1$$

This vector description of line segments and rays enables us to give simple proofs of many theorems of geometry. For example, the terminal point of $\bar{w} = \frac{1}{2}\bar{v} + \frac{1}{2}\bar{u}$ bisects the line segment connecting the terminal

Figure 6

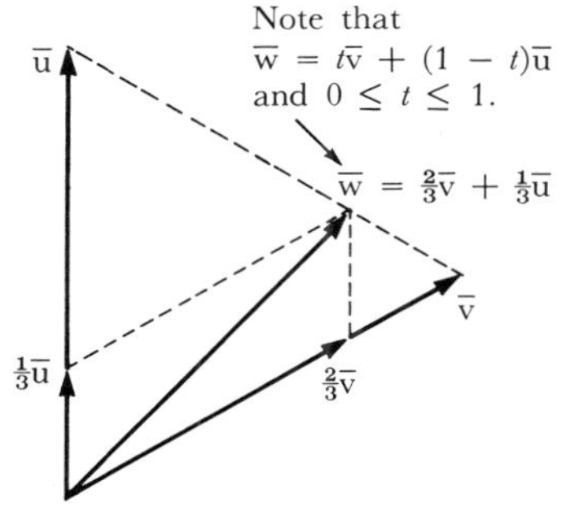

Figure 7

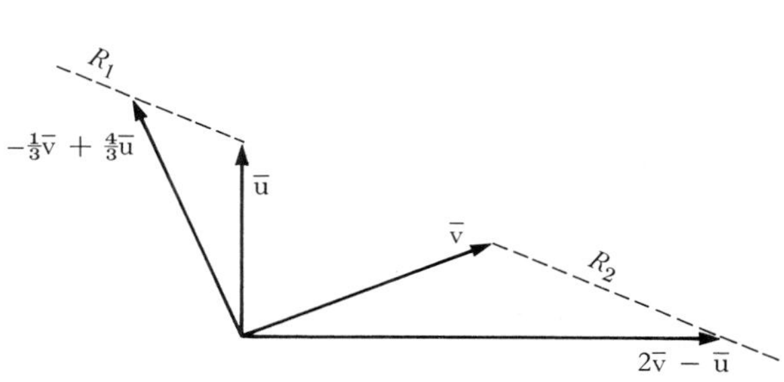

points of $\bar{u}$ and $\bar{v}$. The vector $\bar{v} + \bar{u}$ is the diagonal of the parallelogram determined by $\bar{v}$ and $\bar{u}$. (See Figure 8.) Since we also have

$$\bar{w} = \tfrac{1}{2}(\bar{v} + \bar{u})$$

we see that the terminal point of $\bar{w}$ also bisects the vector $\bar{v} + \bar{u}$. This shows that the *diagonals of a parallelogram bisect each other*.

Figure 8

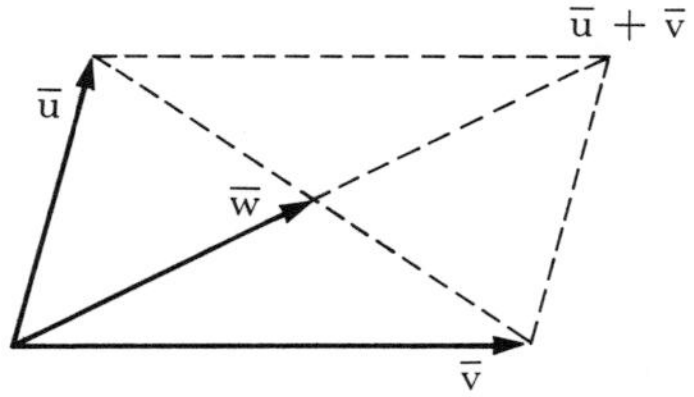

PARALLELOGRAMS AND PARALLELOPIPEDS

If $\bar{u}$ and $\bar{v}$ are not parallel, they determine a parallelogram, as shown in Figure 9.

The points inside or on this parallelogram are precisely the vectors of the form

10 $\quad s\bar{u} + t\bar{v} \quad$ where $0 \leq s \leq 1,\ 0 \leq t \leq 1$

as shown in Figure 10. Thus relation **10** gives a description of the points of this parallelogram.

By the same token, the points of the parallelopiped determined by three independent vectors $\bar{u}$, $\bar{v}$, $\bar{w}$ (see Figure 11) are just the terminal points of

11 $\quad r\bar{u} + s\bar{v} + t\bar{w} \quad$ with $0 \leq r \leq 1,\ 0 \leq s \leq 1,\ 0 \leq t \leq 1$

Figure 9

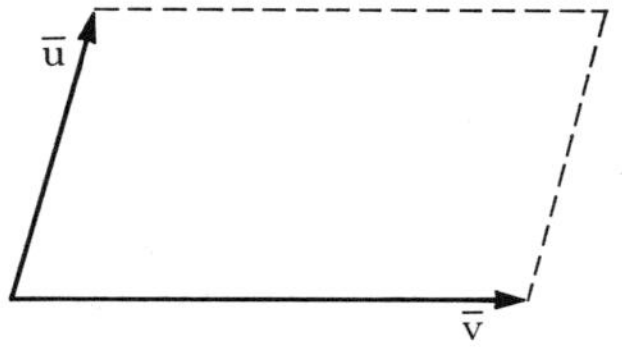

Figure 10 $\overline{w} = s\overline{u} + t\overline{v}$ *with* $0 < s < 1$. $0 < t < 1$, *while* $\overline{w}_1 = s\overline{u} + t\overline{v}$ *with* $t > 1$.

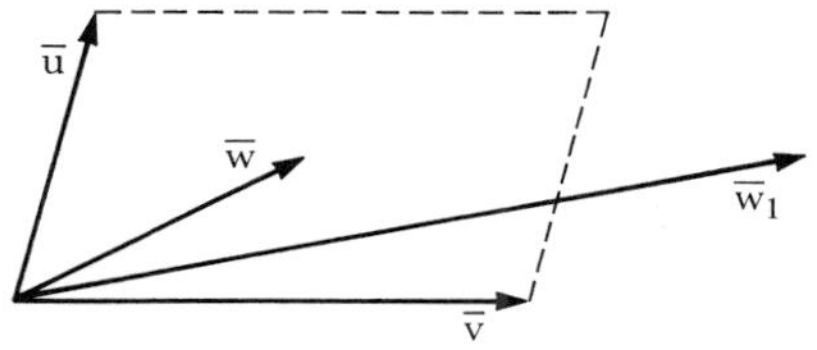

Figure 11

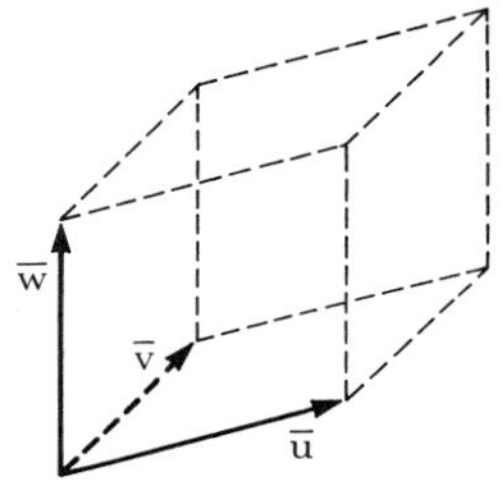

EXERCISES

Equations of Lines

1 Find a vector form for the line through (2,1) parallel to $\bar{i} - 3\bar{j}$.

2 Find a vector form for the line $2x + 5y = 7$. (Hint: Express this equation in the form **4**.)

3 Find a vector form for the line through $(1,-5)$ which is perpendicular to $2\bar{i} - \bar{j}$.

4 Find Cartesian equations for the line through (0,1,0) parallel to $\bar{i} - 2\bar{j} + \bar{k}$.

5 Find a parametric description of the line through (0,1,2) and $(-1,5,1)$.

Line Segments

6 Describe in vector form the line segment from $(3,-1,2)$ to (5,7,7).

7 What does $\overline{w} = t(1,2) + (1 - t)(3,-1)$ with $t \geq 5$ describe?

8 Describe the vectors of the form $s\overline{u} + t\overline{v}$, where $s + t = 1$.

9 Find the points that divide the segment from $(-1,2,1)$ to $(3,1,-1)$ into five equal parts.

Parallelograms and Parallelopipeds

10 Describe in vector form the points of the parallelogram determined by the four vertices (1,0,0), (0,1,0), (0,0,1), (−2,1,1).

11 What point set does

$$s\bar{u} + t\bar{v} \qquad 0 \leq s \leq 1 \qquad 0 \leq t \leq 1 \qquad s + t \leq 1$$

describe?

12 Give a vector description of the triangle with vertices (1,2,1), (2, −1,1), (1, −1,0).

13 Use vector methods to prove each of the following:

a The diagonals of a parallelogram have equal length if and only if the parallelogram is a rectangle.

b The lines from each vertex of a triangle to the midpoint of the opposite side meet in a point.

CHAPTER 2

INDEPENDENCE AND DIMENSION

The concept of vector space was introduced in Chapter 1 as a convenient framework for discussing collections of n-tuples (or column matrices) which are the solutions to a homogeneous linear system (or matrix equation). Our interest now will shift to the study of these and other vector spaces. In this chapter our goal is to obtain a suitable algebraic formulation of the concept of dimension. Full use is made of ideas drawn from systems of equations, and the elimination method will be very useful for specific calculations. The emphasis, however, will be on formal algebraic and geometric aspects of vectors and collections of vectors.

The first three sections introduce the concepts of independence and dimension. These are applied in the fourth section to obtain another description of the rank of matrix. The last two sections explore some of the algebraic relations associated with a strong geometric form of independence, that of orthogonality.

SECTION 1 Independence

A plane in R^3 through the origin can be expressed as the collection of sums of multiples of two vectors, which are not parallel, as shown, for example, in Figure 1.

Figure 1 *The vectors $\bar{w}_1$, $\bar{w}_2$, and $\bar{w}_3$ are sums of multiples of $\bar{u}$ and $\bar{v}$ and lie in the plane of $\bar{u}$ and $\bar{v}$. Any vector in the plane of $\bar{u}$ and $\bar{v}$ is a sum of multiples of $\bar{u}$ and $\bar{v}$.*

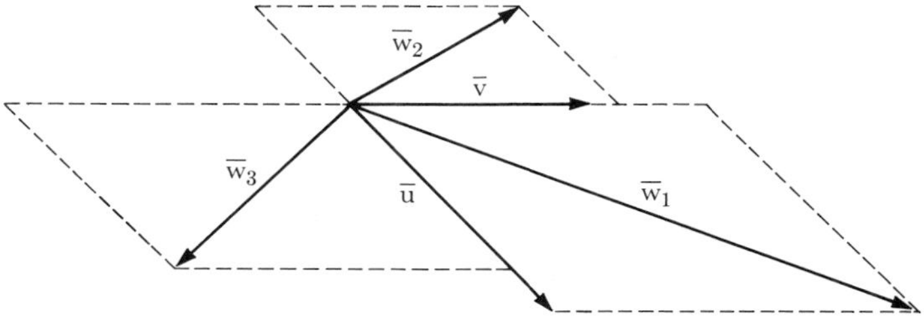

Such a plane is a subspace of R^3, for it certainly contains $\bar{0}$, which is the sum of $0 \cdot \bar{u}$ and $0 \cdot \bar{v}$, and it is closed under addition and scalar multiplication. The statement that $\bar{u}$ and $\bar{v}$ are not parallel is a geometric statement. We shall in this section give an algebraic formulation of the concept of parallelism which is also very useful in studying those vector spaces which have no convenient geometric representation. It will be seen that the simplest way to do this is to discuss the properties of equations of the form

1 $$c_1\bar{u}_1 + c_2\bar{u}_2 + \cdots + c_n\bar{u}_n = \bar{0}$$

where $\bar{u}_1, \bar{u}_2, \ldots, \bar{u}_n$ are vectors in a vector space V and $c_1, c_2, \ldots, c_n$ are numbers.

If we are given the vectors $\bar{u}_1, \bar{u}_2, \ldots, \bar{u}_n$ we can consider the $c_1, c_2, \ldots, c_n$ as variables and ask whether we can find solutions to equation **1**. Since

$$0\bar{u}_1 + 0\bar{u}_2 + \cdots + 0\bar{u}_n = \bar{0}$$

is always true, we know that

$$c_1 = c_2 = \cdots = c_n = 0$$

is always a solution to equation **1**. Thus the relevant question to ask is whether this is the *only* solution to equation **1**.

We give the following two definitions:

> A given set of vectors $\bar{u}_1, \bar{u}_2, \ldots, \bar{u}_n$ is **independent** if the only solution to $c_1\bar{u}_1 + c_2\bar{u}_2 + \cdots + c_n\bar{u}_n = \bar{0}$ is $c_1 = c_2 = \cdots = c_n = 0$.

A set which is not independent is said to be **dependent**. A useful formulation of this is:

> A given set of vectors $\bar{u}_1, \bar{u}_2, \ldots, \bar{u}_n$ is *dependent* if there are numbers $c_1, c_2, \ldots, c_n$, *not* all zero such that $c_1\bar{u}_1 + c_2\bar{u}_2 + \cdots + c_n\bar{u}_n = \bar{0}$.

In many vector spaces, rather simple methods can be applied to determine independence or dependence of a set of vectors. The following examples indicate some of these methods.

EXAMPLE 1 It will again be convenient to express vectors in R^n as column matrices, just as was done in Chapter 1. We shall show that $\begin{bmatrix} 1 \\ 0 \end{bmatrix}$ and $\begin{bmatrix} 0 \\ 1 \end{bmatrix}$ are independent in R^2. In this case, equation **1** has the form

2 $$c_1\begin{bmatrix} 1 \\ 0 \end{bmatrix} + c_2\begin{bmatrix} 0 \\ 1 \end{bmatrix} = \begin{bmatrix} 0 \\ 0 \end{bmatrix}$$

The left-hand side can be rewritten as

$$\begin{bmatrix} c_1 \\ 0 \end{bmatrix} + \begin{bmatrix} 0 \\ c_2 \end{bmatrix} = \begin{bmatrix} c_1 \\ c_2 \end{bmatrix} \quad \text{so that} \quad \begin{bmatrix} c_1 \\ c_2 \end{bmatrix} = \begin{bmatrix} 0 \\ 0 \end{bmatrix}$$

Thus the only solution to equation **2** is $c_1 = c_2 = 0$ and, from the definition of independence,

$$\begin{bmatrix} 1 \\ 0 \end{bmatrix} \quad \text{and} \quad \begin{bmatrix} 0 \\ 1 \end{bmatrix}$$

are independent.

EXAMPLE 2 Let us show that

$$\bar{u}_1 = \begin{bmatrix} 1 \\ 2 \end{bmatrix} \quad \text{and} \quad \bar{u}_2 = \begin{bmatrix} -1 \\ 1 \end{bmatrix}$$

are independent in R^2. In this case, equation **1** is $c_1\bar{u}_1 + c_2\bar{u}_2 = \bar{0}$; that is,

3 $$c_1\begin{bmatrix} 1 \\ 2 \end{bmatrix} + c_2\begin{bmatrix} -1 \\ 1 \end{bmatrix} = \begin{bmatrix} 0 \\ 0 \end{bmatrix}$$

The left-hand side can be rewritten as

$$c_1\begin{bmatrix} 1 \\ 2 \end{bmatrix} + c_2\begin{bmatrix} -1 \\ 1 \end{bmatrix} = \begin{bmatrix} c_1 - c_2 \\ 2c_1 + c_2 \end{bmatrix}$$

Note that this is just the same as

$$\begin{bmatrix} 1 & -1 \\ 2 & 1 \end{bmatrix} \begin{bmatrix} c_1 \\ c_2 \end{bmatrix}$$

Equation **3** therefore takes the form

4 $$\begin{bmatrix} 1 & -1 \\ 2 & 1 \end{bmatrix} \begin{bmatrix} c_1 \\ c_2 \end{bmatrix} = \begin{bmatrix} 0 \\ 0 \end{bmatrix}$$

Now apply our previous methods to this equation, to determine whether it has a unique solution. Since

$$\begin{bmatrix} 1 & -1 \\ 2 & 1 \end{bmatrix} \quad \text{reduces to} \quad \begin{bmatrix} 1 & 0 \\ 0 & 1 \end{bmatrix}$$

it has rank 2; hence equation **3** has the unique solution $c_1 = c_2 = 0$, so that $\bar{u}_1$ and $\bar{u}_2$ are indeed independent.

This example illustrates a standard procedure for determining independence. From the vectors $\bar{u}_1$ and $\bar{u}_2$ we created the equation **4**, where $\bar{u}_1$ and $\bar{u}_2$ are the columns of the left-hand matrix. The rank of this matrix was then determined and we were able to conclude that equation **4** had a unique solution. This told us that if

$$c_1\bar{u}_1 + c_2\bar{u}_2 = \bar{0}$$

then c_1 and c_2 must both be zero, which was exactly what was needed to show that $\bar{u}_1$ and $\bar{u}_2$ are independent.

EXAMPLE 3 The statement that $\bar{u}$ and $\bar{v}$, in R^2 or R^3, are dependent is equivalent to saying that their arrow representations are parallel. For suppose $\bar{u}$ and $\bar{v}$ are dependent; we can then apply the definition of dependence and conclude that there must be numbers c_1 and c_2 *not both zero* such that

$$c_1\bar{u} + c_2\bar{v} = \bar{0}$$

Suppose $c_1 \neq 0$; we can then solve for $\bar{u}$:

$$\bar{u} = -\frac{c_2}{c_1}\bar{v}$$

If $c_2 \neq 0$, we can solve for $\bar{v}$:

$$\bar{v} = -\frac{c_1}{c_2}\bar{u}$$

One of these two equations must hold if $\bar{u}$ and $\bar{v}$ are dependent. Either equation expresses the fact that the arrow representations of $\bar{u}$ and $\bar{v}$ are *parallel*. This shows that

5 If $\bar{u}$ and $\bar{v}$ are dependent, their arrow representations are parallel.

Conversely, suppose the arrow representations of $\bar{u}$ and $\bar{v}$ are parallel. Thus either $\bar{u}$ is a multiple of $\bar{v}$ or $\bar{v}$ is a multiple of $\bar{u}$; that is

$$\bar{u} = a\bar{v} \qquad \text{or} \qquad \bar{v} = b\bar{u} \qquad \text{(or both)}$$

These two equations can be rewritten as

$$1 \cdot \bar{u} + (-a)\bar{v} = \bar{0} \qquad \text{or} \qquad b\bar{u} + (-1)\bar{v} = \bar{0}$$

We conclude that

$$c_1\bar{u} + c_2\bar{v} = \bar{0}$$

must have a solution in which either $c_1 \neq 0$ or $c_2 \neq 0$, for it must have at least one of the two solutions

$$c_1 = 1 \quad \text{and} \quad c_2 = -a \quad \text{or} \quad c_1 = b \quad \text{and} \quad c_2 = -1$$

From the definition of dependence we must therefore have:

6 If the arrow representations of $\bar{u}$ and $\bar{v}$ are parallel, $\bar{u}$ and $\bar{v}$ are dependent.

One can restate facts **5** and **6** in terms of independence as follows:

Two vectors $\bar{u}$ and $\bar{v}$ in R^2 or R^3 are independent if and only if their arrow representations are not parallel.

It follows that a plane in R^3 through the origin can be described as the collection of sums of multiples of two *independent* vectors, for the set of linear combinations of two nonparallel vectors is such a plane.

DISCUSSION These examples indicate a method for determining independence in coordinate spaces as well as an alternative description of independence. Let us first discuss the method for determining independence indicated in Example **2**. In order not to confuse the subscripts with the notation for our vector space we shall write R^m rather than R^n.

Suppose $\bar{u}_1, \bar{u}_2, \ldots, \bar{u}_n$ are vectors in R^m, written as column matrices as in previous sections. Then equation **1**, that is,

$$c_1\bar{u}_1 + c_2\bar{u}_2 + \cdots + c_n\bar{u}_n = \bar{0}$$

can be rewritten as the matrix equation

7 $A\bar{c} = \bar{0}$, where $\bar{c} = \begin{bmatrix} c_1 \\ c_2 \\ \vdots \\ c_n \end{bmatrix}$ and the columns of A are the vectors $\bar{u}_1, \bar{u}_2, \ldots, \bar{u}_n$.

Equation **1** therefore has a unique solution (which must be $c_1 = c_2 = \cdots = c_n = 0$) if and only if the equation $A\bar{c} = \bar{0}$ has the unique solution $\bar{c} = \bar{0}$. The condition that equation **1** have a unique solution is just the statement that the vectors $\bar{u}_1, \bar{u}_2, \ldots, \bar{u}_n$ are independent. In summary, we have

THEOREM 3 The equation $A\bar{c} = \bar{0}$ has a unique solution if and only if the *columns* of A are independent vectors.

Equivalent formulations of the fact that the columns of A are independent are:

8 If A has m rows and n columns, then A can be reduced to a reduced matrix whose first n rows have nonzero entries; that is, rank $A = n$.

9 The null space of A is the zero subspace, that is, contains only $\bar{0}$.

The above are consequences of Theorem **1**, as is the fact that

10 Any set $\bar{u}_1, \bar{u}_2, \ldots, \bar{u}_n$ of vectors in R^m with $m < n$ must be dependent.

The matrix A, whose columns are $\bar{u}_1, \bar{u}_2, \ldots, \bar{u}_n$, must then have fewer rows than columns, and hence its rank is less than n.

EXAMPLE 4

a The vectors (1,2,0), (1,−1,1), (2,1,1), and (3,1,2) must be dependent (from statement **10**).

b The matrix

$$\begin{bmatrix} 2 & 1 \\ 1 & 2 \\ 3 & 0 \end{bmatrix} \quad \text{reduces to} \quad \begin{bmatrix} 1 & 0 \\ 0 & 1 \\ 0 & 0 \end{bmatrix}$$

so it has rank 2. Therefore the vectors (2,1,3), and (1,2,0) must be independent.

c The matrix

$$\begin{bmatrix} 1 & 3 & 15 \\ 2 & 1 & 10 \\ 1 & 0 & 3 \end{bmatrix}$$

reduces to a matrix with a row of zeros; therefore the columns of this matrix are dependent.

DISCUSSION In Example **3** it was noted that a plane in R^3 is the set of all linear combinations of two independent vectors. We showed in that example that if two vectors are dependent, one is a multiple of the other. (See statement **5**.) We then established the converse, that if one vector is a multiple of the other, the vectors are dependent. (See statement **6**.) This pair of results is a special case of the following (the general proof is omitted):

11 If one of the vectors $\bar{u}_1, \bar{u}_2, \ldots, \bar{u}_n$ is a linear combination of the others, these vectors are dependent. Conversely, if the vectors are dependent, at least one of them is a linear combination of the others.

The next two examples give some methods for using the idea of linear combination, while the final example gives a method for showing independence in function spaces.

EXAMPLE 5 The vectors $(1,-2)$, $(3,1)$, and $(6,4)$ must be dependent because any set of three vectors in R^2 must be dependent, from statement **10**. We shall show how to express one of them as a combination of the other two. First we find c_1, c_2, and c_3, not all zero, such that $c_1(1,-2) + c_2(3,1) + c_3(6,4) = (0,0)$. Equation **7** in this case is

$$\begin{bmatrix} 1 & 3 & 6 \\ -2 & 1 & 4 \end{bmatrix} \begin{bmatrix} c_1 \\ c_2 \\ c_3 \end{bmatrix} = \begin{bmatrix} 0 \\ 0 \end{bmatrix}$$

By reducing the coefficient matrix we see that one solution is $c_1 = 6$, $c_2 = -16$, and $c_3 = 7$, and, therefore, $6(1,-2) - 16(3,1) + 7(6,4) = (0,0)$.

Since the coefficient of $(1,-2)$ is not zero, we can solve for $(1,-2)$. We have

$$(1,-2) = \frac{16}{6}(3,1) - \frac{7}{6}(6,4)$$

which expresses $(1,-2)$ as a linear combination of $(3,1)$ and $(6,4)$.

EXAMPLE 6 In order to express a vector $\bar{u}$ as a linear combination of $\bar{u}_1, \bar{u}_2, \ldots, \bar{u}_n$ we must find a solution $c_1, c_2, \ldots, c_n$ to $\bar{u} = c_1\bar{u}_1 + c_2\bar{u}_2 + \cdots + c_n\bar{u}_n$. If the vectors are in R^m and are written as columns, the equation is just $A\bar{c} = \bar{u}$, where the columns of A are $\bar{u}_1, \bar{u}_2, \ldots, \bar{u}_n$, and

$$\bar{c} = \begin{bmatrix} c_1 \\ c_2 \\ \vdots \\ c_n \end{bmatrix}$$

For example, suppose we wish to know whether $(15,10,3)$ is a linear combination of $(1,2,1)$ and $(3,1,0)$. Writing these as columns we seek to solve the equation

$$\begin{bmatrix} 15 \\ 10 \\ 3 \end{bmatrix} = c_1 \begin{bmatrix} 1 \\ 2 \\ 1 \end{bmatrix} + c_2 \begin{bmatrix} 3 \\ 1 \\ 0 \end{bmatrix}$$

which can be rewritten as

$$\begin{bmatrix} 1 & 3 \\ 2 & 1 \\ 1 & 0 \end{bmatrix} \begin{bmatrix} c_1 \\ c_2 \end{bmatrix} = \begin{bmatrix} 15 \\ 10 \\ 3 \end{bmatrix}$$

We apply reduction to find a solution $c_1 = 3$, $c_2 = 4$ so that $(15,10,3) = 3(1,2,1) + 4(3,1,0)$.

We conclude also that the vectors (1,2,1), (3,1,0), and (15,10,3) are dependent. In fact, $3(1,2,1) + 4(3,1,0) - (15,10,3) = (0,0,0)$.

EXAMPLE 7 □ The definition of independence is the same for other vector spaces. For example, let us consider the vector space $C[0,1]$ of all functions f which are defined and continuous for $0 \leq x \leq 1$, discussed in Example **6**, pages 39ff. In this case we say that the functions $f_1, f_2, \ldots, f_k$ are *independent* if the only solution $c_1, c_2, \ldots, c_k$ to the equation

12 $$c_1 f_1 + c_2 f_2 + \cdots + c_k f_k = 0$$

is $c_1 = c_2 = \cdots = c_k = 0$.

The "=" symbol in the expression **12** means function equality, that is, "equal for all x, $0 \leq x \leq 1$." Thus to show that $f_1, f_2, \ldots, f_k$ are dependent one needs to find $c_1, c_2, \ldots, c_k$, *not all zero*, such that

$$c_1 f_1(x) + c_2 f_2(x) + \cdots + c_k f_k(x) = 0 \qquad \text{for all } x,\ 0 \leq x \leq 1$$

One method for treating independence questions for functions is to create, if possible, a system of equations in the $c_1, c_2, \ldots, c_k$, much as was done in Example **3**. For example, suppose we wish to know if the functions 1, x, and x^2 are independent in $C[0,1]$. Suppose c_1, c_2, and c_3 are numbers such that

13 $$c_1 \cdot 1 + c_2 x + c_3 x^2 = 0 \qquad 0 \leq x \leq 1$$

One way to create a set of equations in the c's is to observe that since the identity **13** is to hold for all x between 0 and 1, one can assign various values to x, each of which will give an equation in c_1, c_2, and c_3. For example, put $x = 0$, then $x = \frac{1}{2}$, then $x = 1$ in the identity **13** to obtain the three equations:

$$
\begin{array}{ll}
c_1 \cdot 1 + c_2 \cdot 0 + c_3 \cdot 0 = 0 & \qquad c_1 = 0 \\
c_1 \cdot 1 + c_2 \cdot \tfrac{1}{2} + c_3 \cdot \tfrac{1}{4} = 0 \qquad \text{that is} & \qquad c_1 + \tfrac{1}{2}c_2 + \tfrac{1}{4}c_3 = 0 \\
c_1 \cdot 1 + c_2 \cdot 1 + c_3 \cdot 1 = 0 & \qquad c_1 + c_2 + c_3 = 0
\end{array} \tag{14}
$$

This system has rank 3, so its only solution is $c_1 = c_2 = c_3 = 0$. In particular, if identity **13** holds then $c_1 = c_2 = c_3 = 0$ so that 1, x, and x^2 are indeed independent.

In this case, as in many other instances, one can use differentiation to obtain an even simpler system of equations than system **14.** If we put $x = 0$ in identity **13** then we obtain $c_1 = 0$. Now differentiate identity **13** to obtain the identity

$$c_2 + 2c_3x = 0 \qquad 0 \leq x \leq 1 \tag{15}$$

Put $x = 0$ in this to obtain $c_2 = 0$. Differentiate identity **15** to obtain $2c_3 = 0$, which certainly implies that $c_3 = 0$.

EXERCISES

1 Is a single nonzero vector independent?

2 Use the methods of Example **4** to determine which of the following sets are independent and which are dependent.
a (1,2), (2,1), (3,0) **b** (4,1,0), (2, −1,1) **c** (3,6,1), (2,1,1), (−1,0,1)
d (1,1,0,1), (1,2,1,1), (2,1,1,1), (0,1,0,1), (1,0,1,1)
e (1,1,1,1), (0,1,1,1), (0,0,1,1), (0,0,0,1)

3 Show that each of the following sets are dependent. Express one of the vectors as a linear combination of the others using the methods of Example **5**.
a (−4,3, −5), (1,3, −1), (2,1,1) **b** (1,3), (1, −1), (4,1)
c (1,2,1,1,0), (1,3, −1,0, −1), (1,0,5,3,2)

4 Use the method of Example **6** to determine whether (1,2,1) is a linear combination of (3,1,0) and (−1,1,1).

5 Show that $\bar{u}_1 = (1,-1,0)$, $\bar{u}_2 = (2,1,3)$, and $\bar{u}_3 = (4,2,6)$ are dependent. Can you express $\bar{u}_1$ as a linear combination of $\bar{u}_2$ and $\bar{u}_3$? Why does this not contradict statement **11**?

6 Determine whether (1,0,3) is a linear combination of (2,1,1), (1, −1,1), and (0,0,1) and find one such combination.

7 Show that $(x - 1)^2$ is a linear combination of 1, x, and x^2.

8 Show that the vectors (a,b) and (c,d) are dependent if and only if they are collinear with (0,0).

9 Show that three vectors in R^3 are dependent if and only if they are coplanar or collinear with (0,0,0).

□ **10** Show that $\sin x$ and $\sin 2x$ are independent.

□ **11** Show that e^x and e^{2x} are independent.

□ **12** Suppose f_1 and f_2 are solutions to the differential equation $f'' + f = 0$ and that $f_1(0) = 1$, $f_1'(0) = 0$, $f_2(0) = 0$, $f_2'(0) = 1$. Show that f_1 and f_2 are independent. Deduce from this that $\sin x$ and $\cos x$ are independent.

13 **a** Can an independent set of vectors contain the zero vector?

b Suppose $\bar{u}_1, \bar{u}_2, \bar{u}_3, \ldots, \bar{u}_n$ are independent. Are $\bar{u}_2, \bar{u}_3, \ldots, \bar{u}_n$ independent?

c Suppose $\bar{u}_1, \bar{u}_2, \ldots, \bar{u}_n$ are dependent. Are $\bar{u}_1, \bar{u}_2, \ldots, \bar{u}_n, \bar{v}_1, \bar{v}_2, \ldots, \bar{v}_k$ dependent?

14 Suppose $\bar{u}_1$, $\bar{u}_2$, and $\bar{u}_3$ are independent.

a Show that if $a \neq 0$, then $a\bar{u}_1$, $\bar{u}_2$, and $\bar{u}_3$ are independent.

b Show that $\bar{u}_1 + a\bar{u}_2, \bar{u}_2, \bar{u}_3$ are independent.

□ **15** The definitions and results of this section extend to the complex case.

a Show that $(1 + i, 2i)$ and $(i, 1 - i)$ are independent.

b Show that $(1,i)$ and $(i, -1)$ are dependent.

c Express $(3 + i, i)$ as a linear combination of $(1 + i, 2i)$ and $(i, 1 - i)$.

SECTION 2 Bases and Coordinates

In the previous section the question of a unique solution to the equation

$$c_1\bar{u}_1 + c_2\bar{u}_2 + \cdots + c_n\bar{u}_n = \bar{0}$$

was discussed. The vectors $\bar{u}_1, \bar{u}_2, \ldots, \bar{u}_n$ are independent if this equation has a unique solution (which must be $c_1 = c_2 = \cdots = c_n = 0$); otherwise they are dependent.

Suppose $\bar{u}_1, \bar{u}_2, \ldots, \bar{u}_n$ are vectors contained in a vector space V. Consider the equation

1 $$\bar{u} = c_1\bar{u}_1 + c_2\bar{u}_2 + \cdots + c_n\bar{u}_n$$

We say that $\bar{u}_1, \bar{u}_2, \ldots, \bar{u}_n$ **span** the vector space V if for *each* $\bar{u}$ in V there is *at least one* solution to equation **1**, that is, if each vector in V is a linear combination of $\bar{u}_1, \bar{u}_2, \ldots, \bar{u}_n$.

We say that $\bar{u}_1, \bar{u}_2, \ldots, \bar{u}_n$ are a **basis** for V if for *each* $\bar{u}$ in V equation **1** has a *unique* solution.

Our first observation is:

2 A basis is an independent set that spans V.

A basis is certainly independent, for, if for each $\bar{u}$ in V equation **1** has a unique solution, then, *in particular*, it has a unique solution when $\bar{u} = \bar{0}$. A basis also spans V, for, if for each $\bar{u}$ in V equation **1** has a unique solution, then such a solution certainly expresses $\bar{u}$ as a linear combination of the basis vectors $\bar{u}_1, \bar{u}_2, \ldots, \bar{u}_n$.

Suppose $\bar{u}_1, \bar{u}_2, \ldots, \bar{u}_n$ are an independent set that spans V. Let us show that this set is a basis for V. For simplicity assume $n = 2$, so that $\bar{u}_1$ and $\bar{u}_2$ are independent and span V. Suppose $\bar{u}$ is in V. Then, since $\bar{u}_1$ and $\bar{u}_2$ span V, there must be scalars c_1 and c_2 so that

3 $$\bar{u} = c_1\bar{u}_1 + c_2\bar{u}_2$$

We shall show that there is *only one pair* c_1, c_2 such that this statement holds. Suppose that, if possible, we also have

$$\bar{u} = d_1\bar{u}_1 + d_2\bar{u}_2$$

Subtract this from equation **3** and regroup as follows:

$$\begin{aligned} \bar{0} &= \bar{u} - \bar{u} \\ &= (d_1\bar{u}_1 + d_2\bar{u}_2) - (c_1\bar{u}_1 + c_2\bar{u}_2) \\ &= (d_1 - c_1)\bar{u}_1 + (d_2 - c_2)\bar{u}_2 \end{aligned}$$

By assumption $\bar{u}_1$ and $\bar{u}_2$ are independent, so we must have

$$d_1 - c_1 = 0 \qquad \text{and} \qquad d_2 - c_2 = 0$$

This forces us to conclude that $d_1 = c_1$, $d_2 = c_2$.

We have shown that if $\bar{u}_1$ and $\bar{u}_2$ are independent and span V, then for each $\bar{u}$ in V, equation **3** has a unique solution. This completes the proof of statement **2** for $n = 2$.

In the next section we shall discuss some useful properties of bases. Since, for a given basis $\bar{u}_1, \bar{u}_2, \ldots, \bar{u}_n$ and a given $\bar{u}$ equation **1** has a unique solution, it is appropriate to give a name to this solution. We say that the numbers $c_1, c_2, \ldots, c_n$ are the **coordinates** of $\bar{u}$ **relative to the basis** $\bar{u}_1, \bar{u}_2, \ldots, \bar{u}_n$ if

$$\bar{u} = c_1\bar{u}_1 + c_2\bar{u}_2 + \cdots + c_n\bar{u}_n$$

EXAMPLE 1 The vectors (1,0) and (0,1) are a basis for R^2. In this case equation **1** is, for a given (x,y),

$$(x,y) = c_1(1,0) + c_2(0,1)$$

Since $c_1(1,0) + c_2(0,1) = (c_1,c_2)$, this equation can be rewritten as

$$(x,y) = (c_1,c_2)$$

which clearly has the unique solution

$$c_1 = x,\ c_2 = y$$

Therefore $(x,y) = x(1,0) + y(0,1)$.

In other words, x and y are the coordinates of (x,y) relative to the basis $(1,0)$ and $(0,1)$. Thus the concept of coordinates relative to a basis is an extension of the usual definition of the coordinates of a point.

A similar argument shows that $(1,0,0)$, $(0,1,0)$, and $(0,0,1)$ are a basis for R^3; that $(1,0,0,0)$, $(0,1,0,0)$, $(0,0,1,0)$, and $(0,0,0,1)$ are a basis for R^4, and so forth.

Let us introduce some useful terminology. The matrix with n rows and n columns given by

$$\begin{bmatrix} 1 & 0 & 0 & \cdots & 0 \\ 0 & 1 & 0 & \cdots & 0 \\ 0 & 0 & 1 & \cdots & 0 \\ \cdots & \cdots & \cdots & \cdots & \cdots \\ 0 & 0 & 0 & \cdots & 1 \end{bmatrix}$$

is called the $n \times n$ **identity matrix**. It is usually denoted by I_n, or by I, if n is understood.

We can summarize our observations with the statement

4 The rows of the n-rowed identity matrix are a basis for R^n.

If we consider column rather than row matrices, then the columns of I_n are a basis for R^n.

Now let us describe coordinates relative to these bases.

The coordinates of (x,y,z) relative to $(1,0,0)$, $(0,1,0)$, and $(0,0,1)$ are just x, y, and z, for we have

$$(x,y,z) = x(1,0,0) + y(0,1,0) + z(0,0,1)$$

Note that the ordering chosen to denote the basis affects the coordinates. For example the coordinates of (x,y), relative to the basis $(0,1)$ and $(1,0)$, are y and x, for $(x,y) = y(0,1) + x(1,0)$.

The basis given by the identity matrix is called the **standard basis** of R^n. When it is written in the usual order (the first row of I first, the second row second, and so on), the coordinates of a vector $(x_1,x_2, \ldots, x_n)$ relative to the standard basis are $x_1, x_2, \ldots, x_n$.

When we use the word "coordinates" without reference to a basis *we shall always mean the coordinates relative to the standard basis in its usual order.*

EXAMPLE 2 We shall show that (1,2,1), (1,−1,3), and (1,1,4) are a basis for R^3, and in the process exhibit a general method for determining whether a set of vectors is a basis for R^n.

Suppose (x,y,z) is a vector in R^3. Then, if we write vectors as column matrices, equation **1** becomes

5 $$\begin{bmatrix} x \\ y \\ z \end{bmatrix} = c_1 \begin{bmatrix} 1 \\ 2 \\ 1 \end{bmatrix} + c_2 \begin{bmatrix} 1 \\ -1 \\ 3 \end{bmatrix} + c_3 \begin{bmatrix} 1 \\ 1 \\ 4 \end{bmatrix}$$

Since the right-hand side is

$$\begin{bmatrix} c_1 \\ 2c_1 \\ c_1 \end{bmatrix} + \begin{bmatrix} c_2 \\ -c_2 \\ 3c_2 \end{bmatrix} + \begin{bmatrix} c_3 \\ c_3 \\ 4c_3 \end{bmatrix} = \begin{bmatrix} c_1 + c_2 + c_3 \\ 2c_1 - c_2 + c_3 \\ c_1 + 3c_2 + 4c_3 \end{bmatrix}$$

which can be rewritten as

$$\begin{bmatrix} 1 & 1 & 1 \\ 2 & -1 & 1 \\ 1 & 3 & 4 \end{bmatrix} \begin{bmatrix} c_1 \\ c_2 \\ c_3 \end{bmatrix}$$

we see that equation **5** can be expressed as

$$\begin{bmatrix} 1 & 1 & 1 \\ 2 & -1 & 1 \\ 1 & 3 & 4 \end{bmatrix} \begin{bmatrix} c_1 \\ c_2 \\ c_3 \end{bmatrix} = \begin{bmatrix} x \\ y \\ z \end{bmatrix}$$

The augmented matrix is

$$\left[\begin{array}{ccc:c} 1 & 1 & 1 & x \\ 2 & -1 & 1 & y \\ 1 & 3 & 4 & z \end{array}\right]$$

which reduces to

$$\left[\begin{array}{ccc:c} 1 & 0 & 0 & x + \frac{1}{7}y - \frac{2}{7}z \\ 0 & 1 & 0 & x - \frac{3}{7}y - \frac{1}{7}z \\ 0 & 0 & 1 & -x + \frac{2}{7}y + \frac{3}{7}z \end{array}\right]$$

Thus, for each (x,y,z) equation **5** has the unique solution

6
$$c_1 = x + \tfrac{1}{7}y - \tfrac{2}{7}z$$
$$c_2 = x - \tfrac{3}{7}y - \tfrac{1}{7}z$$
$$c_3 = -x + \tfrac{2}{7}y + \tfrac{3}{7}z$$

We conclude that (1,2,1), (1,−1,3), and (1,1,4) are a basis for R^3. Formulas **6** give the coordinates c_1, c_2, c_3 of (x,y,z) relative to this basis. After further discussion of the theory of bases, we shall obtain in the next section a somewhat shorter method for determining whether a given set of vectors is a basis.

As was noted above, the formulas **6** give the coordinates of (x,y,z) with respect to the basis

$$\bar{u}_1 = (1,2,1) \qquad \bar{u}_2 = (1,-1,3) \qquad \bar{u}_3 = (1,1,4)$$

That is,

$$(x,y,z) = c_1\bar{u}_1 + c_2\bar{u}_2 + c_3\bar{u}_3$$

where c_1, c_2, and c_3 are calculated in terms of x, y, and z by using formulas **6**. Thus, for example, the coordinates of (6,−2,5) with respect to the standard basis

$$(1,0,0) \qquad (0,1,0) \qquad (0,0,1)$$

are 6, −2, and 5, while the coordinates of this vector relative to the basis $\bar{u}_1$, $\bar{u}_2$, $\bar{u}_3$ are c_1, c_2, c_3, given by formulas **6**:

$$c_1 = 6 + \tfrac{1}{7}(-2) - \tfrac{2}{7}(5) = \tfrac{30}{7}$$
$$c_2 = 6 - \tfrac{3}{7}(-2) - \tfrac{1}{7}(5) = \tfrac{43}{7}$$
$$c_3 = -6 + \tfrac{2}{7}(-2) + \tfrac{3}{7}(5) = -\tfrac{31}{7}$$

A thorough discussion of the nature of these formulas for finding coordinates of a vector with respect to bases other than the standard basis, will be given in Chapter 4, Section 4, and in Chapter 5.

EXAMPLE 3 A careful look at the situation in the plane will clarify further the nature of a basis and of coordinates with respect to that basis. Suppose $\bar{u}_1$ and $\bar{u}_2$ are independent vectors in R^2. As we noted in Example **3** (page 70) this implies that the arrow representations of $\bar{u}_1$ and $\bar{u}_2$ are not parallel.

Let $\bar{v}$ be a third vector. Draw a line segment parallel to $\bar{u}_1$, from the terminal point of $\bar{v}$ to the line along $\bar{u}_2$ (line segment AC of Figure 2). Also draw the segment parallel to $\bar{u}_2$ from the terminal point of $\bar{v}$ to the line along $\bar{u}_1$ (line segment AB of Figure 2).

Figure 2

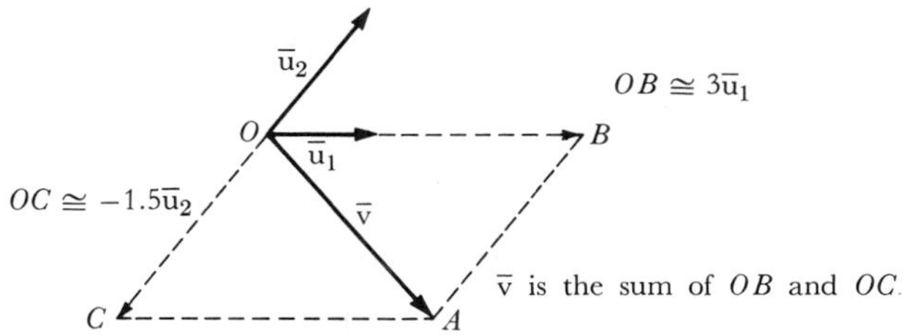

The vector from O to B is parallel to $\bar{u}_1$, and hence is a multiple of $\bar{u}_1$, say c_1; that is,

$$\text{Vector from } O \text{ to } B = c_1\bar{u}_1$$

Likewise, the segment OC is parallel to $\bar{u}_2$ so there is a number c_2 such that

$$\text{Vector from } O \text{ to } C = c_2\bar{u}_2$$

In Figure 2, $c_1 = 3$ and $c_2 = -1.5$. As Figure 2 shows, $\bar{v}$ is the sum of OB and OC; that is,

$$\bar{v} = c_1\bar{u}_1 + c_2\bar{u}_2$$

so that c_1 and c_2 are the coordinates of $\bar{v}$ relative to $\bar{u}_1$ and $\bar{u}_2$.

In Sections 5 and 6 below we shall develop simple formulas for these coordinates in the case when $\bar{u}_1$ and $\bar{u}_2$ are perpendicular. A more general discussion will be given in Section 4, Chapter 4.

EXAMPLE 4 In Example 1, page 53, we discussed a method whereby one can express the null space of a matrix as the set of linear combinations of certain vectors. This construction actually gives a *basis* for the null space of a matrix. To see that this is so let us reexamine the method. Let

$$A = \begin{bmatrix} 1 & 2 & 1 & 1 & 1 \\ 2 & 4 & -1 & 0 & 1 \end{bmatrix}$$

This reduces to

$$R = \begin{bmatrix} 1 & 2 & 0 & \frac{1}{3} & \frac{2}{3} \\ 0 & 0 & 1 & \frac{2}{3} & \frac{1}{3} \end{bmatrix}$$

After transferring the nonleading variables to the right side, the reduced system is

7
$$\begin{aligned} x_1 &= -2x_2 \quad -\tfrac{1}{3}x_4 \quad -\tfrac{2}{3}x_5 \\ x_3 &= \qquad\qquad -\tfrac{2}{3}x_4 \quad -\tfrac{1}{3}x_5 \end{aligned}$$

Then, as in Example **1**, page 53, we determine three vectors $\bar{u}_1$, $\bar{u}_2$, $\bar{u}_3$. The vector $\bar{u}_1$ is obtained by putting $x_2 = 1$, $x_4 = x_5 = 0$, and then calculating x_1 and x_3 from system **7**. The vector $\bar{u}_2$ is obtained by putting $x_2 = 0$, $x_4 = 1$, $x_5 = 0$ and calculating x_1 and x_3 from system **7**. The vector $\bar{u}_3$ is obtained by putting $x_2 = x_4 = 0$, $x_5 = 1$, and calculating x_1 and x_3 from system **7**. This gives

8
$$\bar{u}_1 = \begin{bmatrix} -2 \\ 1 \\ 0 \\ 0 \\ 0 \end{bmatrix} \qquad \bar{u}_2 = \begin{bmatrix} -\frac{1}{3} \\ 0 \\ -\frac{2}{3} \\ 1 \\ 0 \end{bmatrix} \qquad \bar{u}_3 = \begin{bmatrix} -\frac{2}{3} \\ 0 \\ -\frac{1}{3} \\ 0 \\ 1 \end{bmatrix}$$

A vector $\bar{x}$ is in the null space of A if and only if it is a linear combination of $\bar{u}_1$, $\bar{u}_2$, and $\bar{u}_3$, that is, there are numbers c_1, c_2, c_3 such that

$$\bar{x} = c_1\bar{u}_1 + c_2\bar{u}_2 + c_3\bar{u}_3$$

as noted in statement **8**, page 55.

This shows that the constructed vectors $\bar{u}_1$, $\bar{u}_2$, and $\bar{u}_3$ *span* the null space of A. They are also independent, which is a consequence of the fact that each of $\bar{u}_1$, $\bar{u}_2$, and $\bar{u}_3$ is determined by the condition that system **7** hold and that one of x_2, x_4, x_5 is 1, the others being 0. A formal proof of independence goes as follows:

Suppose

9
$$c_1\bar{u}_2 + c_2\bar{u}_2 + c_3\bar{u}_3 = \bar{0}$$

Substitute for $\bar{u}_1$, $\bar{u}_2$, and $\bar{u}_3$ from the expressions **8**, and then express as a single vector:

$$c_1\begin{bmatrix} -2 \\ 1 \\ 0 \\ 0 \\ 0 \end{bmatrix} + c_2\begin{bmatrix} -\frac{1}{3} \\ 0 \\ -\frac{2}{3} \\ 1 \\ 0 \end{bmatrix} + c_3\begin{bmatrix} -\frac{2}{3} \\ 0 \\ -\frac{1}{3} \\ 0 \\ 1 \end{bmatrix} = \begin{bmatrix} -2c_1 & -\frac{1}{3}c_2 & -\frac{2}{3}c_3 \\ c_1 & & \\ & -\frac{2}{3}c_2 & -\frac{1}{3}c_3 \\ & c_2 & \\ & & c_3 \end{bmatrix}$$

The condition that this be the zero vector then means that all its coordinates are zero. In particular its second, fourth, and fifth coordinates are zero; that is $c_1 = c_2 = c_3 = 0$. This shows that the only solution c_1, c_2, c_3 to equation **9** is the trivial solution $c_1 = c_2 = c_3 = 0$, so that $\bar{u}_1$, $\bar{u}_2$, and $\bar{u}_3$ are independent.

We have now shown that the constructed vectors $\bar{u}_1$, $\bar{u}_2$, and $\bar{u}_3$ are independent and span the null space of A; hence, from statement **2**, they must form a basis for A. Note that the number of columns of A (which is five) is the same as the sum of the rank of A (which is 2, since its reduced matrix R has two nonzero rows) and the number of vectors in a basis for the null space of A. This relationship will be discussed further in the next two sections.

EXAMPLE 5 Except for some optional examples and exercises we have thus far confined our discussion to vector spaces of n-tuples, that is, collections of row matrices or collections of column matrices. At this point we introduce another example of a vector space as it will help clarify the notions of independence and coordinates relative to a basis. Let V denote the collection of all real polynomial functions of degree not exceeding 3; that is, a function f is a member of V if there are real numbers c_0, c_1, c_2, and c_3 such that

10 $$f(x) = c_0 + c_1x + c_2x^2 + c_3x^3 \qquad \text{for all real numbers } x$$

This collection V has properties analogous to vector spaces of n-tuples and will be called a *vector space of polynomial functions.* V contains the zero function, since

$$0 = 0 + 0x + 0x^2 + 0x^3$$

Furthermore if f and g are polynomials of degree not exceeding 3, so is their sum $f + g$ and any scalar multiple of either, tf or tg.

Let us show that 1, x, x^2, and x^3 form a basis for V.

By the definition of V any f in V has the form **10**; that is, the polynomials 1, x, x^2, and x^3 *span* V. Suppose, if possible, a given polynomial f had two expressions of the form **10**; that is:

11 $$f(x) = c_0 + c_1x + c_2x^2 + c_3x^3 \qquad \text{for all } x$$

and

12 $$f(x) = d_0 + d_1x + d_2x^2 + d_3x^3 \qquad \text{for all } x$$

Subtracting the latter from the former gives

13 $$0 = (c_0 - d_0) + (c_1 - d_1)x + (c_2 - d_2)x^2 + (c_3 - d_3)x^3$$

If the two expressions **11** and **12** are to be distinct, at least one of the coefficients in the expression **13** must be nonzero. We shall show that this cannot happen by proving the following:

14 The polynomials 1, x, x^2, and x^3 are independent; that is, if $a_0 + a_1x + a_2x^2 + a_3x^3 = 0$ for all x then $a_0 = a_1 = a_2 = a_3 = 0$.

Suppose that for all x

$$a_0 + a_1x + a_2x^2 + a_3x^3 = 0$$

We can create a system of equations in a_0, a_1, a_2, a_3 by assigning various values to x. For example, setting $x = 0, 1, 2,$ and 3 gives the four equations

15

$$\begin{aligned} a_0 \qquad\qquad\qquad\qquad &= 0 \\ a_0 + a_1 + a_2 + a_3 &= 0 \\ a_0 + 2a_1 + 4a_2 + 8a_3 &= 0 \\ a_0 + 3a_1 + 9a_2 + 27a_3 &= 0 \end{aligned}$$

Our usual reduction method shows that the only solution to this is

$$a_0 = a_1 = a_2 = a_3 = 0$$

In summary, if $a_0 + a_1x + a_2x^2 + a_3x^3 = 0$ is to hold for all x, then (a_0,a_1,a_2,a_3) must be a solution to system **15** and hence $a_0 = a_1 = a_2 = a_3 = 0$. This proves statement **14**. Note that we established a result about vector spaces of polynomials by converting it to a question about systems of equations.

We have shown that the polynomials 1, x, x^2, and x^3 span V and are independent; hence they must form a basis for V. Note that the coordinates c_0, c_1, c_2, c_3 of a polynomial f with respect to the basis 1, x, x^2, x^3 are just the coefficients corresponding to these powers of x. This is just another way of saying that a polynomial of degree 3 or less has a unique expression in the form

$$c_0 + c_1x + c_2x^2 + c_3x^3$$

EXERCISES

1 Show that each of the following sets of vectors is a basis for R^3 by using the methods of Example **2**.
  - **a** (3,6,1), (2,1,1), (−1,0,1)
  - **b** (1,1,1), (0,1,1), (0,0,1)
  - **c** (3,7,0), (0,2,4), (5,0,−5)

2 Show that each of the following sets of vectors is not a basis for R^3.
  - **a** (−4,3,−5), (1,3,−1), (2,1,1)
  - **b** (2,1,0), (1,2,−1), (6,7,12), (300,703,427)

3 Show that each of the following sets of vectors is a basis for R^4.

a $(1,1,1,-1)$, $(1,1,-1,1)$, $(1,-1,1,1)$, $(-1,1,1,1)$

b $(1,2,3,4)$, $(0,1,2,3)$, $(0,0,1,2)$, $(0,0,0,1)$

4 Using the methods of Example **4** find a basis for the null space of each of the following matrices.

a $\begin{bmatrix} 1 & 0 & 3 \\ 2 & 1 & 1 \end{bmatrix}$ **b** $\begin{bmatrix} 3 & 1 & 1 & 0 \\ 6 & 2 & 2 & 1 \end{bmatrix}$ **c** $\begin{bmatrix} 2 & 1 & 0 & 6 & 1 & 1 & 2 \\ 4 & 1 & 0 & 6 & 2 & 2 & 1 \end{bmatrix}$

5 Find the coordinates of (x,y,z) relative to each of the bases of Exercise **1**.

6 Find the coordinates of (x,y,z,w) relative to each of the bases of Exercise **3**.

7 Find a basis for the set of vectors of the form $(0,y,z)$.

8 Show that (a,b) and (c,d) form a basis for R^2 if and only if $ad - bc \neq 0$.

9 **a** Show that $1, x, x^2$ are independent. (See Example **5**.)

b Show that $1, x, x^2, x^3, x^4$ are a basis for the vector space of polynomials of degree 4 or less.

c Show that for any n, the polynomials $1, x, x^2, \ldots, x^n$ are independent.

10 Suppose V is the vector space of polynomials of degree ≤ 1.

a Show that $1 + x$ and $1 - x$ are a basis for V.

b Calculate the coordinates of $a + bx$ relative to the basis $1 + x$, $1 - x$.

☐ 11 Show that no matter how large n is, the set $1, x, x^2, \ldots, x^n$ is *not* a basis for $C[0,1]$, the space defined in Example **6**; page 39. (Hint: Show that x^{n+1} is not a linear combination of $1, x, x^2, \ldots, x^n$.)

☐ 12 **a** Show that if f_1 and f_2 are solutions to $f'' + af = 0$ such that $f_1(0)f_2'(0) - f_1'(0)f_2(0) \neq 0$, f_1 and f_2 are then independent.

b It can be shown that if f_1 and f_2 satisfy the conditions of part **a**, they form a basis for the space of solutions to $f'' + af = 0$. Deduce from this that every solution to $f'' + f = 0$ can be uniquely expressed as $c_1 \sin x + c_2 \cos x$.

c Show that every solution to $f'' - f = 0$ can be uniquely expressed as $c_1 e^x + c_2 e^{-x}$.

☐ 13 The results of this section extend immediately to the complex case. Show that each of the following sets of vectors is a basis for C^2.

a $(1,0)$, $(0,1)$

b $(i,1)$, $(-1,-i)$

c $(2 + i, 3)$, $(1 - i, 1 - 2i)$

SECTION 3 Dimension

We present two theorems about bases in this section. The proofs of these results are somewhat technical and can be found in the Appendix.

Theorem **4** tells us that any two bases for a finite-dimensional vector space must contain the same number of vectors.

THEOREM 4 Suppose $\bar{u}_1, \bar{u}_2, \ldots, \bar{u}_n$ are a basis for V. Any other basis for V must contain exactly n vectors.

A vector space V that has a basis consisting of a finite number of vectors is said to be **finite-dimensional**. The number of vectors in such a basis is called the **dimension** of V.

EXAMPLE 1 Since the rows of the n-rowed identity matrix are a basis for R^n, we see that R^n has dimension n. (See Example **1**, page 77.)

A line through the origin in R^2 or R^3 consists of all multiples of a single nonzero vector; therefore, such a line has dimension *one*. A plane in R^3 through (0,0,0) consists of all linear combinations of two independent vectors and hence is *two*-dimensional. Thus, our algebraic concept of dimension coincides with our geometric intuition (as it should).

EXAMPLE 2 To determine the dimension of the null space of a matrix one can use the method of Example **4**, page 81, to find a basis for this null space. The dimension will then be the number of vectors in this basis. This dimension is easily calculated from knowledge of the rank and number of columns of the matrix, for, as was noted in Example **4**, page 81,

$$\text{Dimension of null space of } A = \text{number of columns of } A - \text{rank of } A$$

Thus, for example,

$$A = \begin{bmatrix} 2 & -1 & 1 \\ -7 & \frac{7}{2} & -\frac{7}{2} \\ 4 & 1 & -2 \end{bmatrix}$$

has rank 2, and three columns; hence its null space has dimension 1.

EXAMPLE 3 One consequence of Theorem **4** is that if the dimension of V is n, then no set containing either *more* or *less* than n vectors can be a basis for V. We have seen in statement **10**, page 72, that *no* set containing more than n vectors can be independent and hence such a set cannot be a basis for

R^n. We now also know that *no* set containing less than n vectors can be a basis for R^n.

Consider the matrix

$$A = \begin{bmatrix} 2 & 1 \\ 1 & 1 \\ 0 & 1 \end{bmatrix}$$

The columns of A are independent. Since A has only two columns they cannot be a basis for R^3. We conclude that the columns of A *cannot* span R^3 (for, otherwise, they would be an independent set that spans R^3 and hence a basis for R^3). Since the columns of A do not span R^3, there must be a vector $\bar{c}$ in R^3 such that

$$\bar{c} = x_1 \begin{bmatrix} 2 \\ 1 \\ 0 \end{bmatrix} + x_2 \begin{bmatrix} 1 \\ 1 \\ 1 \end{bmatrix}$$

has *no* solution x_1, x_2. In other words, there must be a vector $\bar{c}$ in R^3 such that

$$A \begin{bmatrix} x_1 \\ x_2 \end{bmatrix} = \bar{c}$$

has no solution.

A general statement of this result would be:

1 If A has more rows than columns, there is a vector $\bar{c}$ such that $A\bar{x} = \bar{c}$ has no solution.

A proof of this is outlined in the exercises.

DISCUSSION Our second result about bases will enable us to deduce spanning from independence or independence from spanning when the number of vectors equals the dimension.

THEOREM 5 Suppose V has dimension n.

a If $\bar{u}_1, \bar{u}_2 \ldots, \bar{u}_n$ are independent, they span V.

b If $\bar{u}_1, \bar{u}_2, \ldots, \bar{u}_n$ span V, they are independent.

The proof will be found in the Appendix. A fundamental consequence of this result for matrix equations is given in

THEOREM 6 Suppose A is a square matrix with n rows and n columns.

a If the columns of A are independent, they span R^n.
b If the columns of A span R^n, they are independent.
In particular,
c If $A\bar{x} = \bar{0}$ has a unique solution, then *for any* $\bar{c}$ in R^n, $A\bar{x} = \bar{c}$ has a unique solution.

PROOF: Parts **a** and **b** are just restatements of Theorem **5a** and **b**. We shall prove part **c**.

If $A\bar{x} = \bar{0}$ has a unique solution, Theorem **3**, page 71, then tells us that the columns of A are independent. Part **a** of this theorem tells us that the columns of A must span R^n. Denote the columns of A by $\bar{u}_1, \bar{u}_2, \ldots, \bar{u}_n$. Since these span R^n and are independent we know that they are a basis for R^n. Hence, for *any* $\bar{c}$ in R^n, the equation

2 $$\bar{c} = x_1\bar{u}_1 + x_2\bar{u}_2 + \cdots + x_n\bar{u}_n$$

has a *unique* solution $x_1, x_2, \ldots, x_n$. This equation can be rewritten as

3 $$A\bar{x} = \bar{c} \qquad \text{where} \qquad \bar{x} = \begin{bmatrix} x_1 \\ x_2 \\ \vdots \\ x_n \end{bmatrix}$$

Since equation **2** has a unique solution we are forced to conclude that equation **3** has a unique solution. This proves part **c** and completes the proof of Theorem **6**.

EXAMPLE 4 Consider the matrix

$$A = \begin{bmatrix} 1 & 1 & 1 \\ 2 & -1 & 1 \\ 1 & 3 & 4 \end{bmatrix}$$

It has rank 3 and three columns; therefore, the columns of A are independent. Theorem **6** shows that (1,2,1), (1,−1,3), (1,1,4) are a basis for R^3.

We know, therefore, that for any vector (c_1,c_2,c_3) the equation

4 $$A\begin{bmatrix} x \\ y \\ z \end{bmatrix} = \begin{bmatrix} c_1 \\ c_2 \\ c_3 \end{bmatrix}$$

has a unique solution. In Example **2** of the previous section we also deduced this by reducing the augmented matrix of equation **4**. As a result of Theorem **6** we are able to deduce *existence* and uniqueness for equation **4** merely by reducing the *coefficient* matrix.

EXAMPLE 5 *Subspaces* of R^2 *and* R^3.

Several other results will be established in the Appendix. Among them are the following facts:

If V has a spanning set containing n vectors, then

a V has a basis containing no more than n vectors.

5 **b** No independent set in V can contain more than n vectors.

c Any nonzero subspace of V has a basis containing no more than n vectors.

Let us use this result to classify the subspaces of R^3. First, there is the subspace consisting of the zero vector only. This is called the *zero* subspace and is said to have dimension *zero*.

A subspace of R^3 which is not the zero subspace, must, because of statement **5c**, be of dimension 1, 2, or 3, since R^3 is spanned by (1,0,0), (0,1,0), and (0,0,1). A one-dimensional subspace has a basis consisting of one independent vector. A single vector is independent if and only if it is nonzero. Thus, a one-dimensional subspace must consist of all multiples of a single nonzero vector. In summary

> The one-dimensional subspaces of R^3 are precisely the lines through the origin.

The two-dimensional subspaces have bases containing two vectors and hence must consist of all linear combinations of *two* independent vectors.

In other words,

> The two-dimensional subspaces of R^3 are precisely the planes through the origin.

A three-dimensional subspace has a basis consisting of three vectors. Since these vectors must be independent, they must (from Theorem **5**) span R^3. Thus

> R^3 has only one three-dimensional subspace, namely, R^3 itself.

In summary

> The nonzero subspaces of R^3 which are *not all* of R^3 are just the lines and planes through (0,0,0).

EXAMPLE 6 Theorem **5** is frequently useful, for it tells us when an independent set also spans. In this example we show how it can be used to obtain an important polynomial formula.

Consider the vector space V of polynomials of degree not exceeding 2. The polynomials 1, x, and x^2 are independent. (See Exercise **9**, page 85.) They also span V, since, by definition, a polynomial of degree not exceeding 2 is a linear combination of 1, x, and x^2. We conclude that V is *three*-dimensional.

Let us construct another basis for V, which is of interest. Put

$$f_1 = \frac{(x-1)(x-2)}{2} \qquad f_2 = \frac{x(x-2)}{-1} \qquad f_3 = \frac{x(x-1)}{2}$$

Note that

6
$$\begin{aligned} f_1(1) &= f_1(2) = 0 & f_1(0) &= 1 \\ f_2(0) &= f_2(2) = 0 & f_2(1) &= 1 \\ f_3(0) &= f_3(1) = 0 & f_3(2) &= 1 \end{aligned}$$

The polynomials f_1, f_2, and f_3 each have degree 2 and thus belong to V. We shall use the relations **6** to show that they are independent.

Suppose that for all x

7
$$c_1f_1(x) + c_2f_2(x) + c_3f_3(x) = 0$$

Substitute $x = 0$ and use relations **6**. The result is

$$f_1(0) = 1 \qquad f_2(0) = 0 \qquad f_3(0) = 0$$

Thus we must have $c_1 = 0$. Substituting $x = 1$ and using relations **6** give $c_2 = 0$, and a similar argument with $x = 2$ shows that $c_3 = 0$. We conclude that identity **7** necessarily entails that $c_1 = c_2 = c_3 = 0$. Consequently, f_1, f_2, and f_3 are independent.

Since V has dimension 3 Theorem **5** says that f_1, f_2, and f_3 are a basis for V. In particular if f is a polynomial of degree not exceeding 2, then we can find c_1, c_2, and c_3 such that *for all* x

8 $$f(x) = c_1 f_1(x) + c_2 f_2(x) + c_3 f_3(x)$$

Setting $x = 0$ and using relations **6** we have $c_1 = f(0)$. Repeating this process with $x = 1$ and $x = 2$ gives

$$c_2 = f(1) \qquad c_3 = f(2)$$

Equation **8** can then be rewritten as

$$f(x) = f(0)f_1(x) + f(1)f_2(x) + f(2)f_3(x)$$

This formula is known as the **Lagrange interpolation formula** and expresses a polynomial of degree 2 or less in terms of its values at 0, 1, and 2, relative to a basis consisting of polynomials of degree exactly 2.

EXERCISES

Additional exercises concerning bases and independence can be found in the Appendix.

1 Use the method of Examples **3** and **4** to decide which of the following sets are bases for R^3.

a $(3,6,1), (2,1,1), (-1,0,1)$
b $(1,1,1), (1,1,0)$
c $(-4,3,-5), (1,3,-1), (2,1,1)$
d $(3,7,0), (0,2,4), (5,0,-5)$

2 Show that there are vectors such that

$$\begin{bmatrix} 2 & 1 & 4 \\ 1 & 2 & 1 \\ 1 & 1 & 0 \\ 2 & 1 & 1 \end{bmatrix} \bar{x} = \bar{c}$$

has *no* solution. Show also that the columns of the coefficient matrix are independent.

3 Find the dimension of the null space of

$$A = \begin{bmatrix} 3 & 1 & 1 & 0 \\ 6 & 2 & 2 & 1 \end{bmatrix}$$

Note that this corresponds to the number of variables which are not leading variables in the reduced form.

4 Find the dimension of the null space of (see Exercise **3**)

$$\begin{bmatrix} 2 & 1 & 0 & 6 & 1 & 1 & 2 \\ 4 & 1 & 0 & 6 & 2 & 2 & 1 \end{bmatrix}$$

5 Show that the columns of

$$A = \begin{bmatrix} 3 & 1 & 2 \\ 2 & 6 & 4 \\ 2 & 1 & 3 \end{bmatrix}$$

span R^3. Do the rows of A also span R^3?

6 Show that if $ad - bc \neq 0$, then for any α and β the system

$$\begin{aligned} ax + by &= \alpha \\ cx + dy &= \beta \end{aligned}$$

has a unique solution.

7 What is the dimension of the subspace of vectors of the form $(0,y,z)$?

8 Classify the subspaces of R^2. The subspaces of R^4.

9 For which of the following matrices A can the equation $A\bar{x} = \bar{c}$ be solved, for any given $\bar{c}$ of appropriate size?

a $A = \begin{bmatrix} 2 & 1 \\ 1 & 2 \end{bmatrix}$ **b** $A = \begin{bmatrix} 2 & 1 & 1 \\ 1 & 2 & 1 \end{bmatrix}$

c $A = \begin{bmatrix} 2 & 1 \\ 1 & 2 \\ 1 & 1 \end{bmatrix}$ **d** $A = \begin{bmatrix} 2 & 1 & 3 \\ 1 & 2 & 3 \\ 1 & 1 & 2 \end{bmatrix}$ **e** $A = \begin{bmatrix} 2 & 1 & 3 & 4 \\ 1 & 2 & 3 & 1 \\ 1 & 1 & 2 & 0 \end{bmatrix}$

10 Suppose V is the space of polynomials of degree not exceeding 3. (See Example **6**.)

a What is the dimension of V?

b Find a basis for V such that the coordinates of f relative to this basis are $f(0), f(1), f(2), f(3)$.

c Find a basis for V such that the coordinates of f relative to this basis are $f(-2), f(-1), f(1), f(2)$.

□ 11 Show that $C[0,1]$ is not finite-dimensional. ($C[0,1]$ was defined in Example **6**, page 90. First show that for any n, the set $1, x, x^2, \ldots, x^n$ is independent, then apply the result **5b**.)

12 Prove statement **1**. (Hint: Adjoin additional columns of zeros to obtain a square matrix B. Then apply Theorem **6b**.)

□ 13 Suppose A has n rows and n columns and has the property that no matter what $\bar{c}$ is chosen in R^n, the equation $A\bar{x} = \bar{c}$ has a solution. Show that the equation $A\bar{x} = \bar{0}$ must have a unique solution. (Hint: Apply Theorem **6b**.)

□ 14 The results of this section are also true for complex vector spaces. Decide which of the following are bases for C^2.

a $(i,1)$ **b** $(i,1), (1,-i)$

c $(i,1), (1,i)$ **d** $(2i + 1, 1), (1 - 3i, 0)$

SECTION 4 Rank

The rank of a matrix was defined in Chapter 1 as the number of nonzero rows in its reduced form. In this section we discuss three important theorems about rank. Each of these theorems states that the rank is given as the dimension of a vector space associated with the matrix.

In Example 4, page 81, a method was given for constructing a basis for the null space of a matrix A. First a reduced matrix R was obtained, and then, corresponding to each variable which is *not* the leading variable of any equation of the reduced system, we constructed a basis vector. (In that example, $\bar{u}_1$, $\bar{u}_2$, $\bar{u}_3$ were constructed from the nonleading variables x_2, x_4, and x_5.) The dimension of the null space of A is usually called the *nullity* of A. Thus

1 Nullity of A = number of nonleading variables of reduced system

The number of leading variables of the reduced system is the rank of A, while the total number of variables is the same as the number of columns of A. Thus we have the following result.

THEOREM 7 The rank of A + the nullity of A is equal to the number of columns of A.

This result connects the rank of A with the dimension of two vector spaces associated with A. The nullity of A is just the dimension of the set of all $\bar{x}$ such that $A\bar{x} = \bar{0}$. The number of columns of A is the same as the number of entries in a column vector $\bar{x}$ for which $A\bar{x}$ is defined; in other words, the number of columns of A is just the dimension of the set of all $\bar{x}$ for which $A\bar{x}$ is defined.

Our next characterization of rank depends upon Theorem 7 and an observation about linear combinations made in Example 6, page 73. If the columns of A are denoted by $\bar{a}_1, \bar{a}_2, \ldots, \bar{a}_n$, then the equation $A\bar{x} = \bar{c}$ can be rewritten as

$$\bar{c} = x_1\bar{a}_1 + x_2\bar{a}_2 + \cdots + x_n\bar{a}_n$$

Thus $A\bar{x} = \bar{c}$ will have a solution if and only if $\bar{c}$ is a linear combination of the columns of A. The set of all linear combinations of the columns of A is usually called the *column space* of A. Thus we have the following result.

2 The column space of A consists of all vectors $\bar{c}$ such that $A\bar{x} = \bar{c}$ has a solution.

EXAMPLE 1 The only new idea involved in the above discussion is that of column space, so let us describe this carefully for a given matrix, say

$$A = \begin{bmatrix} 1 & 5 \\ 0 & 2 \\ -1 & 3 \end{bmatrix}$$

The column space of A consists of all vectors of the form

$$b_1 \begin{bmatrix} 1 \\ 0 \\ -1 \end{bmatrix} + b_2 \begin{bmatrix} 5 \\ 2 \\ 3 \end{bmatrix}$$

where b_1 and b_2 are any two real numbers. For example, the column space of A contains each of the following vectors:

$$0 \begin{bmatrix} 1 \\ 0 \\ -1 \end{bmatrix} + 0 \begin{bmatrix} 5 \\ 2 \\ 3 \end{bmatrix} = \begin{bmatrix} 0 \\ 0 \\ 0 \end{bmatrix}$$

$$\tfrac{3}{2} \begin{bmatrix} 1 \\ 0 \\ -1 \end{bmatrix} - \tfrac{6}{5} \begin{bmatrix} 5 \\ 2 \\ 3 \end{bmatrix} = \begin{bmatrix} -\frac{9}{2} \\ -\frac{12}{5} \\ -\frac{51}{10} \end{bmatrix}$$

$$-\begin{bmatrix} 1 \\ 0 \\ -1 \end{bmatrix} + \begin{bmatrix} 5 \\ 2 \\ 3 \end{bmatrix} = \begin{bmatrix} 4 \\ 2 \\ 4 \end{bmatrix}$$

The two columns of A happen to be independent so that they are a basis for the column space of A.

The result **2** tells us that the equation $A\bar{x} = \bar{c}$ can be solved when and only when $\bar{x}$ is a linear combination of the columns of A. For example, if

$$\bar{c} = \begin{bmatrix} -\frac{9}{2} \\ -\frac{12}{5} \\ -\frac{51}{10} \end{bmatrix} \quad \text{then} \quad A \begin{bmatrix} \frac{3}{2} \\ -\frac{6}{5} \end{bmatrix} = \bar{c}$$

The vector $\bar{w} = \begin{bmatrix} 1 \\ 0 \\ 0 \end{bmatrix}$ is *not* in the column space of A. This can be verified by showing that $\overline{w}$ is independent of the columns of A, or by showing directly that the equation $A\overline{x} = \overline{w}$ cannot be solved.

DISCUSSION We now prove the following theorem.

THEOREM 8 The rank of a matrix is the dimension of its column space.

PROOF: This theorem is a consequence of the following result.

3 The columns of A which correspond to *leading* variables of the reduced system are a basis for the column space of A.

This will certainly prove Theorem 8, for the rank of A is the same as the number of leading variables in the reduced system. We sketch here a proof of result 3. It is suggested that the reader read this proof in conjunction with Example 2.

Suppose A has rank equal to r and reduces to a reduced matrix R. Delete from A and R those columns corresponding to nonleading variables of the reduced system, giving A' and R', respectively. The matrices A' and R' each have r columns and R' has r nonzero rows. It is not hard to convince oneself that R' is the reduced matrix of A' so that rank $A' = r$. This proves that the columns of A' are independent (Theorem 3).

If A has more than r columns, we consider the equation $A'\overline{x} = \bar{a}$, where $\bar{a}$ is some column of A not included in A'. The augmented matrix of this system reduces to a matrix with R' as its first r columns, and the column of R corresponding to $\bar{a}$ as its last column, so that no entry of this last column below the rth entry can be nonzero (for this is true of every column of R). Thus result 6, page 19, implies that $A'\overline{x} = \bar{a}$ has a solution, and we have shown that any column of A not included among those of A' must be a linear combination of the columns of A'. This completes the proof of result 3 and of Theorem 8.

EXAMPLE 2 To assist in understanding the proof of Theorem 8, we look at an example. The matrix

$$A = \begin{bmatrix} 4 & 8 & 1 & -1 & 1 \\ 2 & 4 & 1 & 1 & 0 \\ -1 & -2 & 1 & 4 & 1 \\ 3 & 6 & 1 & 0 & 1 \end{bmatrix}$$

reduces to

$$R = \begin{bmatrix} 1 & 2 & 0 & -1 & 0 \\ 0 & 0 & 1 & 3 & 0 \\ 0 & 0 & 0 & 0 & 1 \\ 0 & 0 & 0 & 0 & 0 \end{bmatrix}$$

The matrix A' is made up of the first, third, and fifth columns of A, so

$$A' = \begin{bmatrix} 4 & 1 & 1 \\ 2 & 1 & 0 \\ -1 & 1 & 1 \\ 3 & 1 & 1 \end{bmatrix} \qquad \text{and} \qquad R' = \begin{bmatrix} 1 & 0 & 0 \\ 0 & 1 & 0 \\ 0 & 0 & 1 \\ 0 & 0 & 0 \end{bmatrix}$$

The sequence of operations which reduces A to R will obviously reduce A' to R' so that rank $A' = 3$ and the columns of A' are independent.

Now let $\bar{a}$ be the *fourth* column of A. The augmented matrix of $A'\bar{x} = \bar{a}$ is

$$\left[\begin{array}{ccc:c} 4 & 1 & 1 & -1 \\ 2 & 1 & 0 & 1 \\ -1 & 1 & 1 & 4 \\ 3 & 1 & 1 & 0 \end{array}\right]$$

Apply the same sequence operations to this which were used to reduce A to R. One then obtains

$$\left[\begin{array}{ccc:c} 1 & 0 & 0 & -1 \\ 0 & 1 & 0 & 3 \\ 0 & 0 & 1 & 0 \\ 0 & 0 & 0 & 0 \end{array}\right]$$

that is, a matrix with R' in its first three columns and the *fourth* column of R as its last column. Clearly, $A'\bar{x} = \bar{a}$, has solutions so that the fourth column of A is a linear combination of the first, third, and fifth columns of A.

The reader may find it instructive to look at further examples to convince himself that these ideas will work for a general matrix A.

EXAMPLE 3 The result **3** provides a means for selecting a basis from a spanning set. For example, if $\bar{u}_1, \bar{u}_2, \ldots, \bar{u}_k$ span V, we can form the matrix A with

$\bar{u}_1, \bar{u}_2, \ldots, \bar{u}_k$ as its columns. Then V will be the column space of A, and principle **3** will select a set of the $\bar{u}_1, \bar{u}_2, \ldots, \bar{u}_k$ which form a basis for V. For example, suppose

$$\bar{u}_1 = (1,0,2) \qquad \bar{u}_2 = (1,1,2) \qquad \bar{u}_3 = (5,3,10)$$
$$\bar{u}_4 = (0,0,1) \qquad \bar{u}_5 = (1,2,8)$$

The matrix

$$A = \begin{bmatrix} 1 & 1 & 5 & 0 & 1 \\ 0 & 1 & 3 & 0 & 2 \\ 2 & 2 & 10 & 1 & 8 \end{bmatrix}$$

has these $\bar{u}$'s as columns. Since A reduces to

$$\begin{bmatrix} 1 & 0 & 2 & 0 & -1 \\ 0 & 1 & 3 & 0 & 2 \\ 0 & 0 & 0 & 1 & 6 \end{bmatrix}$$

we conclude from principle **3** that $\bar{u}_1$, $\bar{u}_2$, and $\bar{u}_4$ form a basis for the space spanned by $\bar{u}_1, \bar{u}_2, \bar{u}_3, \bar{u}_4, \bar{u}_5$.

DISCUSSION Our final result about rank expresses the idea that the elimination method is designed to eliminate redundant equations, finally yielding a set of "independent" equations. To make this precise, we first define the **row space** of A as the set of all linear combinations of the rows of A. We then have the following theorem.

THEOREM 9 The rank of A is the dimension of its row space.

PROOF: The basic idea of the proof is that row operations *do not* change the row space. Example **5** below indicates why this is so. Thus if A reduces to a reduced matrix R, then

4 A and R have the same row space.

Since each nonzero row of R has its first nonzero entry equal to 1, and all other entries in the column in which that 1 appears are zero, it follows (see Example **6**) that

5 The nonzero rows of R are a basis for the row space of R.

The results **4** and **5** immediately establish Theorem **9**.

Note that Theorem **9** (or Theorem **8**) establishes that the rank of a matrix does not depend upon the particular sequence of operations used to reduce it to reduced form.

EXAMPLE 4 Let us describe the row space for the matrix

$$A = \begin{bmatrix} 1 & -1 & 3 \\ 2 & 1 & 0 \end{bmatrix}$$

The row space is the set of all *triples* which can be obtained from

6 $$c_1(1,-1,3) + c_2(2,1,0)$$

by assigning arbitrary values to c_1 and c_2. For example,

$$0 \cdot (1,-1,3) + 0 \cdot (2,1,0) = (0,0,0)$$
$$5 \cdot (1,-1,3) - 4(2,1,0) = (-3,-9,15)$$

and

$$\frac{-4}{7}(1,-1,3) + \frac{2}{9}(2,1,0) = \left(\frac{-8}{63}, \frac{50}{63}, \frac{-12}{7}\right)$$

are all members of the row space of A, being obtained from expression **6** by putting $c_1 = 0$, $c_2 = 0$, then $c_1 = 5$, $c_2 = -4$, then $c_1 = -\frac{4}{7}$, $c_2 = \frac{2}{9}$. In other words:

The row space of A consists of all triples which are

A multiple of $(1,-1,3)$ *plus* a multiple of $(2,1,0)$

The row space is thus the subspace of R^3 spanned by the two vectors

$$\bar{u}_1 = (1,-1,3) \quad \text{and} \quad \bar{u}_2 = (2,1,0)$$

EXAMPLE 5 It will be shown here that applying a row operation to the matrix

$$A = \begin{bmatrix} 1 & -1 & 3 \\ 2 & 1 & 0 \end{bmatrix}$$

changes the description of its row space but does not change the row space itself. Let B be obtained from A by adding to the second row -2 times the first row:

$$B = \begin{bmatrix} 1 & -1 & 3 \\ 0 & 3 & -6 \end{bmatrix}$$

Denote the rows of A by $\bar{u}_1$ and $\bar{u}_2$ and those of B by $\bar{v}_1$ and $\bar{v}_2$, so that

$$\bar{u}_1 = (1,-1,3) \qquad \bar{v}_1 = (1,-1,3)$$
$$\bar{u}_2 = (2,1,0) \qquad \bar{v}_2 = (0,3,-6)$$

The row space of A consists of all vectors of the form

7 $$c_1\bar{u}_1 + c_2\bar{u}_2 \qquad \text{for arbitrary } c_1 \text{ and } c_2$$

while the row space of B consists of all vectors of the form

8 $$c_1\bar{v}_1 + c_2\bar{v}_2 \qquad \text{for arbitrary } c_1 \text{ and } c_2$$

The manner in which B is obtained from A enables us to show that anything of the form **8** also has the form **7** (of course, different c's will appear in each expression). To show this we first observe that

9 $$\bar{v}_1 = \bar{u}_1 \qquad \text{and} \qquad \bar{v}_2 = \bar{u}_2 - 2\bar{u}_1$$

(for this is how B was obtained from A). Consider the sum $c_1\bar{v}_1 + c_2\bar{v}_2$ and replace $\bar{v}_1$ and $\bar{v}_2$ by the expressions **9**:

$$c_1\bar{v}_1 + c_2\bar{v}_2 = c_1\bar{u}_1 + c_2(\bar{u}_2 - 2\bar{u}_1)$$

Now regroup to obtain $(c_1 - 2c_2)\bar{u}_1 + c_2\bar{u}_2$. This has the form **7**, for it is

A multiple of $\bar{u}_1$ *plus* a multiple of $\bar{u}_2$

In summary

Each vector in the row space of B is in the row space of A.

Likewise

Each vector in the row space of A is in the row space of B

for we can substitute using expressions **9** to replace $\bar{u}_1$ and $\bar{u}_2$ in an expression

$$c_1\bar{u}_1 + c_2\bar{u}_2$$

to obtain

$$(c_1 + 2c_2)\bar{v}_1 + c_2\bar{v}_2$$

This establishes that A and B indeed have the same row space. Similar arguments can be used to prove this result for other row operations.

EXAMPLE 6 Let us show that the nonzero rows of the reduced matrix

$$R = \begin{bmatrix} 1 & 0 & 2 & 0 \\ 0 & 1 & -1 & 0 \\ 0 & 0 & 0 & 1 \\ 0 & 0 & 0 & 0 \end{bmatrix}$$

are a basis for the row space of R. Denote these rows by $\bar{u}_1$, $\bar{u}_2$, $\bar{u}_3$, and $\bar{u}_4$, respectively. The row space of R consists of all sums of the form

$$c_1\bar{u}_1 + c_2\bar{u}_2 + c_3\bar{u}_3 + c_4\bar{u}_4$$

Since $\bar{u}_4 = \bar{0}$, any sum of this type is a combination of $\bar{u}_1$, $\bar{u}_2$, and $\bar{u}_3$ only. This shows that the nonzero rows $\bar{u}_1$, $\bar{u}_2$, and $\bar{u}_3$ of R span the row space of R.

Now it will be shown that $\bar{u}_1$, $\bar{u}_2$, and $\bar{u}_3$ are independent and hence are indeed a basis for the row space of R.

Suppose

$$c_1\bar{u}_1 + c_2\bar{u}_2 + c_3\bar{u}_3 = \bar{0}$$

Replacing $\bar{u}_1$, $\bar{u}_2$, and $\bar{u}_3$ by the first three rows of R gives

$$c_1(1,0,2,0) + c_2(0,1,-1,0) + c_3(0,0,0,1) = (0,0,0,0)$$

The left-hand side is

$$(c_1, c_2, 2c_1 - c_2, c_3)$$

so that the equation can only hold if

$$c_1 = c_2 = c_3 = 0$$

and hence $\bar{u}_1$, $\bar{u}_2$, and $\bar{u}_3$ are independent.

For a general reduced matrix these same phenomena occur. If its nonzero rows are $\bar{u}_1, \bar{u}_2, \ldots, \bar{u}_k$, then for each i some coordinate of the sum $c_1\bar{u}_1 + c_2\bar{u}_2 + \cdots + c_k\bar{u}_k$ consists of c_i only. It follows that if such a sum is 0, then, since each c_i is a coordinate of the sum, each c_i must be 0.

EXERCISES

1 Find three vectors in the column space of

a $\begin{bmatrix} 3 & 4 \\ 1 & 2 \\ -1 & 6 \end{bmatrix}$ **b** $\begin{bmatrix} 3 & 1 & -1 \\ 4 & 2 & 6 \end{bmatrix}$

2 Show that $\bar{c}$ is in the column space of A, where

$$\bar{c} = \begin{bmatrix} -6 \\ -4 \\ -20 \end{bmatrix} \qquad A = \begin{bmatrix} 3 & 4 \\ 1 & 2 \\ -1 & 6 \end{bmatrix}$$

3 The matrix

$$A = \begin{bmatrix} 1 & 4 & 0 & -6 & 5 & 1 \\ 2 & 1 & 1 & 1 & 3 & 0 \\ -1 & 2 & 1 & -7 & 1 & -1 \\ 1 & 1 & 2 & -2 & 2 & 1 \\ 3 & 0 & 3 & 3 & 3 & 2 \end{bmatrix} \quad \text{reduces to} \quad R = \begin{bmatrix} 1 & 0 & 0 & 2 & 1 & 0 \\ 0 & 1 & 0 & -2 & 1 & 0 \\ 0 & 0 & 1 & -1 & 0 & 0 \\ 0 & 0 & 0 & 0 & 0 & 1 \\ 0 & 0 & 0 & 0 & 0 & 0 \end{bmatrix}$$

Each of the following is obtained by deleting some columns of A. Determine their reduced forms without using any row operations.

a $\begin{bmatrix} 1 & 4 & 0 & 1 \\ 2 & 1 & 1 & 0 \\ -1 & 2 & 1 & -1 \\ 1 & 1 & 2 & 1 \\ 3 & 0 & 3 & 2 \end{bmatrix}$ **b** $\begin{bmatrix} 1 & 4 & 5 & 1 \\ 2 & 1 & 3 & 0 \\ -1 & 2 & 1 & -1 \\ 1 & 1 & 2 & 1 \\ 3 & 0 & 3 & 2 \end{bmatrix}$

c $\begin{bmatrix} 1 & 4 & 0 & 1 & -6 \\ 2 & 1 & 1 & 0 & 1 \\ -1 & 2 & 1 & -1 & -7 \\ 1 & 1 & 2 & 1 & -2 \\ 3 & 0 & 3 & 2 & 3 \end{bmatrix}$

4 Let A', R' be obtained from A and R by deleting the second and fourth columns of each. Suppose R is a reduced matrix.

a Give an example to show that R' may *not* be a reduced matrix.

b What conditions on R will guarantee that R' is a reduced matrix?

5 Find the dimension of the column space of each of the following by row reducing and using Theorem **8**.

a $\begin{bmatrix} 2 & 1 & 0 \\ 1 & 2 & 1 \end{bmatrix}$ **b** $\begin{bmatrix} -1 & 1 & 4 \\ -5 & 0 & 7 \\ 3 & 2 & 1 \end{bmatrix}$ **c** $\begin{bmatrix} 2 & 1 & 1 \\ 0 & 0 & 0 \\ 3 & 1 & 2 \end{bmatrix}$

6 Use the method of Example **3** to choose a basis for the space spanned by

a $\bar{u}_1 = (1,1,1,1,-1)$, $\bar{u}_2 = (1,1,-1,1,-1)$
$\bar{u}_3 = (1,1,0,1,0)$, $\bar{u}_4 = (2,2,-1,2,-1)$

b $\bar{v}_1 = (1,2,4,7)$, $\bar{v}_2 = (1,-1,1,1)$
$\bar{v}_3 = (1,1,3,5)$, $\bar{v}_4 = (1,1,1,1)$

7 If A has m rows and n columns, is the column space of A a subspace of R^m or R^n?

8 Find four vectors in the row space of

$$A = \begin{bmatrix} 2 & 1 & 0 & 3 & 1 \\ 1 & -1 & 2 & 1 & 0 \\ 1 & 1 & 2 & 1 & 0 \end{bmatrix}$$

9 **a** Suppose C is obtained from the matrix

$$A = \begin{bmatrix} 1 & -1 & 3 \\ 2 & 1 & 0 \end{bmatrix}$$

by multiplying row one by 10. Let $\bar{u}_1$ and $\bar{u}_2$ be the first and second rows of A and $\bar{v}_1$ and $\bar{v}_2$ the first and second rows of C. Show that any linear combination $c_1\bar{u}_1 + c_2\bar{u}_2$ is a multiple of $\bar{v}_1$ plus a multiple of $\bar{v}_2$. (See Example **5**.)

b Suppose D is obtained from the matrix A by interchanging the rows of A. Let $\bar{u}_1$ and $\bar{u}_2$ be the first and second rows of A and $\bar{v}_1$ and $\bar{v}_2$ the first and second rows of C. Show that any sum of the form $c_1\bar{u}_1 + c_2\bar{u}_2$ can be expressed as a linear combination of $\bar{v}_1$ and $\bar{v}_2$. (See Example **5**.)

10 Show that the nonzero rows of the reduced matrix

$$R = \begin{bmatrix} 1 & 0 & 0 & 2 & 1 & 0 \\ 0 & 1 & 0 & -2 & 1 & 0 \\ 0 & 0 & 1 & -1 & 0 & 0 \\ 0 & 0 & 0 & 0 & 0 & 1 \\ 0 & 0 & 0 & 0 & 0 & 0 \end{bmatrix}$$

are independent. (See Example **6**.)

11 Use the result **5** to find a basis for the row space of

$$A = \begin{bmatrix} 3 & -1 & 1 & 3 \\ 7 & 1 & -1 & 6 \\ 2 & 1 & -1 & 1 \end{bmatrix}$$

12 Suppose

$$A = \begin{bmatrix} 4 & 1 & 2 \\ 8 & 2 & 4 \\ 3 & 1 & 0 \end{bmatrix}$$

a Find a basis for the row space of A.
b Find a basis for the column space of A.

13 Show that $\bar{u}_1 = (1,1,1)$, $\bar{u}_2 = (1,2,-1)$, $\bar{u}_3 = (-1,0,1)$ are independent by determining the rank of the matrix with these as columns. Show the same thing by determining the rank of the matrix with these as rows.

14 Let A' be obtained from A by deleting some of the columns of A. Show that rank $A' \leq$ rank A.

15 Let A' be obtained from A by deleting some of the rows of A. What is the relation between rank A' and rank A? Prove it.

16 Suppose A has rank 5. Give reasons for your answers to each of the following.
a If A has six columns, can they be independent?
b If A has six columns and $A\bar{u} = A\bar{v} = \bar{0}$, what can you say about $\bar{u}$ and $\bar{v}$?
c Can A have exactly four rows?
d If the first and third rows of A are dependent, what is the minimum possible number of rows A can have?

17 Suppose every matrix A' obtained from A by deleting some columns of A has the same rank on A. What is A?

□ **18** **a** Find a basis for the column space of

$$A = \begin{bmatrix} i & 2i & 3i \\ 0 & 1 & 1 \\ 1+i & -i & 1 \end{bmatrix}$$

b Find a basis for the row space of

$$A = \begin{bmatrix} 1 & i & 1-i & -i \\ i & -1 & 1+i & 1 \\ 1+i & -1+i & 2 & 1-i \end{bmatrix}$$

SECTION 5 The Dot Product

It is frequently possible to use geometric ideas to show independence of a set of vectors and to calculate coordinates with respect to a basis. For example, if two nonzero vectors are mutually perpendicular in the arrow representation of R^2, they are certainly not parallel. Therefore they are independent and hence are a basis for R^2. The coordinates of a vector with respect to such a basis can be calculated by projecting onto the basis vectors.

In this section the concepts of perpendicularity and projection are given an algebraic formulation. These ideas can then be adapted to the study of vector spaces that have no convenient geometric description, such as R^4 and R^5, and to spaces of functions.

If $\bar{u} = (x_1, x_2, \ldots, x_n)$ and $\bar{v} = (y_1, y_2, \ldots, y_n)$, then the **dot product** is defined by

$$\bar{u} \cdot \bar{v} = x_1 y_1 + x_2 y_2 + \cdots + x_n y_n$$

The dot product has the following properties, all of which are simple consequences of the definition

a $\bar{u} \cdot \bar{v}$ *is a real number*

b $\bar{u} \cdot \bar{v} = \bar{v} \cdot \bar{u}$

1 **c** $\bar{u} \cdot (\bar{v} + \bar{w}) = \bar{u} \cdot \bar{v} + \bar{u} \cdot \bar{w}$

d $\bar{u} \cdot (a\bar{v}) = a(\bar{u} \cdot \bar{v})$ *for any real number a*

e $\bar{u} \cdot \bar{u} > 0$ *if* $\bar{u} \neq \bar{0}$

All these rules also hold for the ordinary product of real numbers. Hence, we can manipulate with the dot product just as though the vectors were numbers. [Of course, $\bar{u} \cdot (\bar{v} \cdot \bar{w})$ makes no sense in this situation since $\bar{v} \cdot \bar{w}$ would be a number, and the dot product of a number with a vector is not defined.]

For example, if a and b are numbers, then

$$(a - b)(a - b) = aa - 2ab + bb$$

This equation is simply a consequence of the distributive and commutative laws of ordinary multiplication. For the dot product, these laws are just properties **1b** and **c**. We therefore conclude that

2 $$(\bar{u} - \bar{v}) \cdot (\bar{u} - \bar{v}) = \bar{u} \cdot \bar{u} - 2\bar{u} \cdot \bar{v} + \bar{v} \cdot \bar{v}$$

must also be true.

The **length** (or **norm**) of $\bar{u}$ is defined by

$$|\bar{u}| = \sqrt{\bar{u} \cdot \bar{u}}$$

Therefore, if $\bar{u} = (x_1, x_2, \ldots, x_n)$, the length of $\bar{u}$ is

$$|\bar{u}| = \sqrt{x_1^2 + x_2^2 + \cdots + x_n^2}$$

Figure 3 shows that when $n = 2$, this definition of length corresponds with the usual definition of length for the arrow used to represent $\bar{u}$.

We can show in this case that

$$\bar{u} \cdot \bar{v} = |\bar{u}|\,|\bar{v}| \cos \theta \tag{3}$$

where θ is the angle between the arrow representations of $\bar{u}$ and $\bar{v}$. For example, if $\bar{u} = (x_1,y_1)$ and $\bar{v} = (x_2,y_2)$, the law of cosines gives

$$|\bar{u} - \bar{v}|^2 = |\bar{u}|^2 + |\bar{v}|^2 - 2|\bar{u}|\,|\bar{v}| \cos \theta \tag{4}$$

as shown in Figure 4.

Figure 3 *The length of* $\bar{u}$ *is* $\sqrt{x^2 + y^2}$.

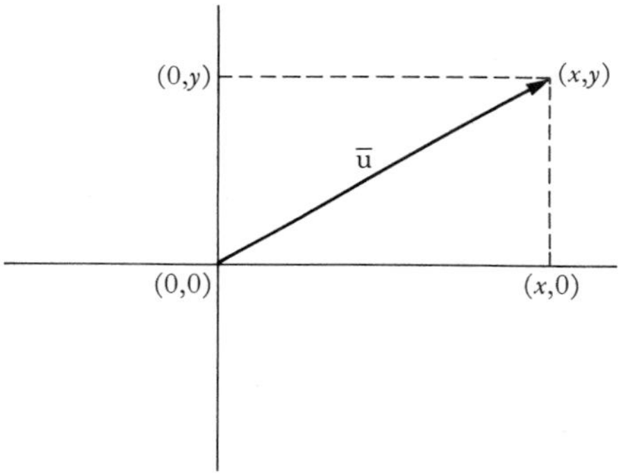

Figure 4 *The law of cosines gives* $|PQ|^2 = |OQ|^2 + |OP|^2 - 2|OQ|\,|OP| \cos \theta$.

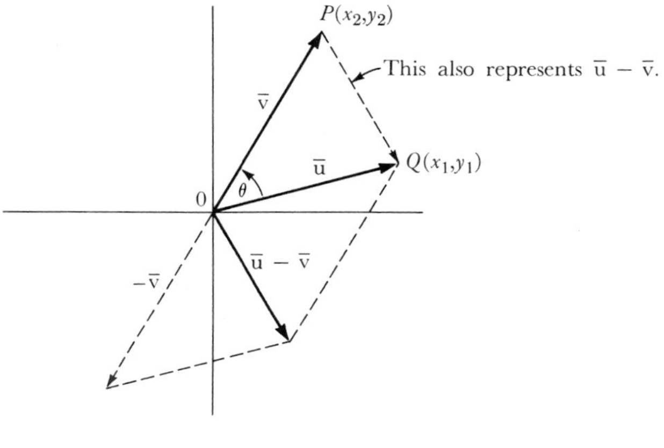

We can also compute $|\bar{u} - \bar{v}^2|$ using the definition of length and result **2** to obtain:

$$|\bar{u} - \bar{v}|^2 = (\bar{u} - \bar{v}) \cdot (\bar{u} - \bar{v})$$
$$= \bar{u} \cdot \bar{u} - 2\bar{u} \cdot \bar{v} + \bar{v} \cdot \bar{v}$$
$$= |\bar{u}|^2 - 2\bar{u} \cdot \bar{v} + |\bar{v}|^2$$

Comparing this with equation **4** we see that we must have

$$\bar{u} \cdot \bar{v} = |\bar{u}|\,|\bar{v}| \cos \theta$$

which establishes statement **3**.

It follows that for $\bar{u}$ and $\bar{v}$ in R^2 we must have

$$\bar{u} \cdot \bar{v} = 0$$

if and only if $\bar{u} = \bar{0}$ or $\bar{v} = \bar{0}$ or θ is an odd multiple of $\pi/2$.

A common convention is that the zero vector is perpendicular to every vector. We have therefore shown that $\bar{u} \cdot \bar{v} = 0$ if and only if $\bar{u}$ and $\bar{v}$ are perpendicular.

This is taken as the definition in R^n when $n > 2$, that is, $\bar{u}$ and $\bar{v}$ in R^n are **perpendicular** or **orthogonal** if $\bar{u} \cdot \bar{v} = 0$.

REMARK This use of such geometric terminology as "length" and "perpendicular" for vectors in R^n is a good example of the way in which geometric names are used for vector spaces that have no convenient geometric description (which is the case for R^5, for example). In general, if we can show that a geometric concept can be given an algebraic formulation, then we can adopt the same terminology whenever the same algebraic formulation holds in a vector space. This enables us to borrow ideas and insight from geometry into more general situations.

EXAMPLE 1 Some examples are given here of the calculation of lengths and angles using dot products. Suppose

$$\bar{u} = (2,-1,3) \qquad \bar{v} = (3,-1,4) \qquad \bar{w} = (6,8,-10)$$

Then

$$\bar{u} \cdot \bar{v} = 2 \cdot 3 + (-1)(-1) + 3 \cdot 4 = 19$$
$$\bar{u} \cdot \bar{w} = 2 \cdot 6 + (-1)8 + 3(-10) = -26$$

so that

$$\bar{u} \cdot (5\bar{v} - 2\bar{w}) = \bar{u} \cdot 5\bar{v} - \bar{u} \cdot 2\bar{w}$$
$$= 5(\bar{u} \cdot \bar{v}) - 2(\bar{u} \cdot \bar{w})$$
$$= 5 \cdot 19 - 2(-26) = 147$$

The angle between $\bar{u}$ and $\bar{v}$ is the angle θ whose cosine is

$$\cos\theta = \frac{\bar{u}\cdot\bar{v}}{|\bar{u}|\,|\bar{v}|}$$

The lengths of $\bar{u}$ and $\bar{v}$ are

$$|\bar{u}| = \sqrt{2^2 + (-1)^2 + 3^2} = \sqrt{14}$$
$$|\bar{v}| = \sqrt{3^2 + (-1)^2 + 4^2} = \sqrt{26}$$

so that

$$\cos\theta = \frac{19}{\sqrt{14}\cdot\sqrt{26}} = \frac{19}{182}\sqrt{91}$$

An approximate value of θ can be found by approximating $\sqrt{91}$ and then using trigonometric tables. Correct to two places this will give $\theta \cong 85°$.

EXAMPLE 2 Dot products in R^n for $n > 3$ are in principle the same as for R^2 and R^3, although requiring more calculation. For example, if

$$\bar{u} = (3,2,1,5,6) \qquad \text{and} \qquad \bar{v} = (-1,2,1,4,-7)$$

then

$$\bar{u}\cdot\bar{v} = 3\cdot(-1) + 2\cdot 2 + 1\cdot 1 + 5\cdot 4 + 6\cdot(-7) = -20$$

The vector $\bar{w} = (1,1,0,-1,0)$ is orthogonal to $\bar{u}$ since

$$\bar{u}\cdot\bar{w} = 3\cdot 1 + 2\cdot 1 + 1\cdot 0 + 5\cdot(-1) + 6\cdot 0 = 0$$

Note that in this case a Pythagorean-type law holds for the orthogonal pair $\bar{u}$ and $\bar{w}$:

$$|\bar{u} + \bar{w}|^2 = |\bar{u}|^2 + |\bar{w}|^2$$

This can be verified directly:

$$\begin{aligned}|\bar{u} + \bar{w}|^2 &= (3+1)^2 + (2+1)^2 + (1+0)^2 + (5-1)^2 + (6+0)^2\\ &= 78\\ |\bar{u}|^2 + |\bar{w}|^2 &= (3^2 + 2^2 + 1^2 + 5^2 + 6^2) + (1^2 + 1^2 + 0^2 + (-1)^2 + 0^2)\\ &= 78\end{aligned}$$

This can also be verified by using the algebraic rules 1 and the fact that $\bar{u}\cdot\bar{w} = 0$:

$$\begin{aligned}|\bar{u} + \bar{w}|^2 &= (\bar{u} + \bar{w})\cdot(\bar{u} + \bar{w})\\ &= \bar{u}\cdot\bar{u} + 2\bar{u}\cdot\bar{w} + \bar{w}\cdot\bar{w}\\ &= |\bar{u}|^2 + |\bar{w}|^2\end{aligned}$$

DISCUSSION The process of projection can be formulated in terms of the dot product. For example, suppose $\bar{u}$ and $\bar{v}$ are given vectors in R^2 or R^3, and $\bar{v} \neq \bar{0}$. Represent these two vectors as arrows issuing from the same point and drop a perpendicular from the terminal point of $\bar{u}$ onto the line along $\bar{v}$. This is segment AB in Figure 5a or b. The vector $\bar{w}_1$ from O to B is called the *projection of* $\bar{u}$ *along* $\bar{v}$.

Figure 5 $\bar{w}_1 = OB$, *the projection of* $\bar{u}$ *along* $\bar{v}$.

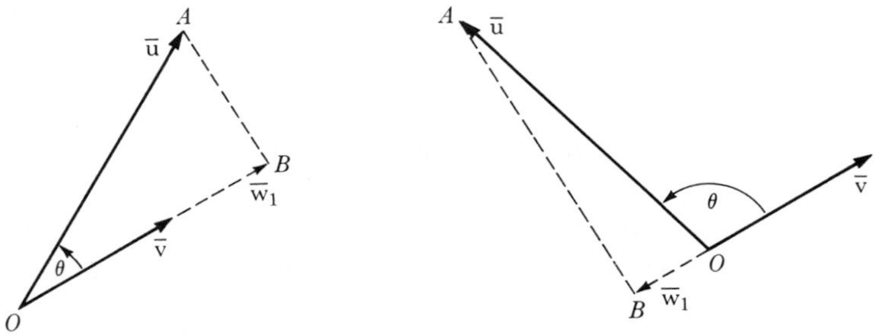

Now construct the vector $\bar{w}_2$ from O to C, so that $\bar{w}_2$ is parallel to AB and has the same length as AB and so that A and C lie on the same side of $\bar{v}$. (See Figure 6a and b.) The vector $\bar{w}_2$ is called the *projection of* $\bar{u}$ *orthogonal to* $\bar{v}$. It is clear from the construction that

$$\bar{u} = \bar{w}_1 + \bar{w}_2$$

That is, $\bar{u}$ is the *sum of its projection along* $\bar{v}$ *and its projection orthogonal to* $\bar{v}$.

Figure 6 $\bar{w}_2 = OC$, *the projection of* $\bar{u}$ *orthogonal to* $\bar{v}$.

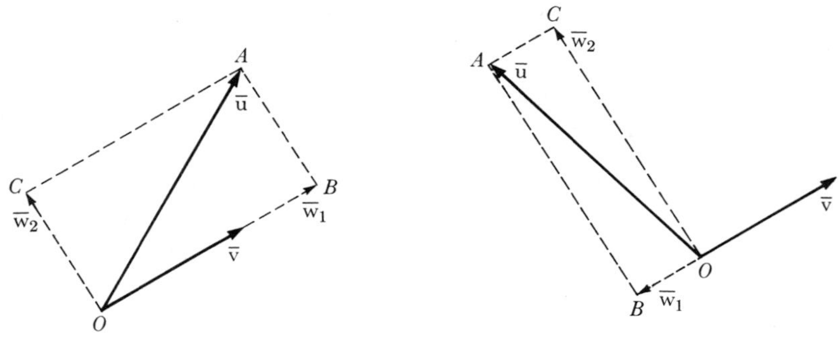

Let us give an explicit algebraic formula for $\bar{w}_1$ and $\bar{w}_2$. These two vectors satisfy the three conditions

5
- a $\bar{w}_1 = a\bar{v}$ (that is $\bar{w}_1$ is parallel to $\bar{v}$)
- b $\bar{w}_2$ is orthogonal to $\bar{v}$
- c $\bar{u} = \bar{w}_1 + \bar{w}_2$

This is all that we need to know to determine $\bar{w}_1$ and $\bar{w}_2$, for if $\bar{w}_1$ and $\bar{w}_2$ satisfy the relations **5**, then

$$\bar{u} = \bar{w}_1 + \bar{w}_2 = a\bar{v} + \bar{w}_2$$

Take the dot product of this expression with $\bar{v}$ and use the fundamental properties **1**:

$$\begin{aligned}\bar{u} \cdot \bar{v} &= (a\bar{v} + \bar{w}_2) \cdot \bar{v} \\ &= (a\bar{v} \cdot \bar{v}) + \bar{w}_2 \cdot \bar{v} \\ &= a(\bar{v} \cdot \bar{v}) + \bar{w}_2 \cdot \bar{v}\end{aligned}$$

Property **5b** implies that $\bar{w}_2 \cdot \bar{v} = \bar{0}$, so we have

$$\bar{u} \cdot \bar{v} = a(\bar{v} \cdot \bar{v}) + \bar{w}_2 \cdot \bar{v} = a(\bar{v} \cdot \bar{v})$$

This can be solved for a to obtain

$$a = \frac{\bar{u} \cdot \bar{v}}{\bar{v} \cdot \bar{v}}$$

Replace this in statements **5a** and **5c** to obtain

6 $$\bar{w}_1 = \frac{\bar{u} \cdot \bar{v}}{\bar{v} \cdot \bar{v}}\bar{v} \qquad \text{the projection of } \bar{u} \text{ along } \bar{v}$$

7 $$\bar{w}_2 = \bar{u} - \bar{w}_1 = \bar{u} - \frac{\bar{u} \cdot \bar{v}}{\bar{v} \cdot \bar{v}}\bar{v} \qquad \text{the projection of } \bar{u} \text{ orthogonal to } \bar{v}$$

These formulas are an algebraic description for the geometric concept of projection. These will be taken as *definitions* for vectors in higher dimensional spaces, that is,

> If $\bar{u}$ and $\bar{v}$ are vectors in R^n and $\bar{v} \neq 0$, then the **projection** $\bar{w}_1$ **of** $\bar{u}$ **along** $\bar{v}$ is defined by formula **6** and the **projection** $\bar{w}_2$ **of** $\bar{u}$ **orthogonal** to $\bar{v}$ is defined by formula **7**.

EXAMPLE 3 The following indicates the calculations of projections for triples, that is, vectors in R^3. Suppose

$$\bar{u} = (1,2,1) \qquad \text{and} \qquad \bar{v} = (3,-1,0)$$

Then

$$\bar{u} \cdot \bar{v} = 1 \cdot 3 + 2 \cdot (-1) + 1 \cdot 0$$
$$= 1$$

and

$$\bar{v} \cdot \bar{v} = 3 \cdot 3 + (-1)(-1) + 0 \cdot 0$$
$$= 10$$

so that formula **6** gives

$$\bar{w}_1 = \frac{\bar{u} \cdot \bar{v}}{\bar{v} \cdot \bar{v}} \bar{v} = \tfrac{1}{10}\bar{v} = \tfrac{1}{10}(3,-1,0)$$
$$= (\tfrac{3}{10},-\tfrac{1}{10},0) \qquad \text{the projection of } \bar{u} \text{ along } \bar{v}$$

The projection $\bar{w}_2$ of $\bar{u}$ orthogonal to $\bar{v}$ is obtained from formula **7**:

$$\bar{w}_2 = \bar{u} - \bar{w}_1 = (1,2,1) - (\tfrac{3}{10},-\tfrac{1}{10},0)$$
$$= (\tfrac{7}{10},\tfrac{21}{10},1)$$

These two projections are indeed perpendicular, for

$$\bar{w}_1 \cdot \bar{w}_2 = (\tfrac{3}{10},-\tfrac{1}{10},0) \cdot (\tfrac{7}{10},\tfrac{21}{10},1)$$
$$= (\tfrac{3}{10})(\tfrac{7}{10}) + (-\tfrac{1}{10})(\tfrac{21}{10}) + 0 \cdot 1$$
$$= \tfrac{21}{100} - \tfrac{21}{100} = 0$$

Furthermore, their sum is $\bar{u}$; that is

$$\bar{w}_1 + \bar{w}_2 = (\tfrac{3}{10},-\tfrac{1}{10},0) + (\tfrac{7}{10},\tfrac{21}{10},1)$$
$$= (1,2,1)$$
$$= \bar{u}$$

EXAMPLE 4 The calculation of projections in R^n for $n > 3$ is a simple extension of the methods of Example **3**. For example, if

$$\bar{u} = (4,0,1,2,0) \qquad \bar{v} = (2,1,-1,1,1)$$

then

$$\bar{u} \cdot \bar{v} = 4 \cdot 2 + 0 \cdot 1 + 1 \cdot (-1) + 2 \cdot 1 + 0 \cdot 1 = 9$$
$$\bar{v} \cdot \bar{v} = 2 \cdot 2 + 1 \cdot 1 + (-1) \cdot (-1) + 1 \cdot 1 + 1 \cdot 1 = 8$$

so the projection $\bar{w}_1$ of $\bar{u}$ along $\bar{v}$ is

$$\bar{w}_1 = \frac{\bar{u} \cdot \bar{v}}{\bar{v} \cdot \bar{v}} \bar{v} = \tfrac{9}{8}\bar{v} = (\tfrac{18}{8},\tfrac{9}{8},-\tfrac{9}{8},\tfrac{9}{8},\tfrac{9}{8})$$

The projection $\bar{w}_2$ of $\bar{u}$ orthogonal to $\bar{v}$ is

$$\bar{w}_2 = \bar{u} - \bar{w}_1 = (\tfrac{14}{8}, -\tfrac{9}{8}, \tfrac{17}{8}, \tfrac{7}{8}, -\tfrac{9}{8})$$

Certainly $\bar{w}_1$ is a multiple of $\bar{v}$ and $\bar{w}_1 + \bar{w}_2 = \bar{u}$. A direct calculation shows that, as we expect, $\bar{w}_1 \cdot \bar{w}_2 = 0$.

EXAMPLE 5 □ *An Inner Product*

It is possible to define for functions a "product" that satisfies all the rules **1**. Since various geometric concepts, such as orthogonality and projection, can be described in terms of the algebraic properties of the dot product we should expect to be able to use the same terminology with this function product.

Suppose $C[0,1]$ is the vector space of functions that are defined and continuous for $0 \leq x \leq 1$. (See Example **6**, page 39.) We shall define a product for this space with properties similar to those of the dot product. This product is usually called an **inner product** and denoted by (f,g). For f and g in $C[0,1]$ we define

$$(f,g) = \int_0^1 f(x)g(x)\, dx$$

The properties of continuous functions and the integral discussed in calculus enables us to conclude that

- **a** (f,g) is a real number
- **b** $(f,g) = (g,f)$
- **c** $(f, (g + h)) = (f,g) + (f,h)$
- **d** $(f, (ag)) = a(f,g)$ for any real number a
- **e** $(f,f) > 0$ if f is not the zero function

Except for the use of different symbols, these properties are the same as those given in statement **1**. We can therefore adopt the language of dot products for this inner product. For example, we say that f and g are *orthogonal* if

$$(f,g) = 0$$

and we define the *length* of f to be

$$|f| = (f,f)^{1/2}$$

Suppose $f(x) = \sin \pi x$. Then

8 $$|f|^2 = \int_0^1 \sin^2 \pi x\, dx = \int_0^1 \frac{1 - \cos 2\pi x}{2}\, dx = \tfrac{1}{2}x - \frac{\sin 2\pi x}{4\pi}\bigg|_0^1 = \frac{1}{2}$$

The functions sin πx and sin $2\pi x$ are orthogonal for

$$\sin \pi x \sin 2\pi x = \tfrac{1}{2}(\cos \pi x - \cos 3\pi x)$$

so that

$$\int_0^1 \sin \pi x \sin 2\pi x \, dx = \frac{1}{2\pi} \sin \pi x - \frac{1}{6\pi} \sin 3\pi x \Big|_0^1 = 0$$

We note that the orthogonality of functions has an interesting geometric interpretation. For example, consider

$$f(x) = \sin \pi x \qquad \text{and} \qquad f_1(x) = \sin 2\pi x$$

As Figure 7 shows, the fact that the inner product $(f,f_1) = 0$ is just the fact that the area above the x-axis of sin πx sin $2\pi x$ for $0 \leq x \leq 1$ equals the area below the x-axis of this function.

Figure 7

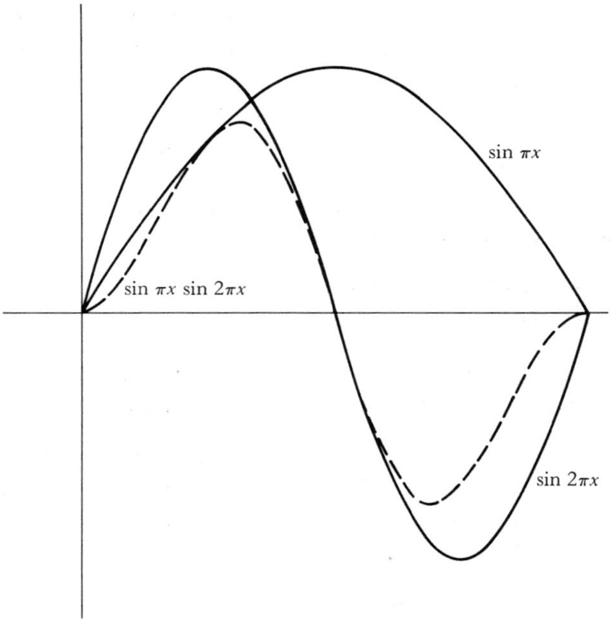

Let us find the projections of g along f and orthogonal to f, where

$$g(x) = x \qquad \text{and} \qquad f(x) = \sin \pi x$$

These projections h_1 and h_2 are defined by the analogs of formulas **6** and **7**:

$$h_1 = \frac{(g,f)}{(f,f)} f \qquad \text{and} \qquad h_2 = g - h_1$$

The result **8** gives $(f,f) = |f|^2 = \frac{1}{2}$ and a calculation gives

$$\begin{aligned}(g,f) &= \int_0^1 x \sin \pi x \, dx \\ &= \frac{-x \cos \pi x}{\pi} + \frac{\sin \pi x}{\pi^2}\Big|_0^1 \\ &= \frac{1}{\pi}\end{aligned}$$

so that

$$h_1(x) = \frac{1/\pi}{1/2} f(x) = \frac{2}{\pi} \sin \pi x$$

$$h_2(x) = g(x) - h_1(x) = x - \frac{2}{\pi} \sin \pi x$$

The projection of x onto $\sin \pi x$ is just *that* multiple of $\sin \pi x$ which gives the "best mean square approximation" in the sense that

9 $$\int_0^1 (x - a \sin \pi x)^2 \, dx \qquad \text{is } \textit{least} \text{ when} \qquad a = \frac{2}{\pi}$$

For we have

$$\begin{aligned}\int_0^1 (x - a \sin \pi x)^2 \, dx &= \int_0^1 \left[\left(x - \frac{2}{\pi} \sin \pi x\right) + \left(\frac{2}{\pi} - a\right) \sin \pi x\right]^2 dx \\ &= \int_0^1 \left(x - \frac{2}{\pi} \sin \pi x\right)^2 \\ &\quad + 2\left(\frac{2}{\pi} - a\right) \int_0^1 \left(x - \frac{2}{\pi} \sin \pi x\right) \sin \pi x \, dx \\ &\quad + \left(\frac{2}{\pi} - a\right)^2 \int_0^1 \sin^2 \pi x \, dx\end{aligned}$$

The first term does not depend upon a, and the third term is least when $a = 2/\pi$; the second term is zero since $x - (2/\pi) \sin \pi x$ is orthogonal to $\sin \pi x$. Therefore, statement **9** is true.

EXAMPLE 6 □ *The Cross Product*

Suppose

$$\bar{v}_1 = x_1\bar{i} + y_1\bar{j} + z_1\bar{k} \qquad \text{and} \qquad \bar{v}_2 = x_2\bar{i} + y_2\bar{j} + z_2\bar{k}$$

are vectors in R^3. These vectors are parallel if and only if they are dependent. They are dependent if and only if

10 $$c_1\bar{v}_1 + c_2\bar{v}_2 = \bar{0}$$

has a solution with at least one of the $c_i \neq 0$. We have

$$c_1\bar{v}_1 + c_2\bar{v}_2 = (c_1x_1 + c_2x_2)\bar{i} + (c_1y_1 + c_2y_2)\bar{j} + (c_1z_1 + c_2z_2)\bar{k}$$

so that equation **10** is the same as the system

$$\begin{aligned} c_1x_1 + c_2x_2 &= 0 \\ c_1y_1 + c_2y_2 &= 0 \\ c_1z_1 + c_2z_2 &= 0 \end{aligned}$$

Using Exercise **14**, page 32, we find that these equations have a nontrivial solution (c_1,c_2) if and only if

$$\begin{aligned} x_1y_2 - x_2y_1 &= 0 \\ y_1z_2 - y_2z_1 &= 0 \\ x_1z_2 - x_2z_1 &= 0 \end{aligned}$$

In other words, the vectors $\bar{v}_1$ and $\bar{v}_2$ are parallel if and only if

$$x_1y_2 - x_2y_1 = y_1z_2 - y_2z_1 = x_1z_2 - x_2z_1 = 0$$

This statement can be rewritten as

11 The vectors $\bar{v}_1$ and $\bar{v}_2$ are nonparallel if and only if at least one of $x_1y_2 - x_2y_1$, $y_1z_2 - y_2z_1$, $x_1z_2 - x_2z_1$ is *not* zero.

We shall show that if $\bar{v}_1$ and $\bar{v}_2$ are not parallel the vectors that are *perpendicular* to *both* $\bar{v}_1$ and $\bar{v}_2$ are precisely the multiples of

12 $$(y_1z_2 - y_2z_1)\bar{i} - (x_1z_2 - x_2z_1)\bar{j} + (x_1y_2 - x_2y_1)\bar{k}$$

This vector is called the **cross product** of $\bar{v}_1$ and $\bar{v}_2$ and is denoted by $\bar{v}_1 \times \bar{v}_2$.

Condition **11** shows that if $\bar{v}_1$ and $\bar{v}_2$ are not parallel, then $\bar{v}_1 \times \bar{v}_2 \neq \bar{0}$; formula **12** therefore gives us a means of finding a vector orthogonal to two given nonparallel vectors. A number of uses for this procedure are given in the supplement to this chapter.

Suppose $\bar{u} = a\bar{i} + b\bar{j} + c\bar{k}$. Then $\bar{u}$ is orthogonal to $\bar{v}_1$ and $\bar{v}_2$ if and only if $\bar{v}_1 \cdot \bar{u} = 0$ and $\bar{v}_2 \cdot \bar{u} = 0$. In terms of coordinates these equations are

13
$$x_1a + y_1b + z_1c = 0$$
$$x_2a + y_2b + z_2c = 0$$

Simple calculation shows that

$$a = y_1z_2 - y_2z_1$$
$$b = -(x_1z_2 - x_2z_1)$$
$$c = x_1y_2 - x_2y_1$$

is a solution to equations **13**. Thus $\bar{v}_1 \times \bar{v}_2$ is indeed orthogonal to $\bar{v}_1$ and $\bar{v}_2$.

The null space of system **13** has dimension 1, for conditions **11** guarantee that the matrix

$$\begin{bmatrix} x_1 & y_1 & z_1 \\ x_2 & y_2 & z_2 \end{bmatrix}$$

has a reduced form with two nonzero rows. (See Exercise **13** below.)

We have shown that if $\bar{v}_1$ and $\bar{v}_2$ are nonparallel the coordinates of $\bar{v}_1 \times \bar{v}_2$ give a nonzero solution to system **13**, and the null space of this system is one-dimensional. We conclude that indeed the vectors that are orthogonal to $\bar{v}_1$ and $\bar{v}_2$ are precisely the multiples of $\bar{v}_1 \times \bar{v}_2$.

EXERCISES

1 Given $\bar{u} = (1,1)$, $\bar{v} = (1,2)$, and $\bar{w} = (3,-1)$ find
a $\bar{u} \cdot (3\bar{v} + \bar{w})$ **b** $|\bar{u} - \bar{v}|^2$ **c** $(\bar{u} \cdot \bar{v})\bar{w}$

2 Given $\bar{u} = (1,-1,1)$, $\bar{v} = (1,1,2)$, and $\bar{w} = (1,3,-1)$ find
a $\bar{u} \cdot (3\bar{v} + \bar{w})$ **b** $|\bar{u} - \bar{v}|^2$ **c** $(\bar{u} \cdot \bar{v})\bar{w}$

3 Given $\bar{u} = (1,1,1,1)$, $\bar{v} = (1,1,-1,1)$, and $\bar{w} = (2,-1,1,3)$ find
a $\bar{u} \cdot (3\bar{v} + \bar{w})$ **b** $|\bar{u} - \bar{v}|^2$ **c** $(\bar{u} \cdot \bar{v})\bar{w}$

4 Find the lengths of each of the following.
a $(1,1)$ **b** $(-1,2,-2)$ **c** $(3,1,0,1,-5)$

5 Show that each of the following pairs of vectors are orthogonal.
a $(1,1)$ and $(1,-1)$ **b** $(2,1,1)$ and $(0,1,-1)$
c $(4,2,1,6,1)$ and $(3,-1,0,-1,-4)$

6 For each of the following vectors $\bar{u}$ find a unit vector parallel to $\bar{u}$.
a $\bar{u} = (1,2)$ **b** $\bar{u} = (1,-1,1)$
c $\bar{u} = (3,1,-1,0)$ **d** $\bar{u} = (1,1,1,1,1,1,1,1,1)$

7 For each of the following pairs $\bar{u}$ and $\bar{v}$ find the projection of $\bar{u}$ onto $\bar{v}$.
a $\bar{u} = (1,1)$, $\bar{v} = (1,2)$ **b** $\bar{u} = (1,1)$, $\bar{v} = (1,-1)$
c $\bar{u} = (1,2,1)$, $\bar{v} = (2,0,-1)$ **d** $\bar{u} = (2,1,0,0,6)$, $\bar{v} = (1,3,-1,1,0)$

8 For each of the pairs $\bar{u}$ and $\bar{v}$ of Exercise **7** find the projection of $\bar{u}$ orthogonal to $\bar{v}$.

9 For each of the pairs $\bar{u}$ and $\bar{v}$ of Exercise **7**, find vectors $\bar{w}_1$ and $\bar{w}_2$ such that $\bar{u} = \bar{w}_1 + \bar{w}_2$, $\bar{w}_1$ is parallel to $\bar{v}$, and $\bar{w}_2$ is orthogonal to $\bar{v}$.

10 Suppose $\bar{u} \cdot \bar{v} = 0$ and $\bar{v} \neq \bar{0}$. Find the projections of $\bar{u}$ onto $\bar{v}$ and of $\bar{u}$ orthogonal to $\bar{v}$.

11 Establish the identities
a $|\bar{u} - \bar{v}|^2 + |\bar{u} + \bar{v}|^2 = 2|\bar{u}|^2 + 2|\bar{v}|^2$
b $\bar{u} \cdot \bar{v} = \frac{1}{4}|\bar{u} + \bar{v}|^2 - \frac{1}{4}|\bar{u} - \bar{v}|^2$
What does part **a** say about the diagonals of a parallelogram?

☐ 12 Use the inner product of Example **5** in each of the following.
a Find the length of $\cos \pi x$ in $C[0,1]$.
b Show that 1 and $\cos \pi x$ are orthogonal in $C[0,1]$.
c Find the projections of x onto 1 and $\cos \pi x$, and orthogonal to 1 and $\cos \pi x$ in $C[0,1]$.
d Show that $a + bx$ and $c + dx$ are orthogonal in $C[0,1]$ if and only if

$$ac + \frac{ad + bc}{2} + \frac{bd}{3} = 0$$

☐ 13 Show that if conditions **11** hold, the coefficient matrix of system **13** reduces to a reduced form with two nonzero rows.

☐ 14 Find a vector perpendicular to $(3,1,0)$ and $(-2,1,-1)$ by using the cross product.

☐ 15 Show that

$$|\bar{u} \times \bar{v}|^2 = |\bar{u}|^2|\bar{v}|^2 - (\bar{u} \cdot \bar{v})^2$$

and use this to deduce that

$$|\bar{u} \times \bar{v}| = |\bar{u}||\bar{v}| \sin \theta$$

where θ is the angle between $\bar{u}$ and $\bar{v}$.

☐ 16 **a** Show that $\bar{u} \times \bar{v} = -\bar{v} \times \bar{u}$.
b Show that $\bar{u} \times (\bar{v} \times \bar{w})$ lies in the plane of $\bar{v}$ and $\bar{w}$.
c Show that $(\bar{u} \times \bar{v}) \times \bar{w}$ lies in the plane of $\bar{u}$ and $\bar{v}$.
d Is it true that $\bar{u} \times (\bar{v} \times \bar{w}) = (\bar{u} \times \bar{v}) \times \bar{w}$?

☐ 17 The complex dot product is somewhat different. If $z = a + ib$, where a and b are real, then the *conjugate* z^* is defined by $z^* = a - ib$. Then $zz^* = a^2 + b^2$. The dot product of two vectors $\bar{u} = (x_1,x_2, \ldots, x_n)$ and $\bar{v} = (y_1,y_2, \ldots, y_n)$ in C^n is defined by $\bar{u} \cdot \bar{v} = x_1y_1^* + x_2y_2^* + \cdots + x_ny_n^*$
a Show that rules **1** all hold for this dot product, except that rules **1b** and **1d** are

$$\bar{u} \cdot \bar{v} = (\bar{v} \cdot \bar{u})^*$$
$$\bar{u} \cdot (a\bar{v}) = a^*(\bar{u} \cdot \bar{v})$$

b Find $\bar{u} \cdot \bar{v}$, $\bar{u} \cdot (i\bar{v})$ and $(3\bar{u} - i\bar{v}) \cdot (1 + i)\bar{u}$ for $\bar{u} = (1,i)$, $\bar{v} = (1 + i, 0)$.

c Show that $\bar{u} = (1,i)$ and $\bar{v} = (1, -i)$ are orthogonal. Find the length of $\bar{u}$ and $\bar{v}$.

d Find the projections of $\bar{u} = (i, 1 + i)$ onto and orthogonal to $\bar{v} = (1,i)$.

e If the conjugate were not used in the modified definition of $\bar{u} \cdot \bar{v}$, which of rules **1** would *not* be true?

SECTION 6 Orthogonal and Orthonormal Sets

In this section it will be seen that a set of nonzero mutually perpendicular vectors is automatically an independent set. A simple formula for calculating coordinates with respect to such a set will be given.

A collection of vectors is an **orthogonal** set if the vectors are mutually perpendicular. In other words, a set $\bar{u}_1, \bar{u}_2, \ldots, \bar{u}_k$ is orthogonal if

$$\bar{u}_i \cdot \bar{u}_j = 0 \qquad \text{whenever } i \neq j$$

Recall that the length of a vector $\bar{u}$ is given by $|\bar{u}| = (\bar{u} \cdot \bar{u})^{1/2}$, so that $|\bar{u}|^2 = \bar{u} \cdot \bar{u}$.

An orthogonal set may contain the zero vector. If an orthogonal set does not contain the zero vector, then one can normalize each vector in the set and obtain an orthogonal set in which the vectors all have length 1. For example, if $\bar{u}_1, \bar{u}_2, \ldots, \bar{u}_k$ is an orthogonal set, and none of the $\bar{u}_i$ are zero, one can then put

$$\bar{v}_1 = \frac{1}{|\bar{u}_1|}\bar{u}_1, \qquad v_2 = \frac{1}{|\bar{u}_2|}\bar{u}_2, \ldots, \qquad \bar{v}_k = \frac{1}{|\bar{u}_k|}\bar{u}_k$$

The vectors $\bar{v}_1, \bar{v}_2, \ldots, \bar{v}_k$ are an orthogonal set, and furthermore each of the vectors $\bar{v}_i$ has length 1. Such a set is usually called an orthonormal set.

A set $\bar{v}_1, \bar{v}_2, \ldots, \bar{v}_k$ is an **orthonormal** set if

$$\bar{v}_i \cdot \bar{v}_j = 0 \qquad \text{if} \qquad i \neq j$$

and

$$\bar{v}_i \cdot \bar{v}_i = 1$$

Formulas are often easier to remember when given in terms of orthonormal, rather than orthogonal sets. Before discussing the basic formula of this section, we present some examples.

EXAMPLE 1 The vectors $\bar{u}_1 = (1,0)$ and $\bar{u}_2 = (0,1)$ are orthonormal, for, clearly, $\bar{u}_1 \cdot \bar{u}_2 = 0$ and $\bar{u}_1 \cdot \bar{u}_1 = \bar{u}_2 \cdot \bar{u}_2 = 1$.

In general, the vectors $\bar{u}_1 = (1,0,0,\ldots,0)$, $\bar{u}_2 = (0,1,0,\ldots,0)$, $\ldots$, $\bar{u}_n = (0,0,0,\ldots,0,1)$ are orthonormal in R^n. In summary, the standard basis for R^n is orthonormal.

EXAMPLE 2 The vectors $\bar{u}_1 = (1,1)$ and $\bar{u}_2 = (1,-1)$ are orthogonal since

$$\bar{u}_1 \cdot \bar{u}_2 = 1 \cdot 1 + 1 \cdot (-1) = 0$$

We have

$$|\bar{u}_1|^2 = \bar{u}_1 \cdot \bar{u}_1 = 2 \qquad |\bar{u}_2|^2 = \bar{u}_2 \cdot \bar{u}_2 = 2$$

so that if we put $\bar{v}_1 = (1/\sqrt{2})\bar{u}_1$ and $\bar{v}_2 = (1/\sqrt{2})\bar{u}_2$, $\bar{v}_1$ and $\bar{v}_2$ are then orthonormal; that is

$$\bar{v}_1 \cdot \bar{v}_1 = \bar{v}_2 \cdot \bar{v}_2 = 1 \qquad \text{and} \qquad \bar{v}_1 \cdot \bar{v}_2 = 0$$

EXAMPLE 3 The vectors $\bar{u}_1 = (1,1,1,1)$, $\bar{u}_2 = (1,-1,1,-1)$, and $\bar{u}_3 = (1,2,-1,-2)$ are orthogonal. The student should check that

$$\bar{u}_1 \cdot \bar{u}_2 = \bar{u}_1 \cdot \bar{u}_3 = \bar{u}_2 \cdot \bar{u}_3 = 0$$

We have

$$\begin{aligned} |\bar{u}_1|^2 &= \bar{u}_1 \cdot \bar{u}_1 = 4 \\ |\bar{u}_2|^2 &= \bar{u}_2 \cdot \bar{u}_2 = 4 \\ |\bar{u}_3|^2 &= \bar{u}_3 \cdot \bar{u}_3 = 10 \end{aligned}$$

so that if

$$\bar{v}_1 = \tfrac{1}{2}\bar{u}_1 \qquad \bar{v}_2 = \tfrac{1}{2}\bar{u}_2 \qquad \bar{v}_3 = (1/\sqrt{10})\bar{u}_3$$

the vectors $\bar{v}_1$, $\bar{v}_2$, and $\bar{v}_3$ are orthonormal.

EXAMPLE 4 □ The definitions of orthogonal and orthonormal were formulated in terms of the dot product and thus can be given for functions in $C[0,1]$ relative to the inner product (f,g). (See Example 5, page 111.) We shall show that the functions 1, $\cos \pi x$, $\cos 2\pi x$, $\ldots$, $\cos n\pi x$ are orthogonal relative to this inner product.

For any integer m

$$(1, \cos m\pi x) = \int_0^1 1 \cos m\pi t \, dt = \frac{1}{m\pi} \sin m\pi t \Big|_0^1 = 0$$

and 1 is therefore orthogonal to each of the functions $\cos \pi x$, $\cos 2\pi x$, $\ldots$, $\cos n\pi x$. To show that these functions are orthogonal to each other we use the formula

1 $$\cos m\pi x \cos k\pi x = \tfrac{1}{2}[\cos (m + k)\pi x + \cos (m - k)\pi x]$$

It follows that if $m \neq k$

$$\begin{aligned}(\cos m\pi x, \cos k\pi x) &= \int_0^1 \cos m\pi t \cos k\pi t \, dt \\ &= \frac{1}{2}\int_0^1 (\cos (m + k)\pi t + \cos (m - k)\pi t) \, dt \\ &= \frac{1}{2}\left(\frac{\sin (m + k)\pi t}{(m + k)\pi} + \frac{\sin (m - k)\pi t}{(m - k)\pi}\right)\Bigg|_0^1 \\ &= 0\end{aligned}$$

In order to normalize this set we need to find the length of each of the functions, recalling that the length of f is $|f| = (f,f)^{1/2}$ or, in other words, $|f|^2 = (f,f)$.

Clearly, since

$$|1|^2 = \int_0^1 1 \, dt = 1$$

the constant function 1 has length 1. For the function $\cos m\pi x$ we use formula **1** with $k = m$. We have

$$\begin{aligned}(\cos m\pi x, \cos m\pi x) &= \int_0^1 \cos^2 m\pi t \, dt \\ &= \frac{1}{2}\int_0^1 (\cos 2m\pi t + 1) \, dt \\ &= \frac{1}{2}\left(\frac{\sin 2m\pi t}{2m\pi} + t\right)\Bigg|_0^1 \\ &= \frac{1}{2}\end{aligned}$$

so that the length of $\cos m\pi x$ is $1/\sqrt{2}$.

Thus, the functions $1, \sqrt{2}\cos \pi x, \sqrt{2}\cos 2\pi x, \ldots, \sqrt{2}\cos n\pi x$ are orthonormal.

DISCUSSION The next result shows that an orthonormal set is independent and provides a formula for calculation of coordinates with respect to an orthonormal basis. Formula **4** below gives the analogous result for orthogonal sets.

2 If $\bar{v}_1, \bar{v}_2, \ldots, \bar{v}_k$ is an orthonormal set and $\bar{u} = c_1\bar{v}_1 + c_2\bar{v}_2 + \cdots + c_k\bar{v}_k$, then $c_1 = \bar{u} \cdot \bar{v}_1$, $c_2 = \bar{u} \cdot \bar{v}_2, \ldots, c_k = \bar{u} \cdot \bar{v}_k$.

We shall give the proof of this statement only for the case $k = 2$. Suppose $\bar{v}_1$ and $\bar{v}_2$ are orthonormal and

$$\bar{u} = c_1\bar{v}_1 + c_2\bar{v}_2$$

Take the dot product of both sides with $\bar{v}_1$ and perform some elementary manipulations:

$$\bar{u} \cdot \bar{v}_1 = (c_1\bar{v}_1 + c_2\bar{v}_2) \cdot \bar{v}_1 = c_1(\bar{v}_1 \cdot \bar{v}_1) + c_2(\bar{v}_2 \cdot \bar{v}_1)$$

By assumption, $\bar{v}_1$ and $\bar{v}_2$ are orthonormal, and we can therefore substitute the relations $\bar{v}_1 \cdot \bar{v}_2 = 0$ and $\bar{v}_1 \cdot \bar{v}_1 = 1$ to obtain $c_1 = \bar{u} \cdot \bar{v}_1$.

A similar argument, using $\bar{v}_2$, shows that $c_2 = \bar{u} \cdot \bar{v}_2$.

Recall (formula **6**, page 109) that the projection of $\bar{u}$ along $\bar{v}$ is

$$\frac{\bar{u} \cdot \bar{v}}{\bar{v} \cdot \bar{v}} \bar{v}$$

In the case where $\bar{v} \cdot \bar{v} = 1$, this simplifies to $(\bar{u} \cdot \bar{v})\bar{v}$. Thus, the result **2** can be restated as:

3 A vector $\bar{u}$ is a linear combination of the orthonormal vectors $\bar{v}_1, \bar{v}_2, \ldots, \bar{v}_k$ if and only if $\bar{u}$ is the sum of its projections onto each of the $\bar{v}_i$.

The basic idea used in this discussion is that $\bar{v}_1$ and $\bar{v}_2$ are orthogonal and have length 1. The assumption that they have length 1 can be relaxed, as long as it is assumed that each is nonzero. This is summarized in the following:

Suppose $\bar{u}_1, \bar{u}_2, \ldots, \bar{u}_k$ is an orthogonal set such that each $\bar{u}_i \neq \bar{0}$:

4 If $\bar{u} = c_1\bar{u}_1 + c_2\bar{u}_2 + \cdots + c_k\bar{u}_k$, then

$$c_1 = \frac{\bar{u} \cdot \bar{u}_1}{\bar{u}_1 \cdot \bar{u}_1}, \; c_2 = \frac{\bar{u} \cdot \bar{u}_2}{\bar{u}_2 \cdot \bar{u}_2}, \ldots, c_k = \frac{\bar{u} \cdot \bar{u}_k}{\bar{u}_k \cdot \bar{u}_k}$$

This can be derived by an argument analogous to that used in deriving the result **2**. Let us instead show that it is a consequence of the result **2**. First divide each $\bar{u}_i$ by its length, that is, put

5 $$\bar{v}_i = \frac{1}{|\bar{u}_i|} \bar{u}_i \qquad i = 1, 2, \ldots, k$$

Then use this to rewrite

$$\bar{u} = c_1\bar{u}_1 + c_2\bar{u}_2 + \cdots + c_k\bar{u}_k$$

as

$$\bar{u} = (c_1|\bar{u}_1|)\bar{v}_1 + (c_2|\bar{u}_2|)\bar{v}_2 + \cdots + (c_k|\bar{u}_k|)\bar{v}_k$$

which expresses $\bar{u}$ as a linear combination of the orthonormal set $\bar{v}_1$, $\bar{v}_2, \ldots, \bar{v}_k$. The result **2** can then be applied to conclude that

$$c_1|\bar{u}_1| = \bar{u} \cdot \bar{v}_1,\ c_2|\bar{u}_2| = \bar{u} \cdot \bar{v}_2, \ldots, c_k|\bar{u}_k| = \bar{u} \cdot \bar{v}_k$$

Now replace $\bar{v}_i$ by its expression **5** to obtain

$$c_1|\bar{u}_1| = \bar{u} \cdot \frac{\bar{u}_1}{|\bar{u}_1|},\ c_2|\bar{u}_2| = \bar{u} \cdot \frac{\bar{u}_2}{|\bar{u}_2|}, \ldots, c_k|\bar{u}_k| = \bar{u} \cdot \frac{\bar{u}_k}{|\bar{u}_k|}$$

Solve for the c_i and use the fact that $\bar{u}_i \cdot \bar{u}_i = |\bar{u}_i|^2$, $i = 1, 2, \ldots, k$, to obtain the desired result **4**:

$$c_1 = \frac{\bar{u} \cdot \bar{u}_1}{\bar{u}_1 \cdot \bar{u}_1},\ c_2 = \frac{\bar{u} \cdot \bar{u}_2}{\bar{u}_2 \cdot \bar{u}_2}, \ldots, c_k = \frac{\bar{u} \cdot \bar{u}_k}{\bar{u}_k \cdot \bar{u}_k}\,.$$

Let us note one important consequence of these results:

6 An orthonormal set (or an orthogonal set of nonzero vectors) is independent.

EXAMPLE 5 The standard basis for R^n is an orthonormal basis for R^n. For example, $\bar{v}_1 = (1,0,0)$, $\bar{v}_2 = (0,1,0)$, and $\bar{v}_3 = (0,0,1)$ are an orthonormal basis for R^3. In this case, formula **2** has a simple form, for we have

$$(x,y,z) = x\bar{v}_1 + y\bar{v}_2 + z\bar{v}_3$$

In other words

$$(x,y,z) \cdot \bar{v}_1 = x \qquad (x,y,z) \cdot \bar{v}_2 = y \qquad (x,y,z) \cdot \bar{v}_3 = z$$

EXAMPLE 6 The vectors $\bar{u}_1 = (1,1)$ and $\bar{u}_2 = (1,-1)$ are orthogonal, and thus statement **6** implies that they are independent. Since R^2 has dimension 2, we see from Theorem **5** that $\bar{u}_1$ and $\bar{u}_2$ are a basis for R^2. To express $\bar{u} = (x,y)$ in terms of this basis we can either use formula **4** directly or use the process by which we deduced formula **4** from formula **2**. It is often easier to remember formula **2** and to reconstruct this process.

We put

$$\bar{v}_1 = \frac{1}{\sqrt{2}}(1,1) \qquad \bar{v}_2 = \frac{1}{\sqrt{2}}(1,-1)$$

so that $\bar{v}_1$ and $\bar{v}_2$ are an orthonormal basis for R^2. For $\bar{u} = (x,y)$ we have

$$\bar{u} \cdot \bar{v}_1 = \frac{x + y}{\sqrt{2}} \qquad \bar{u} \cdot \bar{v}_2 = \frac{x - y}{\sqrt{2}}$$

and

$$(\bar{u} \cdot \bar{v}_1)\bar{v}_1 = \left[\frac{x + y}{\sqrt{2}}\right]\left[\frac{1}{\sqrt{2}}\right](1,1) = \frac{x + y}{2}\,\bar{u}_1$$

$$(\bar{u} \cdot \bar{v}_2)\bar{v}_2 = \left[\frac{x - y}{\sqrt{2}}\right]\left[\frac{1}{\sqrt{2}}\right](1,-1) = \frac{x - y}{2}\,\bar{u}_2$$

so that

$$\bar{u} = \frac{x + y}{2}\,\bar{u}_1 + \frac{x - y}{2}\,\bar{u}_2$$

Therefore, the coordinates of $\bar{u} = (x,y)$ with respect to $\bar{u}_1$ and $\bar{u}_2$ are

$$\frac{x + v}{2}, \frac{x - y}{2}.$$

EXAMPLE 7 We shall show that $\bar{u} = (1,2,3,4)$ is *not* a linear combination of the orthogonal vectors

$$\bar{u}_1 = (1,1,1,1) \qquad \bar{u}_2 = (1,-1,1,-1) \qquad \bar{u}_3 = (1,2,-1,-2)$$

by showing that $\bar{u}$ is not the sum of its projections onto $\bar{u}_1$, $\bar{u}_2$, and $\bar{u}_3$. Put

$$\bar{v}_1 = \tfrac{1}{2}\bar{u}_1 \qquad \bar{v}_2 = \tfrac{1}{2}\bar{u}_2 \qquad \bar{v}_3 = \left(\frac{1}{\sqrt{10}}\right)\bar{u}_3$$

so that the projections of $\bar{u}$ onto $\bar{u}_1$, $\bar{u}_2$, and $\bar{u}_3$ are, respectively,

$$(\bar{u} \cdot \bar{v}_1)\bar{v}_1 = \left(\frac{10}{2}\right)\left(\frac{1}{2}\right)\bar{u}_1 = \frac{5}{2}(1,1,1,1)$$

$$(\bar{u} \cdot \bar{v}_2)\bar{v}_2 = \left(\frac{-2}{2}\right)\left(\frac{1}{2}\right)\bar{u}_2 = \frac{-1}{2}(1,-1,1,-1)$$

$$(\bar{u} \cdot \bar{v}_3)\bar{v}_3 = \left(\frac{-6}{\sqrt{10}}\right)\left(\frac{1}{\sqrt{10}}\right)\bar{u}_3 = \frac{-3}{5}(1,2,-1,-2)$$

The sum of these projections is

$$\bar{v} = \tfrac{1}{5}(7,9,13,21)$$

which is not equal to $\bar{u}$, so that from result **4**, $\bar{u}$ is *not* a linear combination of $\bar{u}_1$, $\bar{u}_2$, and $\bar{u}_3$.

Put

$$\bar{w} = \bar{u} - \bar{v} = \tfrac{1}{5}(-2,1,2,-1)$$

By the usual computations we have

$$\bar{w} \cdot \bar{u}_1 = \bar{w} \cdot \bar{u}_2 = \bar{w} \cdot \bar{u}_3 = 0$$

The set $\bar{u}_1$, $\bar{u}_2$, $\bar{u}_3$, and $\bar{w}$ therefore is an orthogonal set. In general, if we subtract from $\bar{u}$ its projections onto each member of an orthogonal set, we obtain a vector which is orthogonal to each member of the orthogonal set. This observation is the key to a general method of obtaining an orthonormal basis from any given basis. The method is called the Gram-Schmidt process and is outlined in Exercise **11**.

EXERCISES

1 Show that each of the following sets of vectors is orthogonal.
 a (1,2), (−2,1)
 b (1,0,1), (1,1,−1), (−1,2,1)
 c (1,0,1,0), (1,1,−1,0), (1,−2,−1,1)

2 Normalize each of the sets in Exercise **1** to obtain an orthonormal set.

3 Find the coordinates of $\bar{u} = (3,1)$ with respect to the orthogonal basis $\bar{u}_1 = (1,2)$, $\bar{u}_2 = (-2,1)$ by using projection methods. (See statement **4**.)

4 Find the coordinates of $\bar{u} = (-1,3,5)$ with respect to the orthogonal basis $\bar{u}_1 = (1,0,1)$, $\bar{u}_2 = (1,1,-1)$, $\bar{u}_3 = (-1,2,1)$ by using projection methods. (See statement **4**.)

5 Express (3,2,1) as a linear combination of (1,0,1), (1,1,−1), and (−1,2,1) by using statement **4**.

6 Show that (1,4,1) is *not* a linear combination of (1,0,1) and (1,1,−1) by using statement **4**. (See Example **7**.)

7 By subtracting from (1,4,1) its projections onto (1,0,1) and (1,1,−1) find a nonzero vector which is orthogonal to (1,0,1) and (1,1,−1).

□ 8 Show that the functions $\sin \pi x$, $\sin 2\pi x$, . . ., $\sin n\pi x$ are orthogonal in $C[0,1]$. By normalizing obtain an orthonormal set of functions.

9 Find an orthogonal basis for R^4 that includes the vectors (1,0,1,0), (1,1,−1,0), and (1,−2,−1,1). (Hint: Find a vector $\bar{u}$ that is not a linear combination of the given vectors and subtract from $\bar{u}$ its projections onto the given vectors.)

10 Find an orthogonal basis for R^3 which includes $\bar{u}_1 = (1,2,3)$. (Hint: Select a vector which is not a multiple of $\bar{u}_1$ and let $\bar{u}_2$ be its projection orthogonal to $\bar{u}_1$. Construct a third vector $\bar{u}_3$ by proceeding as in Exercise **9**.)

□ 11 Suppose $\bar{w}_1$, $\bar{w}_2$, and $\bar{w}_3$ are a basis for R^3. Find an orthogonal basis $\bar{u}_1$, $\bar{u}_2$, and $\bar{u}_3$ such that
 a $\bar{w}_1$ is a multiple of $\bar{u}_1$.

b $\overline{w}_2$ is a linear combination of $\overline{u}_1$ and $\overline{u}_2$. (Hint: Put $\overline{u}_1 = \overline{w}_1$, $\overline{u}_2 =$ projection of $\overline{w}_2$ orthogonal to $\overline{u}_1$, and $\overline{u}_3 = \overline{w}_3$ minus the projections of $\overline{w}_3$ onto $\overline{u}_1$ and $\overline{u}_2$. This process is called the *Gram-Schmidt process*.)

c Apply this process to the basis (1,1,1), (1,2,1), and (1,1,2).

d Apply this process to the functions 1, x, and x^2 to find three orthogonal polynomials of degree not exceeding 2, using the inner product of Example **4**.

☐ **12** Using the complex dot product of Exercise **17** of the previous section:

a Show that $(1,i,1)$, $(i,0,-i)$, and $(i,2,i)$ are orthogonal.

b Normalize the vectors of part **a** to obtain an orthonormal set of vectors.

c Express $(1,1,i)$ as a linear combination of the vectors of part **a** by using formula **2**.

Supplement on Planes and Projection

In this section we use vector methods to describe planes and show how to find distances by using projection methods.

PLANES

Suppose $\overline{n} = a\overline{i} + b\overline{j} + c\overline{k}$ is a nonzero vector. The plane through (x_0,y_0,z_0) which is *perpendicular* to $\overline{n}$ can be described as follows: A point (x,y,z) lies in this plane if and only if

$$(x - x_0)\overline{i} + (y - y_0)\overline{j} + (z - z_0)\overline{k}$$

is *orthogonal* to $\overline{n}$. (See Figure 8.)

Figure 8

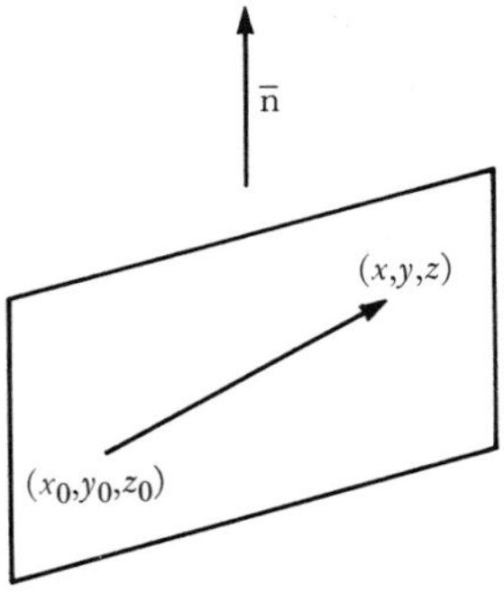

In other words, this plane consists of all points (x,y,z) such that

$$[(x - x_0)\bar{\text{i}} + (y - y_0)\bar{\text{j}} + (z - z_0)\bar{\text{k}}] \cdot \bar{\text{n}} = 0$$

Calculating this dot product gives the equation

1 $$a(x - x_0) + b(y - y_0) + c(z - z_0) = 0$$

In summary, this must be a *Cartesian equation of the plane through* (x_0,y_0,z_0) *with normal vector* $a\bar{\text{i}} + b\bar{\text{j}} + c\bar{\text{k}}$.

This is often the most useful description of a plane. We can, however, also describe this plane as a translation of a subspace (which is a plane through the origin). For clearly the plane of equation 1 is parallel to the plane

2 $$ax + by + cz = 0$$

since these two planes have the same normal vector. The plane of equation 2 is just the null space of the matrix

$$[a \quad b \quad c]$$

Since, by assumption, $\bar{\text{n}} \neq \bar{0}$, at least one of a, b, and c is *not* zero. Thus, since this null space is two-dimensional, it is the subspace spanned by $\bar{\text{u}} = (x_1,y_1,z_1)$ and $\bar{\text{v}} = (x_2,y_2,z_2)$, where the vectors $\bar{\text{u}}$ and $\bar{\text{v}}$ are independent solutions to equation 2. In other words, the plane 2 consists of all vectors of the form $s\bar{\text{u}} + t\bar{\text{v}}$.

Put $\bar{\text{w}}_0 = (x_0,y_0,z_0)$. Then the plane of equation 1 consists of all vectors $\bar{\text{w}}$ of the form

3 $$\bar{\text{w}} = s\bar{\text{u}} + t\bar{\text{v}} + \bar{\text{w}}_0$$

With $\bar{\text{w}} = (x,y,z)$ we can express this equation in coordinate form:

4 $$\begin{aligned} x &= sx_1 + tx_2 + x_0 \\ y &= sy_1 + ty_2 + y_0 \\ z &= sz_1 + tz_2 + z_0 \end{aligned}$$

Equation 3 is known as a *vector equation* of the plane 1, while the set of equations 4 is known as a *parametric form* of this plane with *parameters s and t*.

The cross product discussed in Example 6, page 113, is a useful tool in discussing planes. For example, if $\bar{\text{u}}$ and $\bar{\text{v}}$ are independent solutions to equation 2, then $\bar{\text{u}} \times \bar{\text{v}}$ is normal to the plane of this equation and hence parallel to $\bar{\text{n}}$.

The plane through three noncollinear points $\bar{u}_1$, $\bar{u}_2$, and $\bar{u}_3$ can be described in several ways. We can directly substitute the coordinates of these three vectors for (x,y,z) in equation **1** and solve for a, b, and c in the resulting three equations; or we can use the fact that this plane passes through $\bar{u}_1$ and is parallel to the plane of $\bar{u}_1 - \bar{u}_2$ and $\bar{u}_1 - \bar{u}_3$ to obtain the vector description

$$\bar{w} = s(\bar{u}_1 - \bar{u}_2) + t(\bar{u}_1 - \bar{u}_3) + \bar{u}_1$$

Using the cross product we can obtain a normal vector and then use equation **1**, for the vectors $\bar{u}_1 - \bar{u}_2$ and $\bar{u}_1 - \bar{u}_3$ are parallel to this plane and consequently

$$\bar{n} = (\bar{u}_1 - \bar{u}_2) \times (\bar{u}_1 - \bar{u}_3)$$

is normal to the plane. For example, to describe the plane through (1,2,1), (−1,1,3), and (2,1,0) we find the normal

$$\begin{aligned}\bar{n} &= [(1,2,1) - (-1,1,3)] \times [(1,2,1) - (2,1,0)]\\ &= (2,1,-2) \times (-1,1,1)\\ &= (3,0,3)\end{aligned}$$

so that the equation of this plane is

$$3(x - 1) + 0(y - 2) + 3(z - 1) = 0$$

PROJECTION METHODS

Projection methods are particularly useful in determining distances. For convenience we list here the projection formulas of page 109:

5 The projection of $\bar{u}$ onto $\bar{v}$ is $\dfrac{\bar{u} \cdot \bar{v}}{\bar{v} \cdot \bar{v}}\,\bar{v}$.

6 The projection of $\bar{u}$ orthogonal to $\bar{v}$ is $\bar{u} - \dfrac{\bar{u} \cdot \bar{v}}{\bar{v} \cdot \bar{v}}\,\bar{v}$.

These formulas are used below, in a number of examples, to solve distance problems.

DISTANCE FROM A POINT TO A LINE: To find the distance from (−5,−3) to the line L:

$$2x - y = 4$$

we first determine a vector parallel to this line. We can write this equation as

$$\frac{x}{1} = \frac{y + 4}{2}$$

so that the vector $\bar{v} = \bar{i} + 2\bar{j}$ is parallel to this line. Furthermore $(0, -4)$ lies on this line.

The vector

$$\bar{u} = [0 - (-5)]\bar{i} + [-4 - (-3)]\bar{j} = 5\bar{i} + \bar{j}$$

connects $(-5, -3)$ to $(0, -4)$. Thus, the projection $\bar{w}$ of $\bar{u}$ orthogonal to $\bar{v}$ is perpendicular to L and can be drawn so as to connect $(-5, -3)$ to L, as shown in Figure 9.

Figure 9

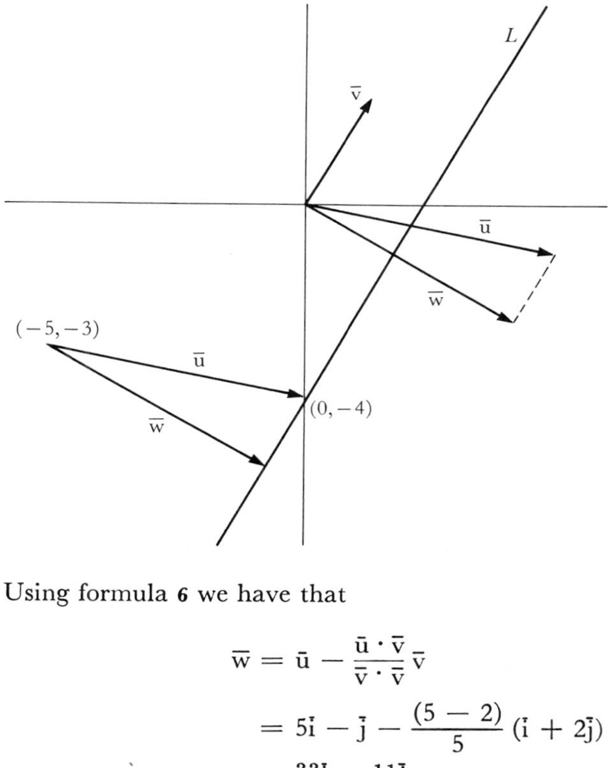

Using formula **6** we have that

$$\begin{aligned}\bar{w} &= \bar{u} - \frac{\bar{u} \cdot \bar{v}}{\bar{v} \cdot \bar{v}}\bar{v} \\ &= 5\bar{i} - \bar{j} - \frac{(5 - 2)}{5}(\bar{i} + 2\bar{j}) \\ &= \tfrac{22}{5}\bar{i} - \tfrac{11}{5}\bar{j}\end{aligned}$$

so our desired distance is

$$|\bar{w}| = \sqrt{(\tfrac{22}{5})^2 + (\tfrac{11}{5})^2} = \tfrac{11}{5}\sqrt{5}$$

DISTANCE FROM A POINT TO A PLANE: To find the distance from (2,1,3) to the plane

$$x + 2y - 2z = 4$$

we note that $\bar{v} = \bar{i} + 2\bar{j} - 2\bar{k}$ is perpendicular to this plane (from equation 1) and that (0,0,−2) lies in the plane. Thus

$$\begin{aligned}\bar{u} &= (0 - 2)\bar{i} + (0 - 1)\bar{j} + (-2 - 3)\bar{k} \\ &= -2\bar{i} - \bar{j} - 5\bar{k}\end{aligned}$$

connects (2,1,3) to (0,0,−2). The projection $\bar{w}$ of $\bar{u}$ *onto* $\bar{v}$ is normal to the plane and can be drawn so as to connect (2,1,3) to the plane. (See Figure 10.)

Figure 10

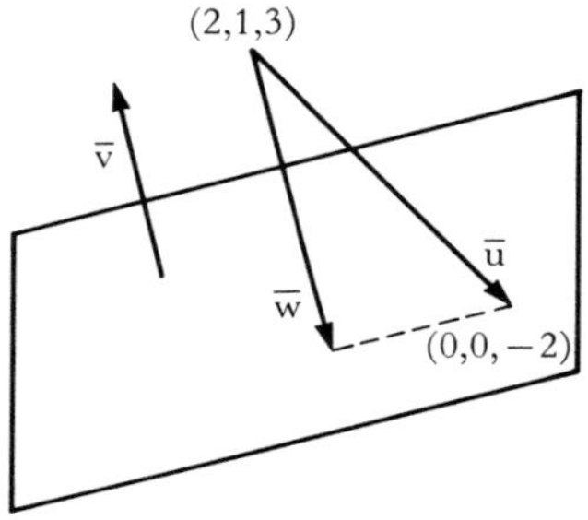

Formula 5 gives

$$\bar{w} = \frac{\bar{u}\cdot\bar{v}}{\bar{v}\cdot\bar{v}}\bar{v} = \tfrac{6}{9}\bar{v} = \tfrac{6}{9}\bar{i} + \tfrac{12}{9}\bar{j} - \tfrac{12}{9}\bar{k}$$

so our desired distance is

$$|\bar{w}| = \sqrt{(\tfrac{6}{9})^2 + (\tfrac{12}{9})^2 + (\tfrac{12}{9})^2} = 2$$

DISTANCE BETWEEN TWO PLANES: To find the distance between the two parallel planes

$$2x - 3y + 2z = 1 \qquad \text{and} \qquad 2x - 3y + 2z = 3$$

we note that $(0,0,\frac{1}{2})$ and $(0,0,\frac{3}{2})$, respectively, lie on these planes, so that

$$\bar{u} = (\tfrac{1}{2} - \tfrac{3}{2})\bar{k} = -\bar{k}$$

connects these two points. The normal to each plane is $\bar{v} = 2\bar{i} - 3\bar{j} + 2\bar{k}$; therefore, the desired distance is the length of the projection of $\bar{u}$ onto $\bar{v}$.

This projection is

$$\bar{w} = \frac{\bar{u} \cdot \bar{v}}{\bar{v} \cdot \bar{v}} \bar{v} = \frac{-2}{17} \bar{v}$$

so our desired distance is

$$|\bar{w}| = \frac{2}{17} |\bar{v}| = \frac{2\sqrt{17}}{17}$$

DISTANCE FROM A POINT TO THE INTERSECTION OF TWO PLANES: The two planes

$$x + 2y - 3z = 0 \qquad x - 3y + z = -1$$

intersect in a line through (1,1,1). To find the distance from (−1,2,−1) to this line we need to find a vector which is parallel to this line so we can use the method we used above to find the distance from a point to a line. The respective normals are

$$\bar{n}_1 = \bar{i} + 2\bar{j} - 3\bar{k} \qquad \text{and} \qquad \bar{n}_2 = \bar{i} - 3\bar{j} + \bar{k}$$

so that $\bar{n}_1 \times \bar{n}_2$ is parallel to our line. Since

$$\begin{aligned} \bar{u} &= [1 - (-1)]\bar{i} + (1 - 2)\bar{j} + [1 - (-1)]\bar{k} \\ &= 2\bar{i} - \bar{j} + 2\bar{k} \end{aligned}$$

connects (−1,2,−1) to (1,1,1), our desired distance is just the length of the projection of $\bar{u}$ orthogonal to $\bar{n}_1 \times \bar{n}_2$.

We have

$$\begin{aligned} \bar{v} &= \bar{n}_1 \times \bar{n}_2 = -7\bar{i} - 4\bar{j} - 5\bar{k} \\ \bar{u} &= \frac{\bar{u} \cdot \bar{v}}{\bar{v} \cdot \bar{v}} \bar{v} = \frac{-4}{9}\bar{i} + \frac{17}{9}\bar{j} - \frac{8}{9}\bar{k} \end{aligned}$$

so the distance is $\frac{1}{3}\sqrt{41}$.

EXERCISES

Planes

1 Find an equation for the plane through (1,2,1) perpendicular to $2\bar{i} - \bar{j} - \bar{k}$.

2 Find a normal vector to the plane

$$2x - 3y + 4z = 1$$

3 Find a vector equation for the plane of Exercise 2.

4 Find in *two* ways a Cartesian equation for the plane through (2,1,0), (−1,1,0), (3,1,1).

5 Find an equation for the plane through (−1,1,1) that is perpendicular to the line of intersection of $x - 2y + 2z = 1$ and $3x - z = 4$.

6 Find a plane through (6,1,2) that is parallel to the plane $x - 3y + 2z = 1$.

7 Show that the line

$$\frac{x+1}{3} = \frac{y}{4} = \frac{z-2}{5}$$

is perpendicular to the plane $3x + 4y + 5z = 10$ and parallel to the plane $5x + 5y - 7z = 1$.

8 Find the point of intersection of the lines

$$\frac{x}{3} = \frac{y}{4} = \frac{z-2}{4} \qquad \text{and} \qquad x - 1 = \frac{y+4}{-4} = \frac{z}{-2}$$

9 Find the plane determined by the lines of Exercise **8**.

Projection Methods

10 Find the distance from (2,1) to $x + 5y = 3$.

11 Find the distance from $(3,1,-1)$ to the line

$$\frac{x-1}{2} = \frac{y+1}{3} = \frac{z-1}{5}$$

12 Find the distance from $(3,1,-1)$ to the line through (0,1,3) which is parallel to the line of intersection of $x - y + z = 1$, $2x + y = 0$.

13 Find the distance from $(-1,1,2)$ to the line $\overline{w} = t(2,1,1) + (3,1,0)$.

14 Find the distance from $(1,2,-1)$ to the line through $(-1,1,1)$, and (0,0,1).

15 Find the distance from the origin to $2x + y - 3z = 4$.

16 Find the distance between the two parallel lines

$$\frac{x-1}{2} = \frac{y+1}{3} = \frac{z}{4} \qquad \frac{x}{2} = \frac{y-2}{3} = \frac{z+1}{4}$$

17 Find the distance between the two *skew* lines

$$\frac{x-1}{2} = \frac{y+1}{3} = \frac{z}{4} \qquad x = y = z - 1$$

(Hint: Find a point on each and a vector orthogonal to each line.)

18 Find the distance between the two planes

$$2x + y + z = 4 \qquad \text{and} \qquad 2x + y + z = 8$$

19 Find the distance from the point $(2,-1,1)$ to the intersection of

$$x + y + z = 1 \qquad \text{and} \qquad x - y + 2z = 0$$

20 Show that the diagonals of a parallelogram are perpendicular if and only if it is a rhombus. Show that they have equal length if and only if it is a rectangle.

21 Show that the altitudes of a triangle meet in a point.

CHAPTER 3

TRANSFORMATIONS AND MATRICES

The process of multiplying a column matrix by a matrix was introduced in Section 5 of Chapter 1. We then saw (Theorem 2) that this process satisfied the laws

$$A(\bar{u} + \bar{v}) = A\bar{u} + A\bar{v} \qquad \text{and} \qquad A(a\bar{u}) = a(A\bar{u})$$

These laws were particularly useful in obtaining the theorems about the solutions to linear systems given in Chapter 1.

Our task in this chapter is to study processes or operations on vectors that satisfy the same laws. These operations are commonly known as linear transformations or linear operations and include such familiar and important geometric operations as rotation, reflection, and projection. These examples will be discussed in Section 1. We shall then show in Section 2 that for the coordinate spaces R^n, every linear transformation is given by the process of multiplying by a matrix. We can then easily derive algebraic formulas for many geometric operations.

The concepts of sum, scalar multiple, and product of linear transformations are introduced in Sections 3 and 4. These concepts will enable us to combine various linear operations to obtain new linear operations, as well as to describe relationships among these processes. They will also lead to an analogous "arithmetic" for matrices, which will be studied further in Section 5.

SECTION 1 Linear Transformations

The simplest kind of real-valued functions of one real variable whose graph passes through the origin is a function f of the form "multiplication by a"; that is, for all x

$$f(x) = ax$$

This function is linear in that its graph is a straight line.

By analogy the simplest kind of vector-valued function of a vector variable is "multiplication by a matrix A"; that is, it is a function defined by the rule

1 $$T(\bar{u}) = A\bar{u}$$

Here we shall use the letter T, for "transformation," and reserve f for functions of one variable. This function T is defined only for vectors $\bar{u}$ in R^n (expressed as column matrices) where n is the number of columns of A. For each $\bar{u}$, the vector $T(\bar{u})$ lies in R^m, where m is the number of rows of A.

The graph of a vector-valued function of a vector variable is difficult to visualize, and hence it is not as useful a concept as it is in the one-variable case and we are forced to rely more on algebraic properties. The basic properties of the function T given by statement **1** are

2 $$T(\bar{u} + \bar{v}) = T(\bar{u}) + T(\bar{v}) \qquad T(a\bar{u}) = aT(\bar{u})$$

These are, of course, just the properties of matrix multiplication summarized in Theorem **2**.

Our purpose in this chapter is to analyze functions which satisfy properties **2**. Let us introduce some definitions.

An operation T which is defined for all vectors $\bar{u}$ in a vector space V, resulting in vectors $T\bar{u}$ in a vector space W, is called a **transformation** from V into W. If this operation satisfies the conditions

2 $$T(\bar{u} + \bar{v}) = T\bar{u} + T\bar{v} \qquad \text{and} \qquad T(a\bar{u}) = a(T\bar{u})$$

it is called a **linear transformation** or **linear operation**.

It is common practice to write $T\bar{u}$ rather than $T(\bar{u})$, using parentheses only where needed for grouping. The primary example of a linear transformation is the operation of multiplication by a matrix. In fact, it will be shown in the next section that every linear transformation from R^n into R^m is really just multiplication by a suitable matrix. This is, however, not always the most convenient way to describe a linear transformation. The examples below give a number of linear transformations defined geometrically.

REMARK We shall usually denote a transformation by a single letter, such as T. In this case, T is the *name* of the operation, while $T\bar{u}$ denotes the *effect* of the operation on the vector $\bar{u}$. Thus, for example, the symbol T of formula **1** is a shorthand expression for the operation "multiply a column matrix by the matrix A," while $T\bar{u}$ is the *vector* obtained by multiplying $\bar{u}$ by A.

This is really a new way to think of a matrix. In Chapter 1 we saw how systems of equations could be solved by reducing a matrix. We also saw how to rewrite a system as a matrix equation, resulting in a simplification of notation, as well as providing a framework for treating the concepts of independence and bases. In this chapter we shall think of a matrix as defining the operation of multiplying by that matrix. Not only does this approach lead us to consider other operations with the same properties (the properties of conditions **2**), but it will also lead us to new concepts about vectors and matrices.

To distinguish between the matrix A and the operation of multiplying by A we shall generally denote the operation by a letter other than A, such as T.

EXAMPLE 1 *Projection.*

The operation of projection discussed in Section 5 of Chapter 2 is a linear operation, for it satisfies conditions **2**. We first give a geometric discussion of this fact and then show how one can use the dot product to obtain the same conclusions.

For convenience let us suppose $\bar{w}$ is a vector of length 1 in R^2 and that P is the name of the operation of projection onto the line through $\bar{w}$. For a given pair $\bar{u}$ and $\bar{v}$, Figure 1 shows that $P(\bar{u} + \bar{v})$ can be calculated in two ways: Find the sum $\bar{u} + \bar{v}$ and then project this along $\bar{w}$; or project $\bar{u}$ and $\bar{v}$ each along $\bar{w}$ and form the sum of these projections. In other words,

$$P(\bar{u} + \bar{v}) = P\bar{u} + P\bar{v}$$

We can also calculate $P(a\bar{u})$ in two ways: Multiply $\bar{u}$ by a and then project; or project $\bar{u}$ and then multiply by a, as shown in Figure 2. In other words, $P(a\bar{u}) = aP\bar{u}$.

We have shown, therefore, that the projection P does satisfy the linearity conditions **2**. Since the operation P is defined for vectors $\bar{u}$ in R^2 and results in vectors $P\bar{u}$ in R^2, we can conclude that the projection P is a *linear transformation* from R^2 into R^2. The same conclusion can be reached by using the dot product properties given in Section 5 of Chapter 2. We recall (see formula **6**, page 109) that $P\bar{u} = (\bar{u} \cdot \bar{w})\bar{w}$, for $\bar{w}$ has length 1.

Figure 1 $P(\bar{u} + \bar{v}) = P\bar{u} + P\bar{v}$.

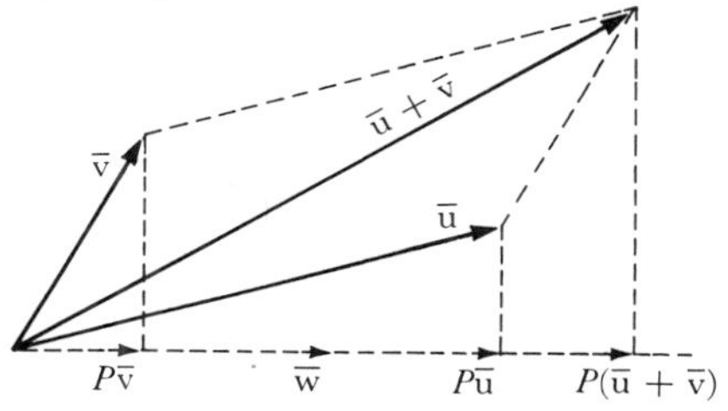

Figure 2 $P(a\bar{u}) = aP(\bar{u})$.

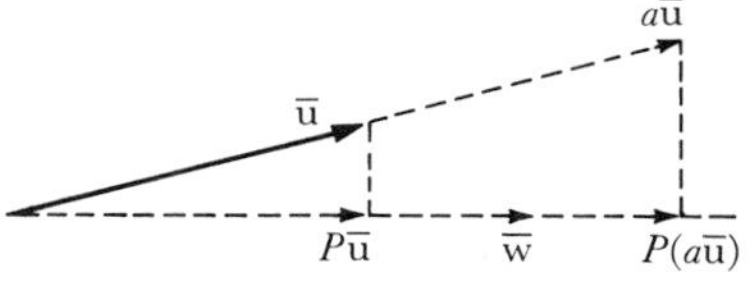

Since $(\bar{u} + \bar{v}) \cdot \bar{w} = \bar{u} \cdot \bar{w} + \bar{v} \cdot \bar{w}$, it follows that

$$\begin{aligned} P(\bar{u} + \bar{v}) &= [(\bar{u} + \bar{v}) \cdot \bar{w}]\bar{w} = (\bar{u} \cdot \bar{w} + \bar{v} \cdot \bar{w})\bar{w} \\ &= (\bar{u} \cdot \bar{w})\bar{w} + (\bar{v} \cdot \bar{w})\bar{w} \\ &= P\bar{u} + P\bar{v} \end{aligned}$$

We also have $(a\bar{u} \cdot \bar{w}) = a(\bar{u} \cdot \bar{w})$; therefore

$$P(a\bar{u}) = (a\bar{u} \cdot \bar{w})\bar{w} = a(\bar{u} \cdot \bar{w})\bar{w} = a(P\bar{u})$$

Thus the fact that the projection P satisfies the linearity conditions 2 is a consequence of properties of the dot product.

EXAMPLE 2 *Transformations of Coordinates.*

Transformations can also be defined by describing their effect on coordinates. For example, suppose T is the operation defined by the formula

$$T(x,y) = (-x,y)$$

In other words, the first coordinate of $T\bar{u}$ is just the negative of the first coordinate of $\bar{u}$, while the second coordinate of $T\bar{u}$ is the second coordinate of $\bar{u}$. This formula can be rewritten in column form as

$$T\begin{bmatrix} x \\ y \end{bmatrix} = \begin{bmatrix} -x \\ y \end{bmatrix}$$

Since

$$\begin{bmatrix} -x \\ y \end{bmatrix} = \begin{bmatrix} -1 & 0 \\ 0 & 1 \end{bmatrix} \begin{bmatrix} x \\ y \end{bmatrix}$$

we see that the transformation T is given by

$$T\bar{u} = A\bar{u} \qquad \text{where} \qquad A = \begin{bmatrix} -1 & 0 \\ 0 & 1 \end{bmatrix}$$

In other words, T is just the operation of multiplication by A; hence T is a linear transformation from R^2 into R^2. As we can see from Figure 3, T can also be described as the operation which sends a vector into its reflection in the y-axis.

EXAMPLE 3 *Reflection.*

Suppose $\bar{w}$ is a vector of length 1 in R^2. The operation T for vectors $\bar{u}$ in R^2 is defined as follows:

> $T\bar{u}$ is the vector such that $\frac{1}{2}(T\bar{u} + \bar{u})$ is the projection $P\bar{u}$ of $\bar{u}$ onto $\bar{w}$.

Figure 3

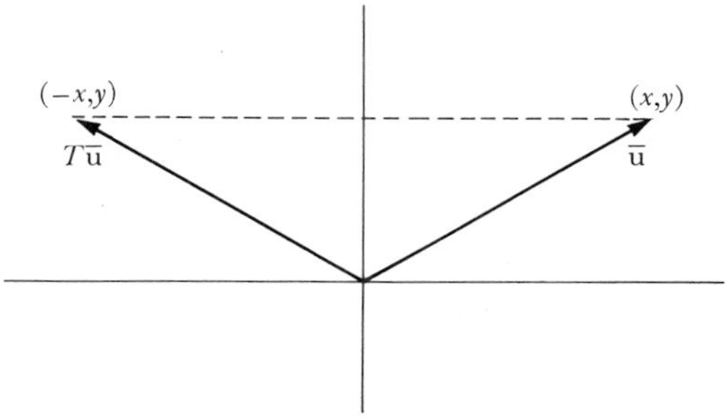

As can be seen in Figure 4, $T\bar{u}$ is just the *reflection* of $\bar{u}$ in the line through $\bar{w}$. We therefore have $T\bar{u} = 2P\bar{u} - \bar{u}$. Arguments similar to those used in Example 1 can be given to show that T is linear, that is, that $T(\bar{u} + \bar{v})$ can be calculated by first calculating $\bar{u} + \bar{v}$ and reflecting; or it can be calculated by reflecting $\bar{u}$, reflecting $\bar{v}$, and then adding. Similar arguments show that $T(a\bar{u})$ can also be calculated in two ways: By first calculating $a\bar{u}$ and then reflecting, or by reflecting $\bar{u}$ and then multiplying by a.

Figure 4 $\frac{1}{2}(T\bar{u} + \bar{u}) = P\bar{u}$.

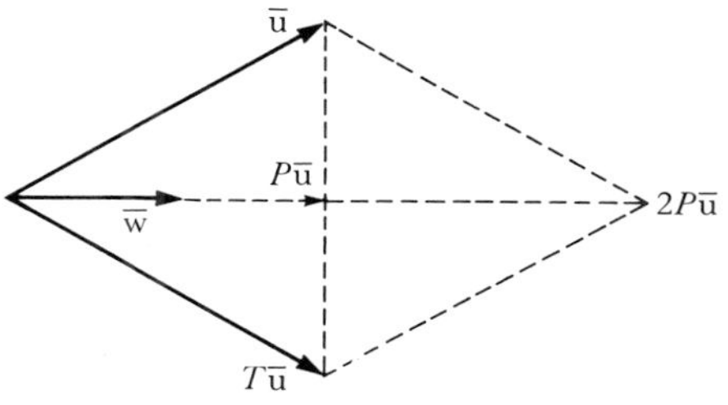

Figure 5 *The operation of counterclockwise rotation through the angle* θ.

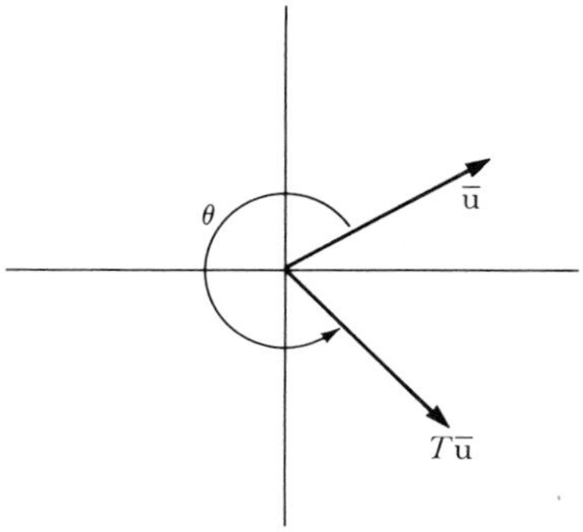

EXAMPLE 4 *Rotation.*

The operation of rotation in a plane is a linear transformation. For example, suppose θ is a fixed number and that we represent vectors in R^2 as arrows issuing from the origin in the usual fashion. Let T denote the operation of counterclockwise rotation through the angle θ. In other words, $T\bar{u}$ is the arrow obtained by rotating $\bar{u}$ around the origin, as shown in Figure 5.

Figure 6a $T(\bar{u} + \bar{v}) = T\bar{u} + T\bar{v}$. *Therefore, the rotation of the sum is the sum of the rotations.*

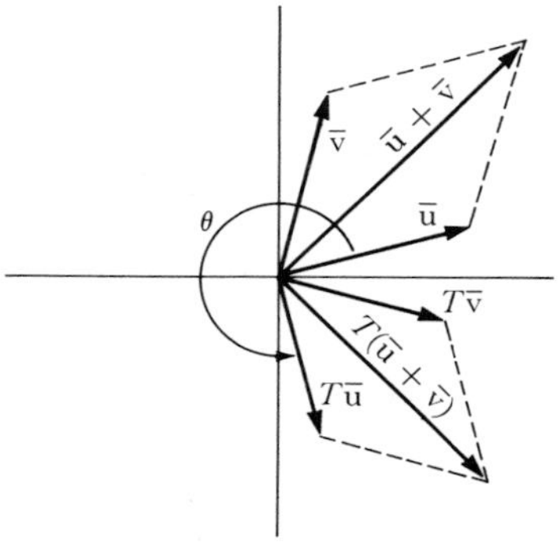

Figure 6b $T(a\bar{u}) = aT\bar{u}$. *Therefore, the multiple of a rotation is the rotation of the multiple.*

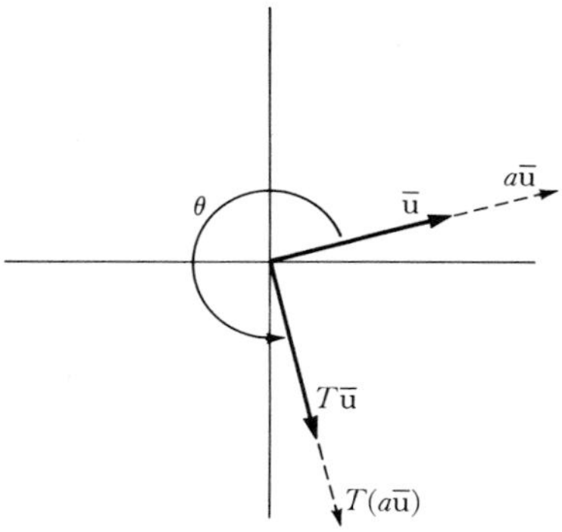

This operation T is a linear transformation, as indicated in Figures 6a and b.

Figure 7 shows that

$$T\bar{i} = \cos\theta\bar{i} + \sin\theta\bar{j}$$

$$T\bar{j} = \cos\left(\theta + \frac{\pi}{2}\right)\bar{i} + \sin\left(\theta + \frac{\pi}{2}\right)\bar{j}$$

where $\bar{i}$ represents (1,0) and $\bar{j}$ represents (0,1). In the next section we shall give a general formula for calculating $T(x,y)$.

Figure 7 $T(1,0) = (\cos\theta, \sin\theta)$
$T(0,1) = [\cos(\theta + \pi/2), \sin(\theta + \pi/2)]$.

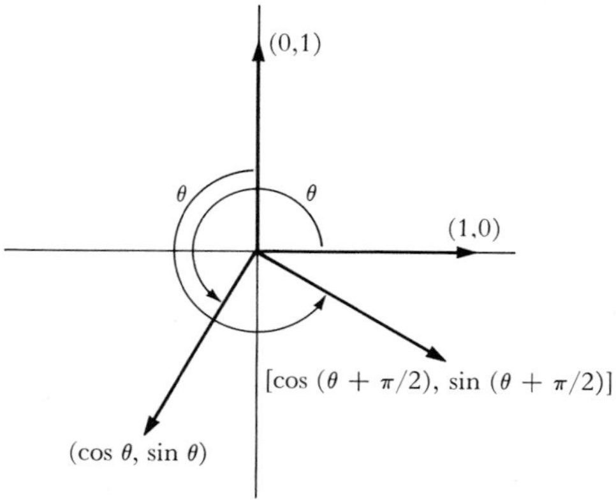

EXAMPLE 5 *Nonlinear Transformations.*

The real-valued functions whose graphs are straight lines through the origin are obviously only a very small class of functions, excluding for example, polynomial functions of degree 2 or higher, trigonometric functions, and exponentials and logarithms. For example, if f is defined by

$$f(x) = x^2$$

then f does not have the linearity properties **2**. For example, $f(1) = 1$ and $f(-1) = 1$. The sum, $1 + (-1)$, is 0, while the sum of the images, $1 + 1$, is 2; that is

$$f(1 + (-1)) \neq f(1) + f(-1)$$

A good example of a nonlinear transformation of the plane is the polar coordinate transformation T defined by

$$T(x,y) = \left(\sqrt{x^2 + y^2}, \arctan \frac{y}{x}\right) \qquad \text{if } (x,y) \neq (0,0)$$

$$T(0,0) = (0,0)$$

This certainly maps the origin into the origin. It does *not* map a multiple of $\bar{u} = (x,y)$ into that same multiple of $T\bar{u}$, for

$$T(tx,ty) = \left(\sqrt{t^2x^2 + t^2y^2}, \arctan \frac{ty}{tx}\right) \neq t\left(\sqrt{x^2 + y^2}, \arctan \frac{y}{x}\right)$$

A very simple class of functions not included in the definition of linear transformation are the translations. A **translation** through $\bar{v}$ is the function T defined by

$$T\bar{u} = \bar{u} + \bar{v}$$

This will be defined for all $\bar{u}$ for which $\bar{u} + \bar{v}$ is defined, that is, all $\bar{u}$ in R^n if $\bar{v}$ is in R^n. If $\bar{v} \neq \bar{0}$, then

$$T\bar{0} = \bar{0} + \bar{v} \neq \bar{0}$$

The linearity properties **2** require however that a linear transformation map $\bar{0}$ into $\bar{0}$ (see Exercise **10** below). In more advanced courses it will be shown that many nonlinear transformations can be approximated by combinations of linear transformations and translations.

EXAMPLE 6 □ *Differentiation and Integration are Linear Transformations.*

Suppose $C[0,1]$ is the vector space of all functions f which are defined and continuous for $0 \leq x \leq 1$. (See Example **6**, page 39.) Suppose further that M is the subspace of $C[0,1]$ consisting of the functions f such that both f and its derivative f' are defined and continuous for $0 \le x \le 1$. In calculus it is shown that

$$(f + g)' = f' + g' \qquad \text{and} \qquad (af)' = af'$$

Thus, if we let D denote the differentiation operation, we have for f and g in M:

$$D(f + g) = Df + Dg \qquad \text{and} \qquad D(af) = aDf$$

In other words, the differentiation operator D is a linear transformation from M into $C[0,1]$.

Suppose T is the integration operation defined by

$$Tf(x) = \int_0^x f(t)\, dt \qquad 0 \le x \le 1$$

For example,

$$T(\sin x) = 1 - \cos x \qquad \text{since} \int_0^x \sin t\, dt = -\cos t \Big|_0^x = 1 - \cos x$$

From calculus, again, we know that

$$\int_0^x \big(f(t) + g(t)\big)\, dt = \int_0^x f(t)\, dt + \int_0^x g(t)\, dt,$$

and

$$\int_0^x af(t)\, dt = a\int_0^x f(t)\, dt$$

or, in other words, that

$$T(f + g) = Tf + Tg \qquad \text{and} \qquad T(af) = aTf$$

If f is continuous, then it also follows that for each x

$$\int_0^x f(t)\, dt \text{ is continuous}$$

These observations can be summarized by stating that the integral operator T is a linear transformation from $C[0,1]$ into $C[0,1]$.

EXAMPLE 7 □ *Linear Differential Operators.*
Suppose $a_1, a_2, \ldots, a_n$ are numbers and f is a function such that $f, f', f'', f''', \ldots, f^{(n)}$ are all defined and continuous for $0 \le x \le 1$. We

define the *linear differential operator* L by

$$Lf = f^{(n)} + a_n f^{(n-1)} + \cdots + a_2 f' + a_1 f$$

An extension of the arguments used in Example **6** can be given to establish that

3
$$L(f+g) = Lf + Lg$$
$$L(af) = aLf$$

Therefore L is a *linear transformation* from the vector space of n times continuously differentiable functions into $C[0,1]$. As the student will see in Chapter 4 this fact allows us to use many of the concepts of linear algebra to discuss linear differential equations, which are equations of the form $Lf = g$. For example, properties **3** can be used to show that

4 The null space of L, that is, the set of all functions f such that $Lf = 0$ is a subspace of $C[0,1]$.

We shall also show that the null space of L has dimension n; that is, there are n independent functions

$$h_1, h_2, \ldots, h_n$$

such that the null space of L consists of all functions of the form

$$c_1h_1 + c_2h_2 + \cdots + c_nh_n$$

where $c_1, c_2, \ldots, c_n$, are arbitrary constants.

EXERCISES

1 Suppose P is projection onto the vector $\bar{w} = \frac{1}{3}(1,2,2)$. For $\bar{u} = (1,3,-1)$, $\bar{v} = (0,-1,2)$ calculate
a $P\bar{u}$ **b** $P\bar{v}$ **c** $P(\bar{u}+\bar{v})$ **d** $P(4\bar{u})$ **e** $4P\bar{u}$

2 For the transformation T defined by $T(x,y) = (-x,y)$ calculate
a $T(3,2)$ **b** $T(-1,-4)$
c $T((3,2) + (-1,-4))$ **d** $T(3,2) + T(-1,-4)$

3 Suppose T and P are as in Example **3**, so that $P\bar{u} = (\bar{u}\cdot\bar{w})\bar{w}$, $T\bar{u} = 2P\bar{u} - \bar{u}$. Show that $T(\bar{u}+\bar{v}) = T\bar{u} + T\bar{v}$ and $T(a\bar{u}) = aT(\bar{u})$. Illustrate these facts with a figure.

4 Suppose T is reflection in the line through $\bar{w} = \frac{1}{3}(1,-2,2)$ and that P is projection onto $\bar{w}$. For $\bar{u} = (2,-3,1)$, $\bar{v} = (2,-1,-1)$, calculate
a $P\bar{u}$ **b** $P\bar{v}$ **c** $T\bar{u}$ **d** $T\bar{v}$
Verify that $T(\bar{u}+\bar{v}) = T\bar{u} + T\bar{v}$.

5 Suppose T is a counterclockwise rotation in R^2 through the angle $\pi/4$. Draw figures similar to Figures 6a and 6b to indicate that T is linear. Calculate $T(1,0)$, $T(0,1)$, and $T(1,1)$.

6 Give a geometric description of each of the following transformations.
a $T(x,y) = (y,x)$ **b** $T(x,y,z) = (0,y,z)$
c $T(x,y) = (x,-y)$ **d** $T(x,y,z,w) = (x,0,0,0)$

7 Suppose P is projection orthogonal to the unit vector $\bar{w}$. Find a formula for $P\bar{u}$ and use this to show that P is linear.

8 Show that if T is linear, then $T(a\bar{u} + b\bar{v}) = aT\bar{u} + bT\bar{v}$.

9 Suppose $T(x,y) = (|x|,y)$. Calculate $T(1,2)$ and $T(-1,2)$. Does $T(1,2) = -T(-1,-2)$? Is T linear?

10 Suppose T is a linear transformation. Show that $T\bar{0} = \bar{0}$. [Hint: $T\bar{0} = T(a\bar{0})$ for any scalar a.]

□ 11 For $Df = f'$ calculate
a $D(\sin x)$ **b** $D(e^x)$ **c** $D(e^x + \sin x)$ **d** $D(ae^x + b \sin x)$

□ 12 For T, the integration operator defined in Example **6**, calculate
a $T(\cos x)$ **b** $T(x^2)$ **c** $T(e^x)$ **d** $T(x^2 + 3 \cos x + 2e^x)$

□ 13 **a** Find a basis for the null space of the differential operator $Lf = f'''$.
b Describe the solutions to $Lf = \sin x$, where L is as in part **a**.

□ 14 Consider the operator $Lf = f'' - f$. Show that e^x and e^{-x} are in the null space of L. Show that these two functions are independent. (Hence, from more advanced techniques, they form a basis for the null space of L.) Show that $Lx = -x$ and describe the solutions to $Lf = x$.

□ 15 Proceed as in Exercise **14** for the operator $Lf = f'' + f$ and the functions $\sin x$, $\cos x$. Describe also the solutions to $Lf = 2 + x^2$. (Hint: $L(x^2) = 2 + x^2$.)

□ 16 Suppose g is in $C[0,1]$ and that Tf is defined to be that function whose value at x is $Tf(x) = f(x)g(x)$. Show that T is a linear transformation from $C[0,1]$ into $C[0,1]$ and find the null space of T. (T is called a *multiplication operator*.)

□ 17 The discussion of this section also extends to complex vector spaces. For the transformation T defined by

$$T(x,y) = (ix + (2 + i)y, \ -3ix)$$

verify that $T[(2,i) + i(3, 1 - i)] = T(2,i) + iT(3, 1 - i)$. Then verify that T is a linear transformation from C^2 into C^2.

SECTION 2 The Matrix of a Linear Transformation

We noted in the previous section that the transformation T defined by $T\bar{u} = A\bar{u}$, where A has m rows and n columns, is a linear transformation from R^n into R^m. In this section we establish the converse of this result, that every linear transformation from R^n into R^m is just multiplication by a suitable matrix. This will give us a useful tool for calculating the effect of a linear transformation, enabling us, for example, to obtain formulas for rotations, reflections, and projections.

In the following theorem we assume that vectors are expressed as column matrices.

THEOREM 10 If T is a linear transformation from R^n into R^m, there is a *unique* matrix A, with m rows and n columns, such that $T\bar{u} = A\bar{u}$, for all $\bar{u}$ in R^n.

PROOF: The matrix A is called the **matrix** of T. This result will be proved for the case when $n = m = 2$ by showing how to determine A.

The basic idea is as follows: If it is known that for all $\bar{u}$, $\bar{v}$, and scalars t

1 $$T(\bar{u} + \bar{v}) = T\bar{u} + T\bar{v} \qquad T(t\bar{u}) = tT\bar{u}$$

then we can calculate $T\bar{u}$ knowing what T does to a basis.

Suppose T satisfies properties **1** and

$$T\begin{bmatrix} 1 \\ 0 \end{bmatrix} = \begin{bmatrix} a \\ b \end{bmatrix} \quad \text{and} \quad T\begin{bmatrix} 0 \\ 1 \end{bmatrix} = \begin{bmatrix} c \\ d \end{bmatrix}$$

To calculate the effect of T upon $\begin{bmatrix} x \\ 0 \end{bmatrix}$, write

$$\begin{bmatrix} x \\ 0 \end{bmatrix} = x\begin{bmatrix} 1 \\ 0 \end{bmatrix} \quad \text{so that} \quad T\begin{bmatrix} x \\ 0 \end{bmatrix} = T\left\{x\begin{bmatrix} 1 \\ 0 \end{bmatrix}\right\}$$

The second of the properties **1** then gives

$$T\left\{x\begin{bmatrix} 1 \\ 0 \end{bmatrix}\right\} = xT\begin{bmatrix} 1 \\ 0 \end{bmatrix} = x\begin{bmatrix} a \\ b \end{bmatrix} = \begin{bmatrix} ax \\ bx \end{bmatrix}$$

In a similar manner

$$T\begin{bmatrix} 0 \\ y \end{bmatrix} = T\left\{y\begin{bmatrix} 0 \\ 1 \end{bmatrix}\right\} = yT\begin{bmatrix} 0 \\ 1 \end{bmatrix} = y\begin{bmatrix} c \\ d \end{bmatrix} = \begin{bmatrix} cy \\ dy \end{bmatrix}$$

It is now a simple matter to calculate the effect of T upon $\begin{bmatrix} x \\ y \end{bmatrix}$. First write

$$\begin{bmatrix} x \\ y \end{bmatrix} = \begin{bmatrix} x \\ 0 \end{bmatrix} + \begin{bmatrix} 0 \\ y \end{bmatrix}$$

then apply the first of the properties **1** to obtain

$$T\begin{bmatrix} x \\ y \end{bmatrix} = T\left\{\begin{bmatrix} x \\ 0 \end{bmatrix} + \begin{bmatrix} 0 \\ y \end{bmatrix}\right\}$$

$$= T\begin{bmatrix} x \\ 0 \end{bmatrix} + T\begin{bmatrix} 0 \\ y \end{bmatrix}$$

$$= \begin{bmatrix} ax \\ bx \end{bmatrix} + \begin{bmatrix} cy \\ dy \end{bmatrix}$$

$$= \begin{bmatrix} ax + cy \\ bx + dy \end{bmatrix}$$

The latter can be written in a convenient matrix form as

$$\begin{bmatrix} a & c \\ b & d \end{bmatrix} \begin{bmatrix} x \\ y \end{bmatrix}$$

It has therefore been shown that if

$$A = \begin{bmatrix} a & c \\ b & d \end{bmatrix} \quad \text{where} \quad T\begin{bmatrix} 1 \\ 0 \end{bmatrix} = \begin{bmatrix} a \\ b \end{bmatrix} \quad \text{and} \quad T\begin{bmatrix} 0 \\ 1 \end{bmatrix} = \begin{bmatrix} c \\ d \end{bmatrix}$$

then for all $\bar{u}$ in R^2, $T\bar{u} = A\bar{u}$. In summary, this proves the existence of such a matrix A for the case $m = n = 2$. To complete the proof of Theorem **10** in this case we need only observe that if B is any matrix with two rows and two columns then the first and second columns of B are, respectively,

$$B\begin{bmatrix} 1 \\ 0 \end{bmatrix} \quad \text{and} \quad B\begin{bmatrix} 0 \\ 1 \end{bmatrix}$$

Hence if $A\bar{u} = B\bar{u}$ for all $\bar{u}$ in R^2, then

$$B\begin{bmatrix} 1 \\ 0 \end{bmatrix} = A\begin{bmatrix} 1 \\ 0 \end{bmatrix} = \begin{bmatrix} a \\ b \end{bmatrix} \quad \text{and} \quad B\begin{bmatrix} 0 \\ 1 \end{bmatrix} = A\begin{bmatrix} 0 \\ 1 \end{bmatrix} = \begin{bmatrix} c \\ d \end{bmatrix}$$

so the columns of B must be the same as the columns of A.

This completes the proof of Theorem **10** for the case $m = n = 2$.

It is possible to define a multiplication so that $T\bar{u}$ is defined directly in terms of row matrices. Our decision to define the matrix of T by expressing vectors as column matrices is arbitrary but does have the virtue of being a long-standing convention. Example 2, below, shows how to convert from rows to columns.

In the general case, similar calculations show that the effect of T on the standard basis determines $T\bar{u}$ for any $\bar{u}$. Thus for all $\bar{u}$ in R^n

$$T\bar{u} = A\bar{u}$$

where A is the matrix whose respective *columns* are:

$$T\begin{bmatrix}1\\0\\0\\\vdots\\0\end{bmatrix}, \quad T\begin{bmatrix}0\\1\\0\\\vdots\\0\end{bmatrix}, \ldots, T\begin{bmatrix}0\\0\\0\\\vdots\\1\end{bmatrix}$$

REMARK The above theorem establishes a one-to-one correspondence between linear operations on coordinate spaces and matrices whereby each such operation can be represented as the operation of multiplication by a matrix. Therefore, each statement about linear operations on coordinate spaces can be translated into a statement about matrices, and each statement about matrices can be translated into a statement about linear operations. We shall make full use of this identification in this chapter and in Chapter 4. In some cases we shall find it easier to discuss linear operations first, and then translate our ideas into matrix form; in other cases we will find it convenient to reverse this procedure.

Note that the construction of the matrix of T was given in terms of the effect of T upon the standard basis vectors. In the next chapter, a similar construction in terms of the effect of T upon other bases will be discussed.

EXAMPLE 1 Let us find the matrix of a projection in R^3. Suppose $\bar{v} = \frac{1}{3}(1,2,2)$ and $P\bar{u}$ is the projection of $\bar{u}$ onto $\bar{v}$. Then $P\bar{u} = (\bar{u} \cdot \bar{v})\bar{v}$, and P is a linear transformation from R^3 into R^3. To find the matrix of P calculate

$$P\begin{bmatrix}1\\0\\0\end{bmatrix} \quad P\begin{bmatrix}0\\1\\0\end{bmatrix} \quad P\begin{bmatrix}0\\0\\1\end{bmatrix}$$

We have

$$\left\{\begin{bmatrix}1\\0\\0\end{bmatrix} \cdot \bar{v}\right\}\bar{v} = \tfrac{1}{3} \cdot \tfrac{1}{3}\begin{bmatrix}1\\2\\2\end{bmatrix} = \tfrac{1}{9}\begin{bmatrix}1\\2\\2\end{bmatrix}$$

$$\left\{\begin{bmatrix}0\\1\\0\end{bmatrix}\cdot\bar{v}\right\}\bar{v} = \tfrac{2}{3}\cdot\tfrac{1}{3}\begin{bmatrix}1\\2\\2\end{bmatrix} = \tfrac{2}{9}\begin{bmatrix}1\\2\\2\end{bmatrix}$$

$$\left\{\begin{bmatrix}0\\0\\1\end{bmatrix}\cdot\bar{v}\right\}\bar{v} = \tfrac{2}{3}\cdot\tfrac{1}{3}\begin{bmatrix}1\\2\\2\end{bmatrix} = \tfrac{2}{9}\begin{bmatrix}1\\2\\2\end{bmatrix}$$

Therefore the matrix of P is

$$P_0 = \begin{bmatrix}\frac{1}{9} & \frac{2}{9} & \frac{2}{9}\\ \frac{2}{9} & \frac{4}{9} & \frac{4}{9}\\ \frac{2}{9} & \frac{4}{9} & \frac{4}{9}\end{bmatrix}$$

and the formula for P can be rewritten as $P\bar{u} = P_0\bar{u}$.

EXAMPLE 2 It is easy to convert from coordinate expressions to column matrices. If T is defined by

$$T(x,y) = (ax + by, cx + dy)$$

rewrite both parts of this expression as column matrices:

$$T\begin{bmatrix}x\\y\end{bmatrix} = \begin{bmatrix}ax + by\\cx + dy\end{bmatrix}$$

Since the right-hand side is the same as

$$\begin{bmatrix}a & b\\c & d\end{bmatrix}\begin{bmatrix}x\\y\end{bmatrix}$$

it follows that

$$T\begin{bmatrix}x\\y\end{bmatrix} = \begin{bmatrix}a & b\\c & d\end{bmatrix}\begin{bmatrix}x\\y\end{bmatrix}$$

We conclude that T is a linear transformation from R^2 into R^2 and that the matrix of T is

$$\begin{bmatrix}a & b\\c & d\end{bmatrix}$$

For example, suppose T is the reflection of Example 2, page 134, so that $T(x,y) = (-x,y)$, which can be rewritten as

$$T\begin{bmatrix}x\\y\end{bmatrix} = \begin{bmatrix}-1 & 0\\0 & 1\end{bmatrix}\begin{bmatrix}x\\y\end{bmatrix}$$

From this we conclude that the matrix of T is

$$\begin{bmatrix} -1 & 0 \\ 0 & 1 \end{bmatrix}$$

EXAMPLE 3 The standard rotation formula is a consequence of Theorem **10**. Suppose T is a counterclockwise rotation in R^2 through the angle θ. Then, as shown in Example **4**, page 135, T is a linear transformation and

$$T(1,0) = (\cos\theta, \sin\theta)$$
$$T(0,1) = \left[\cos\left(\theta + \frac{\pi}{2}\right), \sin\left(\theta + \frac{\pi}{2}\right)\right]$$

Writing these as columns, we have

$$T\begin{bmatrix} 1 \\ 0 \end{bmatrix} = \begin{bmatrix} \cos\theta \\ \sin\theta \end{bmatrix} \qquad T\begin{bmatrix} 0 \\ 1 \end{bmatrix} = \begin{bmatrix} \cos\left(\theta + \frac{\pi}{2}\right) \\ \sin\left(\theta + \frac{\pi}{2}\right) \end{bmatrix}$$

Since $\cos(\theta + \pi/2) = -\sin\theta$ and $\sin(\theta + \pi/2) = \cos\theta$, we have shown that the matrix of T is

$$A = \begin{bmatrix} \cos\theta & -\sin\theta \\ \sin\theta & \cos\theta \end{bmatrix}$$

Therefore

$$T\begin{bmatrix} x \\ y \end{bmatrix} = A\begin{bmatrix} x \\ y \end{bmatrix}$$

This gives the standard formula

$$T\begin{bmatrix} x \\ y \end{bmatrix} = \begin{bmatrix} x\cos\theta - y\sin\theta \\ x\sin\theta + y\cos\theta \end{bmatrix}$$

which enables us to calculate the effect of rotating

$$\begin{bmatrix} x \\ y \end{bmatrix}$$

through the angle θ.

EXAMPLE 4 Formulas for reflection can be obtained by finding the matrix of the reflection. Suppose $\bar{w} = \frac{1}{3}(1,2,2)$ and T is reflection in the line through $\bar{w}$. As shown in Example **3**, page 134, we have

$$T\bar{u} = 2P\bar{u} - \bar{u}$$

where P is the projection onto $\overline{w}$. In Example 1 it was shown that

$$P\begin{bmatrix}1\\0\\0\end{bmatrix}=\begin{bmatrix}\frac{1}{9}\\\frac{2}{9}\\\frac{2}{9}\end{bmatrix},\qquad P\begin{bmatrix}0\\1\\0\end{bmatrix}=\begin{bmatrix}\frac{2}{9}\\\frac{4}{9}\\\frac{4}{9}\end{bmatrix},\qquad P\begin{bmatrix}0\\0\\1\end{bmatrix}=\begin{bmatrix}\frac{2}{9}\\\frac{4}{9}\\\frac{4}{9}\end{bmatrix}$$

so that

$$T\begin{bmatrix}1\\0\\0\end{bmatrix}=\begin{bmatrix}-\frac{7}{9}\\\frac{4}{9}\\\frac{4}{9}\end{bmatrix},\qquad T\begin{bmatrix}0\\1\\0\end{bmatrix}=\begin{bmatrix}\frac{4}{9}\\-\frac{1}{9}\\\frac{8}{9}\end{bmatrix},\qquad T\begin{bmatrix}0\\0\\1\end{bmatrix}=\begin{bmatrix}\frac{4}{9}\\\frac{8}{9}\\-\frac{1}{9}\end{bmatrix}$$

Thus we have the formula

$$T\begin{bmatrix}x\\y\\z\end{bmatrix}=\begin{bmatrix}-\frac{7}{9} & \frac{4}{9} & \frac{4}{9}\\\frac{4}{9} & -\frac{1}{9} & \frac{8}{9}\\\frac{4}{9} & \frac{8}{9} & -\frac{1}{9}\end{bmatrix}\begin{bmatrix}x\\y\\z\end{bmatrix}$$

which can be rewritten in coordinate form as

$$T(x,y,z) = \tfrac{1}{9}(-7x + 4y + 4z,\ 4x - y + 8z,\ 4x + 8y - z)$$

EXAMPLE 5 Suppose T is a linear transformation from R^1 into R^1. The matrix of T is a matrix with one row and one column; that is, the matrix of T is a number. Theorem 10 then tells us that for all x in R^1, $Tx = ax$, where $a = T1$.

In other words, if $f(x)$ is a real-valued function of a real variable such that $f(x + y) = f(x) + f(y)$ and $f(bx) = bf(x)$, for all real numbers x, y, b, then f is of the form $f(x) = ax$, where $a = f(1)$.

EXAMPLE 6 All the above examples discuss transformations from R^n into R^m where $m = n$, in which case the matrix of the transformation is square. In general, the matrix of a linear transformation from R^n into R^m has n columns and m rows. For example, consider

$$T(x,y,z) = (x - y,\ 2y + z)$$

Since

$$T\begin{bmatrix}1\\0\\0\end{bmatrix}=\begin{bmatrix}1\\0\end{bmatrix}\qquad T\begin{bmatrix}0\\1\\0\end{bmatrix}=\begin{bmatrix}-1\\2\end{bmatrix}\qquad T\begin{bmatrix}0\\0\\1\end{bmatrix}=\begin{bmatrix}0\\1\end{bmatrix}$$

we see that

$$T\begin{bmatrix} x \\ y \\ z \end{bmatrix} = \begin{bmatrix} 1 & -1 & 0 \\ 0 & 2 & 1 \end{bmatrix} \begin{bmatrix} x \\ y \\ z \end{bmatrix}$$

EXERCISES

1 Find the matrix of each of the following linear transformations.
a $T(x,y,z,w) = (x, x + y, x + y + z, x + y + z + w)$
b $T(x,y,z) = (x - y, z)$
c $T(x,y,z,w) = (x,x,x,x,x)$

2 Find the matrix of a counterclockwise rotation in R^2 through the angle
a $\pi/4$ **b** $\pi/2$ **c** π **d** $-\pi$ **e** 0

3 For each of the rotations T of Exercise **2**, calculate $T(1,1)$ and $T(2,-3)$.

4 Suppose P is projection onto $\bar{w} = \frac{1}{\sqrt{2}}(1,1)$. Find the matrix of P.

5 Suppose T is a reflection in the line through $\bar{w} = \frac{1}{\sqrt{2}}(1,1)$. Find the matrix of T.

6 Suppose P is projection orthogonal to $\bar{w} = \frac{1}{3}(1,2,2)$. Find the matrix of P.

7 Suppose T is the linear transformation from R^2 into R^2 defined by $T\bar{u} = \bar{0}$. What is the matrix of T?

8 Suppose T is the linear transformation from R^2 into R^2 defined by $T\bar{u} = \bar{u}$. What is the matrix of T?

9 Suppose $T\bar{u} = A\bar{u}$, where $A = \begin{bmatrix} 2 & 1 & 3 \\ 0 & -1 & 1 \end{bmatrix}$. What is $T(x,y,z)$?

10 Give a geometric description of the linear transformation T if the matrix of T is

a $\begin{bmatrix} 0 & 1 \\ 1 & 0 \end{bmatrix}$ **b** $\begin{bmatrix} 1 & 0 \\ 0 & 0 \end{bmatrix}$ **c** $\begin{bmatrix} 0 & -1 \\ -1 & 0 \end{bmatrix}$ **d** $\begin{bmatrix} -1 & 0 \\ 0 & -1 \end{bmatrix}$

11 We noted in the proof of Theorem **10** that $A\begin{bmatrix} 1 \\ 0 \end{bmatrix}$ and $A\begin{bmatrix} 0 \\ 1 \end{bmatrix}$ are the columns of A. This fact enables quick calculation of certain products.
a Find

$$\begin{bmatrix} 3 & 1 & 2 & 1 \\ 0 & 1 & 1 & -1 \\ 0 & 0 & 0 & 1 \end{bmatrix} \begin{bmatrix} 0 \\ 1 \\ 0 \\ 0 \end{bmatrix} \quad \text{and} \quad \begin{bmatrix} 2 & 1 \\ 0 & 1 \end{bmatrix} \begin{bmatrix} 0 \\ 1 \end{bmatrix}$$

b Suppose A has five columns and $\bar{u} = \begin{bmatrix} 0 \\ 0 \\ 1 \\ 0 \\ 0 \end{bmatrix}$. What is $A\bar{u}$?

c Suppose A has five columns. Find a vector $\bar{u}$ such that $A\bar{u}$ is the fifth column of A.

12 The *null space* of a linear transformation T is the set of all $\bar{u}$ such that $T\bar{u} = \bar{0}$. The *range* of T is the set of all $\bar{u}$ for which the equation $T\bar{v} = \bar{u}$ can be solved. The dimension of the range of T is called the *rank* of T. Let T be a linear transformation from R^n into R^m.

- **a** Is the null space of T a subspace of R^n or R^m?
- **b** Is the range of T a subspace of R^n or R^m?

Let A be the matrix of T.

- **c** What is the relation between the null spaces of T and A, the range of T and the column space of A, and the rank of T and the rank of A?
- **d** If $n < m$, can the range of T be R^m?
- **e** If $n > m$, what can you say about the null space of T?

13 Suppose T is a linear transformation from R^2 in R^2 and that

$$T\begin{bmatrix} 3 \\ 1 \end{bmatrix} = \begin{bmatrix} -2 \\ 6 \end{bmatrix} \qquad T\begin{bmatrix} -1 \\ 4 \end{bmatrix} = \begin{bmatrix} 0 \\ 5 \end{bmatrix}$$

Find the matrix of T. (Hint: Express the standard basis vectors as linear combinations of

$$\begin{bmatrix} 3 \\ 1 \end{bmatrix} \text{and} \begin{bmatrix} -1 \\ 4 \end{bmatrix} \text{in order to find } T\begin{bmatrix} 1 \\ 0 \end{bmatrix} \text{and } T\begin{bmatrix} 0 \\ 1 \end{bmatrix}.\Bigg)$$

14 Suppose $\bar{u}_1$ and $\bar{u}_2$ are a basis for R^2 and that S and T are linear transformations from R^2 into R^2 such that $S\bar{u}_1 = T\bar{u}_1$ and $S\bar{u}_2 = T\bar{u}_2$. Show that S and T must be the same. (Hint: By expressing $\bar{u}$ in terms of $\bar{u}_1$ and $\bar{u}_2$ show that $S\bar{u} = T\bar{u}$.)

☐ **15** The concepts of this section extend to complex vector spaces. Find the matrix of each of the following linear transformations.

- **a** $T(x,y) = (ix - y,\ (1 + i)x)$
- **b** $T(x,y) = (0,\ 0,\ ix,\ (i - 1)x + 2y)$
- **c** T is projection onto $\frac{1}{\sqrt{2}}(i,1)$

SECTION 3 Sums and Scalar Multiples

Our next task is to define the sum of two matrices (or two transformations) and the scalar multiple of a matrix (or a transformation). These definitions, along with the definition of a product given in the next section, will provide us with a formal algebra of matrices (and transformations). This transformation (or matrix) algebra is of central importance in modern mathematics. For example, we shall use these formal concepts in Chapter 6 to show how the study of certain differential equations can be converted into the study of roots of polynomials.

If

$$A = \begin{bmatrix} a_1 & b_1 \\ c_1 & d_1 \end{bmatrix} \quad \text{and} \quad B = \begin{bmatrix} a_2 & b_2 \\ c_2 & d_2 \end{bmatrix}$$

then the **sum** $A + B$ is defined by

$$A + B = \begin{bmatrix} a_1 + a_2 & b_1 + b_2 \\ c_1 + c_2 & d_1 + d_2 \end{bmatrix}$$

Note that we merely add together the corresponding entries of A and B. For larger matrices, the definition of sum is analogous. For example

$$\begin{bmatrix} 2 & 1 & -1 & 3 \\ 2 & 1 & 14 & 2 \\ -3 & 0 & -2 & 1 \end{bmatrix} + \begin{bmatrix} -1 & 0 & 2 & -3 \\ -6 & 7 & 0 & -9 \\ 5 & -1 & 0 & 0 \end{bmatrix} = \begin{bmatrix} 1 & 1 & 1 & 0 \\ -4 & 8 & 14 & -7 \\ 2 & -1 & -2 & 1 \end{bmatrix}$$

Note that the sum $A + B$ is defined only when A and B are of the same size, that is, when the number of rows of A equals the number of rows of B and the number of columns of A equals the number of columns of B. It is evident from our definition of matrix sum that the following rules hold:

1 $$A + B = B + A, \qquad A + (B + C) = (A + B) + C$$

A matrix, all of whose entries are zeros, is called a **zero matrix** and usually denoted by 0. Observe that if A and the zero matrix 0 have the same size then

2 $$A + 0 = A$$

If

$$A = \begin{bmatrix} a_1 & b_1 \\ c_1 & d_1 \end{bmatrix}$$

and a is any scalar, the **scalar multiple** aA is defined by

$$aA = \begin{bmatrix} aa_1 & ab_1 \\ ac_1 & ad_1 \end{bmatrix}$$

In other words merely multiply each entry of A by the scalar a. The definition for larger matrices is analogous. For example

$$5\begin{bmatrix} 2 & 1 & -1 & 3 \\ 2 & 1 & 14 & 2 \\ -3 & 0 & -2 & 1 \end{bmatrix} = \begin{bmatrix} 10 & 5 & -5 & 15 \\ 10 & 5 & 70 & 10 \\ -15 & 0 & -10 & 5 \end{bmatrix}$$

The following rules should be evident.

3
$$a(A + B) = aA + aB \qquad (a + b)A = aA + bA$$
$$(ab)A = a(bA)$$
$$1A = A \qquad 0A = 0$$

Also observe that the matrix $-A$ [defined by the rule $-A = (-1)A$] satisfies the relation

4
$$A + (-A) = 0$$

The above rules are just like some of the rules of ordinary arithmetic. Note also that a matrix A with m rows and n columns is just an mn-tuple written in rectangular form. Thus such a matrix can be thought of as a vector in R^{mn}. The above addition and scalar multiple correspond to the sum and multiple in R^{mn}.

EXAMPLE 1 Some examples illustrating the arithmetic rules **1**, **2**, **3**, and **4** are given below.

a
$$\begin{bmatrix} 3 & 1 & 1 \\ 2 & -1 & 1 \end{bmatrix} + \begin{bmatrix} 4 & 2 & -1 \\ 0 & 0 & 2 \end{bmatrix} = \begin{bmatrix} 7 & 3 & 0 \\ 2 & -1 & 3 \end{bmatrix}$$

and

$$\begin{bmatrix} 4 & 2 & -1 \\ 0 & 0 & 2 \end{bmatrix} + \begin{bmatrix} 3 & 1 & 1 \\ 2 & -1 & 1 \end{bmatrix} = \begin{bmatrix} 7 & 3 & 0 \\ 2 & -1 & 3 \end{bmatrix}$$

Illustrate the general law $A + B = B + A$.

b
$$\begin{bmatrix} 2 & 1 \\ 0 & 3 \end{bmatrix} + \begin{bmatrix} -1 & 0 \\ 7 & 2 \end{bmatrix} = \begin{bmatrix} 1 & 1 \\ 7 & 5 \end{bmatrix}$$

and

$$\begin{bmatrix} -2 & 6 \\ 2 & 1 \end{bmatrix} + \begin{bmatrix} 1 & 1 \\ 7 & 5 \end{bmatrix} = \begin{bmatrix} -1 & 7 \\ 9 & 6 \end{bmatrix}$$

We also have

$$\begin{bmatrix} -2 & 6 \\ 2 & 1 \end{bmatrix} + \begin{bmatrix} 2 & 1 \\ 0 & 3 \end{bmatrix} = \begin{bmatrix} 0 & 7 \\ 2 & 4 \end{bmatrix}$$

and

$$\begin{bmatrix} 0 & 7 \\ 2 & 4 \end{bmatrix} + \begin{bmatrix} -1 & 0 \\ 7 & 2 \end{bmatrix} = \begin{bmatrix} -1 & 7 \\ 9 & 6 \end{bmatrix}$$

which is an example of the law $(A + B) + C = A + (B + C)$.

c $$\begin{bmatrix} 2 & 1 \\ 1 & 2 \end{bmatrix} + \begin{bmatrix} 0 & 0 \\ 0 & 0 \end{bmatrix} = \begin{bmatrix} 2 & 1 \\ 1 & 2 \end{bmatrix}$$

and

$$\begin{bmatrix} 2 & 1 \\ 1 & 2 \end{bmatrix} + \begin{bmatrix} -2 & -1 \\ -1 & -2 \end{bmatrix} = \begin{bmatrix} 0 & 0 \\ 0 & 0 \end{bmatrix}$$

These illustrate principles **2** and **4**.

d $$3\begin{bmatrix} 6 & -1 & 0 \\ 1 & 2 & 1 \end{bmatrix} = \begin{bmatrix} 18 & -3 & 0 \\ 3 & 6 & 3 \end{bmatrix}$$

and

$$-5\begin{bmatrix} 18 & -3 & 0 \\ 3 & 6 & 3 \end{bmatrix} = \begin{bmatrix} -90 & 15 & 0 \\ -15 & -30 & -15 \end{bmatrix}$$

We also have

$$-15\begin{bmatrix} 6 & -1 & 0 \\ 1 & 2 & 1 \end{bmatrix} = \begin{bmatrix} -90 & 15 & 0 \\ -15 & -30 & -15 \end{bmatrix}$$

This is just a special case of the principle $(ab)A = a(bA)$.

e If $A = \begin{bmatrix} -1 & 3 \\ 0 & 0 \end{bmatrix}$ and $C = \begin{bmatrix} -2 & -1 \\ -1 & 1 \end{bmatrix}$

then the solution B to $2A - 3B + C = 0$ is

$$B = \frac{2}{3}A + \frac{1}{3}C = \begin{bmatrix} -\frac{4}{3} & \frac{5}{3} \\ -\frac{1}{3} & \frac{1}{3} \end{bmatrix}$$

DISCUSSION We now give the corresponding definitions for linear transformations from V into W. In order to define $S + T$ and aT we need to show what effect each has on a given vector in V. We define $S + T$ as that transformation whose value at $\bar{u}$ is $S\bar{u} + T\bar{u}$. We define aT as that transformation whose value at $\bar{u}$ is $a(T\bar{u})$, more formally:

If S and T are linear transformations from V into W then

> The **sum** $S + T$ is that transformation from V into W whose value at each $\bar{u}$ is given by $S\bar{u} + T\bar{u}$.
>
> The **scalar multiple** aT is that transformation from V into W whose value at each $\bar{u}$ is given by $aT\bar{u}$.

These are just the multidimensional versions of the ordinary function sum and multiple. For example, the function f defined by

$$f(x) = x^2 + 2x$$

is the sum of the two functions g and h defined by

$$g(x) = x^2 \qquad \text{and} \qquad h(x) = 2x$$

Furthermore, h is just $2k$ where $k(x) = x$.

The proof that the sum $S + T$ and multiple aT are each linear if S and T are linear is left to Exercise **10** below. If S and T are both linear transformations from R^n into R^m then

5 The matrix of $S + T$ is the sum of the matrix of S and the matrix of T. The matrix of aT is a times the matrix of T.

In order to establish this we need to show that if A and B are matrices such that $S\bar{u} = A\bar{u}$ and $T\bar{u} = B\bar{u}$ for all $\bar{u}$ in R^n, then $(S + T)\bar{u} = (A + B)\bar{u}$ for all $\bar{u}$ in R^n. From the definition of transformation sum we have $(S + T)\bar{u} = S\bar{u} + T\bar{u}$, which is equal to $A\bar{u} + B\bar{u}$. To complete the proof of property **5** we need to show that

6 $$A\bar{u} + B\bar{u} = (A + B)\bar{u}$$

This is proved by direct calculation. For example, if

$$A = \begin{bmatrix} a_1 & b_1 \\ c_1 & d_1 \end{bmatrix} \qquad B = \begin{bmatrix} a_2 & b_2 \\ c_2 & d_2 \end{bmatrix} \qquad \bar{u} = \begin{bmatrix} x \\ y \end{bmatrix}$$

it follows that

$$
\begin{aligned}
A\bar{u} + B\bar{u} &= \begin{bmatrix} a_1 & b_1 \\ c_1 & d_1 \end{bmatrix} \begin{bmatrix} x \\ y \end{bmatrix} + \begin{bmatrix} a_2 & b_2 \\ c_2 & d_2 \end{bmatrix} \begin{bmatrix} x \\ y \end{bmatrix} \\
&= \begin{bmatrix} a_1x + b_1y \\ c_1x + d_1y \end{bmatrix} + \begin{bmatrix} a_2x + b_2y \\ c_2x + d_2y \end{bmatrix} \\
&= \begin{bmatrix} a_1x + b_1y + a_2x + b_2y \\ c_1x + d_1y + c_2x + d_2y \end{bmatrix} \\
&= \begin{bmatrix} (a_1 + a_2)x + (b_1 + b_2)y \\ (c_1 + c_2)x + (d_1 + d_2)y \end{bmatrix} \\
&= \begin{bmatrix} a_1 + a_2 & b_1 + b_2 \\ c_1 + c_2 & d_1 + d_2 \end{bmatrix} \begin{bmatrix} x \\ y \end{bmatrix} \\
&= (A + B)\bar{u}
\end{aligned}
$$

which establishes statement **6** in this case. The proof that the matrix of aT is aB is even easier and will be omitted.

EXAMPLE 2 Suppose S and T are defined by

$$S(x,y) = (x - 2y,\ 2x + 7y) \qquad \text{and} \qquad T(x,y) = (3x + 2y,\ x - y)$$

Then

$$
\begin{aligned}
(S + T)(x,y) &= S(x,y) + T(x,y) \\
&= (x - 2y,\ 2x + 7y) + (3x + 2y,\ x - y) \\
&= (4x,\ 3x + 6y)
\end{aligned}
$$

and

$$
\begin{aligned}
(3T)(x,y) &= 3[T(x,y)] \\
&= 3(3x + 2y,\ x - y) \\
&= (9x + 6y,\ 3x - 3y)
\end{aligned}
$$

To express these facts in matrix form rewrite as column matrices:

$$
\begin{aligned}
S\begin{bmatrix} x \\ y \end{bmatrix} &= \begin{bmatrix} x - 2y \\ 2x + 7y \end{bmatrix} = \begin{bmatrix} 1 & -2 \\ 2 & 7 \end{bmatrix} \begin{bmatrix} x \\ y \end{bmatrix} \\
T\begin{bmatrix} x \\ y \end{bmatrix} &= \begin{bmatrix} 3x + 2y \\ x - y \end{bmatrix} = \begin{bmatrix} 3 & 2 \\ 1 & -1 \end{bmatrix} \begin{bmatrix} x \\ y \end{bmatrix}
\end{aligned}
$$

so that the matrices of S and T are, respectively,

$$A = \begin{bmatrix} 1 & -2 \\ 2 & 7 \end{bmatrix} \quad \text{and} \quad B = \begin{bmatrix} 3 & 2 \\ 1 & -1 \end{bmatrix}$$

Thus

$$A + B = \begin{bmatrix} 4 & 0 \\ 3 & 6 \end{bmatrix} \quad \text{and} \quad 3B = \begin{bmatrix} 9 & 6 \\ 3 & -3 \end{bmatrix}$$

are the matrices of $S + T$ and $3T$, respectively. This can also be verified by observing that

$$(S + T)\begin{bmatrix} x \\ y \end{bmatrix} = \begin{bmatrix} 4x \\ 3x + 6y \end{bmatrix} = \begin{bmatrix} 4 & 0 \\ 3 & 6 \end{bmatrix} \begin{bmatrix} x \\ y \end{bmatrix}$$

$$(3T)\begin{bmatrix} x \\ y \end{bmatrix} = \begin{bmatrix} 9x + 6y \\ 3x - 3y \end{bmatrix} = \begin{bmatrix} 9 & 6 \\ 3 & -3 \end{bmatrix} \begin{bmatrix} x \\ y \end{bmatrix}$$

EXAMPLE 3 *The Zero and Identity Transformations.*

The **zero transformation** 0 from V into W is the transformation defined by

$$0\bar{u} = \bar{0} \qquad \text{for each } \bar{u} \text{ in } V$$

where $\bar{0}$ denotes the zero vector in W. If V and W are coordinate spaces then the matrix of 0 is the zero matrix 0 of appropriate size.

The **identity transformation** I from V into V is defined by $I\bar{u} = \bar{u}$, for each $\bar{u}$ in V. The matrix of the identity transformation from R^n into R^n is the identity matrix with n rows and n columns. For example, if $n = 2$ and $I\bar{u} = \bar{u}$, for each $\bar{u}$ in R^2, then, certainly we have

$$I\begin{bmatrix} 1 \\ 0 \end{bmatrix} = \begin{bmatrix} 1 \\ 0 \end{bmatrix} \quad \text{and} \quad I\begin{bmatrix} 0 \\ 1 \end{bmatrix} = \begin{bmatrix} 0 \\ 1 \end{bmatrix}$$

so that the matrix of I is

$$\begin{bmatrix} 1 & 0 \\ 0 & 1 \end{bmatrix}$$

We note that for any transformation T,

$$T + (-1)T = 0$$

This follows from the definitions of sum, scalar multiple, and the zero transformation, for

$$[T + (-1)T]\bar{u} = T\bar{u} + (-1)T\bar{u} = T\bar{u} - T\bar{u} = \bar{0}$$
$$= 0\bar{u}$$

The various rules **1**, **2**, and **3**, that is,

$$S + T = T + S \qquad S + (T + U) = (S + T) + U$$
$$S + 0 = S$$
$$a(S + T) = aS + aT \qquad (a + b)S = aS + bS$$
$$(ab)S = a(bS)$$
$$1S = S \qquad 0S = 0$$

also can be shown to be true for transformations by using the appropriate definitions. In particular, the collection of all linear transformations from V into W is a vector space with these definitions. In more advanced courses, the student will study many of the properties of this vector space.

EXAMPLE 4 If we project $\bar{u}$ along $\bar{w}$ and orthogonal to $\bar{w}$, then add these two projections together, we get $\bar{u}$ back again. (See page 109.) Let us see what this means in terms of our transformation sum.

Suppose $\bar{w}$ is a vector of length 1 and P is projection onto $\bar{w}$. Suppose further that Q is projection orthogonal to $\bar{w}$. Then (see formulas **6** and **7**, page 109)

$$P\bar{u} = (\bar{u} \cdot \bar{w})\bar{w}$$
$$Q\bar{u} = \bar{u} - (\bar{u} \cdot \bar{w})\bar{w}$$

so that $P\bar{u} + Q\bar{u} = \bar{u}$. In other words, $P + Q = I$.

For example, suppose $\bar{w} = \frac{1}{3}(1,2,2)$, so that the matrix of P is (see Example **1**, page 143)

$$P_0 = \begin{bmatrix} \frac{1}{9} & \frac{2}{9} & \frac{2}{9} \\ \frac{2}{9} & \frac{4}{9} & \frac{4}{9} \\ \frac{2}{9} & \frac{4}{9} & \frac{4}{9} \end{bmatrix}$$

Denote the matrix of Q by Q_0. Since $P + Q = I$, we have $Q = I - P$; therefore

$$Q_0 = I - P_0 = \begin{bmatrix} \frac{8}{9} & -\frac{2}{9} & -\frac{2}{9} \\ -\frac{2}{9} & \frac{5}{9} & -\frac{4}{9} \\ -\frac{2}{9} & -\frac{4}{9} & \frac{5}{9} \end{bmatrix}$$

EXAMPLE 5 Let us extend the result of Example **4** and show that the sum of projections in three orthogonal directions in R^3 is the identity transformation.

Suppose $\overline{w}_1$, $\overline{w}_2$, and $\overline{w}_3$ are an orthonormal basis for R^3 and P_1, P_2, and P_3 are the projections onto $\overline{w}_1$, $\overline{w}_2$, and $\overline{w}_3$, respectively. We then have the formulas

$$P_1\bar{u} = (\bar{u} \cdot \overline{w}_1)\overline{w}_1$$
$$P_2\bar{u} = (\bar{u} \cdot \overline{w}_2)\overline{w}_2$$
$$P_3\bar{u} = (\bar{u} \cdot \overline{w}_3)\overline{w}_3$$

Since $\bar{u} = (\bar{u} \cdot \overline{w}_1)\overline{w}_1 + (\bar{u} \cdot \overline{w}_2)\overline{w}_2 + (\bar{u} \cdot \overline{w}_3)\overline{w}_3$ (see formula **2**, page 120) we have

$$P_1 + P_2 + P_3 = I.$$

Thus, for example, $P_1 + P_2 = I - P_3$. Example **4** shows that $I - P_3$ is the projection orthogonal to $\overline{w}_3$; therefore, the sum of the projections onto $\overline{w}_1$ and $\overline{w}_2$ is the projection orthogonal to $\overline{w}_3$.

EXAMPLE 6 Reflection through a line and projection along the line are closely related. (See Figure 4, page 135.) This relationship is expressed by a transformation equation.

Suppose T is reflection in the line through $\overline{w}$ and P is projection onto $\overline{w}$. Then, as shown in Example **3**, page 134, we have $T\bar{u} = 2P\bar{u} - \bar{u}$, and, consequently, $T = 2P - I$. Solving for P we have

$$P = \tfrac{1}{2}(T + I)$$

EXAMPLE 7 □ The differential operator L defined by $Lf = f'' - f$ is the difference of the two operators D_1 and I, where $D_1 f = f''$. In fact, a linear differential operator is simply a linear transformation that is a sum of transformations of the form L_i, where

$$L_i f = a_i f^{(i)}$$

(See Example **7**, page 138.)

EXERCISES

1 Express as a single matrix

a $\begin{bmatrix} 2 & 0 \\ 1 & 1 \end{bmatrix} + 3\begin{bmatrix} -1 & 1 \\ 0 & 1 \end{bmatrix}$

b $-\begin{bmatrix} 1 & 2 & 1 & 0 \\ 1 & 1 & 1 & 0 \end{bmatrix} + 4\begin{bmatrix} -1 & 0 & 0 & 1 \\ 1 & 2 & 0 & 0 \end{bmatrix}$

c $-2\begin{bmatrix} 1 & 1 \\ 1 & 1 \\ 1 & 1 \end{bmatrix} - \tfrac{1}{2}\begin{bmatrix} -4 & -4 \\ -4 & -4 \\ -4 & -4 \end{bmatrix}$

2 For

$$A = \begin{bmatrix} 2 & 1 & 3 \\ 1 & -1 & 2 \end{bmatrix} \quad B = \begin{bmatrix} -1 & 0 & -2 \\ 4 & 1 & 2 \end{bmatrix} \quad \bar{u} = \begin{bmatrix} x \\ y \\ z \end{bmatrix}$$

verify that

a $A\bar{u} + B\bar{u} = (A + B)\bar{u}$ **b** $(-5A)\bar{u} = (-5)A\bar{u}$

c $(aB)\bar{u} = a(B\bar{u})$.

3 Suppose $T(x,y) = (3x - y, 2x)$ and $S(x,y) = (2x + y, 0)$. Find

a $(S + T)(3,4)$

b $(2T - 4S)(1,1)$

c The matrix of S, the matrix of T, and the matrix of $S + T$

4 Suppose $T(x,y,z) = (4x, 0, x + y + z, z)$, $S(x,y,z) = (x,x,x,x)$. Find the matrix of $S + T$, $S - T$, $2T$ and $3S - 2T$.

5 Suppose A has m rows and n columns and B has p rows and q columns. What relations must hold among m, n, p, and q for $A + B$ to be defined?

6 Suppose M is a plane through $(0,0,0)$ perpendicular to the unit vector $\bar{w}$. Describe projection onto M and reflection in M in terms of projection onto $\bar{w}$, as in Examples **5** and **6**.

7 Suppose $Df = f'$ and $Tf = \int_0^x f(t)\,dt$. Calculate

a $(D + T)(\cos x)$ **b** $(3D)x^2$ **c** $(-2T + 4D)e^{2x}$

Find a nonzero function in the null space of $T + D$.

8 Suppose R, S, and T are linear transformations which satisfy $3R + 2S - (4S + 2T + I) = 0$. Express T in terms of R, S, and I.

9 Suppose $A = \begin{bmatrix} a_{11} & a_{12} \\ a_{21} & a_{22} \end{bmatrix}$. Find $\lambda I - A$. If $\bar{u}$ is in the null space of $\lambda I - A$, what is $A\bar{u}$?

10 Show that if S and T are linear transformations from V into W, then $S + T$ and aT are linear. [Hint: Calculate $(S + T)(\bar{u} + \bar{v})$, and so on.]

□ 11 Suppose V is the set of all matrices with m rows and n columns. What is the dimension of V?

□ 12 Suppose P is projection onto $\bar{w}$ and Q is projection orthogonal to $\bar{w}$. Are P, Q, and I independent?

□ 13 Suppose S and T are the linear transformations from the complex space C^2 into C^2 defined by

$$S(x,y) = [(i + 2)y, -x - iy]$$
$$T(x,y) = [(3i + 1)x + iy, 0]$$

a What is $(S + T)(x,y)$?

b Verify that the matrix of $S + T$ is the sum of the matrix of S and the matrix of T.

c Find the matrix of iT.

SECTION 4 Transformation and Matrix Products

We continue our discussion of transformation and matrix algebra by defining a product. In Section 3 we first defined matrix sums and scalar multiples and then extended our definitions to transformations. In defining the product we find it more natural to begin with transformations.

Suppose T is a transformation from V into W and S is a transformation from W into U. For each $\bar{u}$ in V, the vector $T\bar{u}$ is in W; therefore, S can be applied to $T\bar{u}$; resulting in a vector $S(T\bar{u})$ which is in U.

The **product** ST is defined as the transformation from V into U whose value at each $\bar{u}$ in V is given by

$$(ST)\bar{u} = S(T\bar{u})$$

Thus, the operation ST is the operation obtained by first applying T, then applying S.

Let us show that ST is linear if S and T are linear. To do this, suppose S and T are linear and $\bar{u}$ and $\bar{v}$ are in V. Then, since T is linear,

$$T(\bar{u} + \bar{v}) = T\bar{u} + T\bar{v}$$

Since S is linear, we can apply S to this and obtain

$$S[T(\bar{u} + \bar{v})] = S(T\bar{u} + T\bar{v}) = S(T\bar{u}) + S(T\bar{v})$$

The definition of transformation product tells us that

$$S[T(\bar{u} + \bar{v})] = (ST)(\bar{u} + \bar{v}) \quad \text{and} \quad S(T\bar{u}) + S(T\bar{v}) = (ST)\bar{u} + (ST)\bar{v}$$

We have therefore shown that

$$(ST)(\bar{u} + \bar{v}) = (ST)\bar{u} + (ST)\bar{v}$$

A similar argument shows that

$$(ST)(a\bar{u}) = a(ST)\bar{u}$$

demonstrating that ST is linear if S and T are linear.

EXAMPLE 1 The product ST is defined by the process of composition, that is, by first performing the operation T, then the operation S. This is an extension of the idea of function composition. For example, if

$$f(x) = \sin x \qquad \text{and} \qquad g(x) = x^2$$

then the composition $f \circ g$ is the function defined by

$$(f \circ g)(x) = f(g(x)) = \sin x^2$$

Note that in this case the composition of two linear functions is linear. For example, if

$$h(x) = ax \qquad \text{and} \qquad k(x) = bx$$

then

$$h(k(x)) = a(k(x)) = a(bx) = (ab)x$$

whose graph is also a straight line through the origin.

EXAMPLE 2 A rotation followed by a rotation is just another rotation; that is, the product of two rotations is again a rotation. To be precise, suppose S is a counterclockwise rotation in R^2 through the angle θ and T is a counterclockwise rotation in R^2 through the angle φ. The product ST is then a counterclockwise rotation through φ followed by a counterclockwise rotation through θ. In other words,

$$ST \text{ is a counterclockwise rotation through } \varphi + \theta.$$

We can also rotate first through θ, then through φ. This is the product TS. Clearly, TS is also a counterclockwise rotation through $\varphi + \theta$, and, therefore, $ST = TS$. The operations S, T, ST, and TS are indicated in Figure 8.

EXAMPLE 3 We say that two operators S and T **commute** if $ST = TS$. Example 2 shows that any two rotations of R^2 must commute. This example shows two operators which *do not commute.*

Suppose S is a counterclockwise rotation in R^2 through the angle $\pi/4$ and that T is reflection in the y-axis. Then ST is reflection in the y-axis followed by rotation through $\pi/4$. The product TS is rotation through

Figure 8 *Rotation product.*

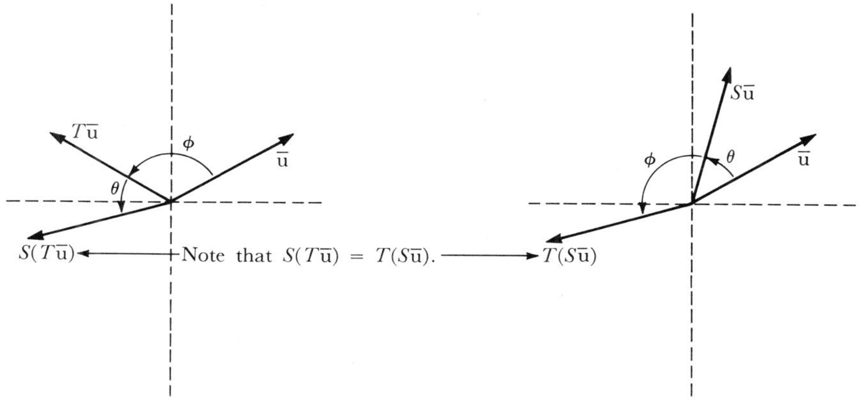

Figure 9

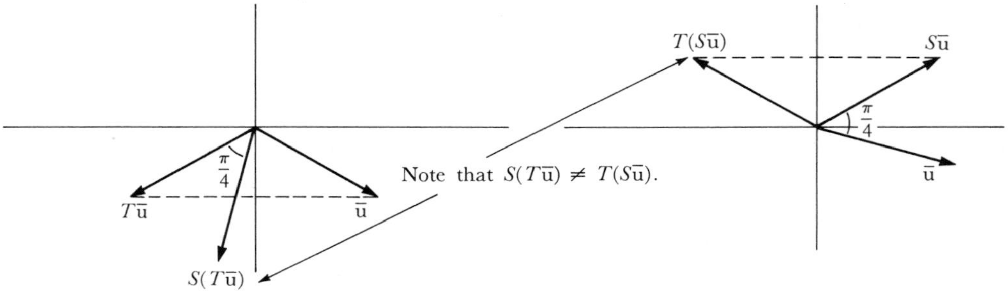

$\pi/4$ followed by reflection in the y-axis. As is shown in Figure 9, the two operations ST and TS are *not* the same.

EXAMPLE 4 Example **3** shows that transformation products differ from number products in that the order in which transformation products are performed may affect the result. Other new phenomena can also occur; for example, for numbers it is true that if $x^2 = x$, then $x = 0$ or $x = 1$. This is not so for transformations, in fact, any projection P satisfies $PP = P$.

Suppose $\bar{v}$ is a vector of length 1 and P is the operation of projection onto $\bar{v}$. As we saw in Example **1**, page 133, $P\bar{u}$ is given by the formula

1 $$P\bar{u} = (\bar{u} \cdot \bar{v})\bar{v}$$

Suppose $\bar{u}_1$ is parallel to $\bar{v}$. We should then expect that $P\bar{u}_1 = \bar{u}_1$. This fact can be established using formula **1**, since saying that $\bar{u}_1$ is parallel to $\bar{v}$ means that $\bar{u}_1 = a\bar{v}$, for some scalar a. Therefore

$$P\bar{u}_1 = P(a\bar{v}) = a(P\bar{v}) = a(\bar{v} \cdot \bar{v})\bar{v}$$

Since $\bar{v} \cdot \bar{v} = 1$ (by assumption, $\bar{v}$ is a unit vector) we have

$$P\bar{u}_1 = a(\bar{v} \cdot \bar{v})\bar{v} = a\bar{v} = \bar{u}_1$$

Now consider the operation of projecting twice, namely, $P^2 = PP$. For each $\bar{u}$ we know that $P\bar{u}$ is parallel to $\bar{v}$ (since formula **1** tells us that $P\bar{u}$ is a multiple of $\bar{v}$). Thus, since $P\bar{u}$ is parallel to $\bar{v}$, the above argument shows that $P(P\bar{u}) = P\bar{u}$.

We know from the definition of product that

$$P^2\bar{u} = (PP)\bar{u} = P(P\bar{u}) = P\bar{u}$$

Consequently, $P^2 = P$. This statement is equivalent to saying that projecting twice in the same direction is the same as projecting once in that direction.

EXAMPLE 5 ☐ Suppose V is the collection of all functions f such that f and *each* of its derivatives $f', f'', \ldots, f^{(n)}, \ldots$, are defined and continuous for $0 \leq x \leq 1$. We denote the differentiation operator by D, so that $Df = f'$.

The operator DD is the operation of differentiating twice. We usually denote this operator by D^2. For example,

$$\begin{aligned} D^2(x^3 + 3x^2) &= D[D(x^3 + 3x^2)] \\ &= D(3x^2 + 6x) \\ &= 6x + 6 \end{aligned}$$

The operator $D^3 = D(DD)$ is just the operation of differentiating three times. A systematic study of the operator D and its powers $D^2, D^3, \ldots$, will be given in Chapter 6.

DISCUSSION The student may wish to do some of the initial exercises at this point to clarify further his understanding of transformation products.

Let us now show how to define a matrix product. It will be defined so that the matrix of ST is the product of the matrix of S and the matrix of T. For simplicity consider matrices with two rows and two columns. Suppose

$$A = \begin{bmatrix} A_{11} & A_{12} \\ A_{21} & A_{22} \end{bmatrix} \quad \text{and} \quad B = \begin{bmatrix} B_{11} & B_{12} \\ B_{21} & B_{22} \end{bmatrix}$$

and that S and T are the linear transformations defined by $S\bar{u} = A\bar{u}$ and $T\bar{u} = B\bar{u}$.

We have shown that ST is linear. Furthermore, the matrix of ST is the matrix with the columns

$$(ST)\begin{bmatrix} 1 \\ 0 \end{bmatrix} \quad \text{and} \quad (ST)\begin{bmatrix} 0 \\ 1 \end{bmatrix}$$

as shown in the discussion of Theorem **10**, page 141ff. Likewise,

$$T\begin{bmatrix} 1 \\ 0 \end{bmatrix} = \begin{bmatrix} B_{11} \\ B_{21} \end{bmatrix} \quad \text{and} \quad T\begin{bmatrix} 0 \\ 1 \end{bmatrix} = \begin{bmatrix} B_{12} \\ B_{22} \end{bmatrix}$$

Now use the definition of ST and the above relations along with the fact that $S\bar{u} = A\bar{u}$ to obtain

$$ST\begin{bmatrix}1\\0\end{bmatrix} = S\left\{T\begin{bmatrix}1\\0\end{bmatrix}\right\} = S\begin{bmatrix}B_{11}\\B_{21}\end{bmatrix} = A\begin{bmatrix}B_{11}\\B_{21}\end{bmatrix}$$

$$ST\begin{bmatrix}0\\1\end{bmatrix} = S\left\{T\begin{bmatrix}0\\1\end{bmatrix}\right\} = S\begin{bmatrix}B_{12}\\B_{22}\end{bmatrix} = A\begin{bmatrix}B_{12}\\B_{22}\end{bmatrix}$$

In other words, the columns of the matrix of ST are given by

$$A\begin{bmatrix}B_{11}\\B_{21}\end{bmatrix} = \begin{bmatrix}A_{11}B_{11} + A_{12}B_{21}\\A_{21}B_{11} + A_{22}B_{21}\end{bmatrix}$$

and

$$A\begin{bmatrix}B_{12}\\B_{22}\end{bmatrix} = \begin{bmatrix}A_{11}B_{12} + A_{12}B_{22}\\A_{21}B_{12} + A_{22}B_{22}\end{bmatrix}$$

The matrix of ST, then, is

$$\begin{bmatrix}A_{11}B_{11} + A_{12}B_{21} & A_{11}B_{12} + A_{12}B_{22}\\A_{21}B_{11} + A_{22}B_{21} & A_{21}B_{12} + A_{22}B_{22}\end{bmatrix}$$

This matrix is called the **product** AB. The product is defined in such a way that

2 If A is the matrix of S and B is the matrix of T, then AB is the matrix of ST.

Follow this same procedure for larger matrices, obtaining the definition: If A has m rows and n columns and B has n rows and k columns, then the **product** AB is the matrix with m rows and k columns such that

3 Each column of AB is A times the corresponding column of B.

This product is so defined that property **2** holds in this general case. Since the student must become adept at computing matrix products before proceeding further, we now give several examples. Also, much of the next section is devoted to practice with matrix products.

EXAMPLE 6 The product AB can be calculated by multiplying A times each column of B. For example,

$$\begin{bmatrix}2 & 1\\3 & 2\end{bmatrix}\begin{bmatrix}1 & 2 & -1\\1 & 1 & 1\end{bmatrix} = \begin{bmatrix}3 & 5 & -1\\5 & 8 & -1\end{bmatrix}$$

since

$$\begin{bmatrix} 2 & 1 \\ 3 & 2 \end{bmatrix}\begin{bmatrix} 1 \\ 1 \end{bmatrix} = \begin{bmatrix} 3 \\ 5 \end{bmatrix}, \quad \begin{bmatrix} 2 & 1 \\ 3 & 2 \end{bmatrix}\begin{bmatrix} 2 \\ 1 \end{bmatrix} = \begin{bmatrix} 5 \\ 8 \end{bmatrix}, \quad \begin{bmatrix} 2 & 1 \\ 3 & 2 \end{bmatrix}\begin{bmatrix} -1 \\ 1 \end{bmatrix} = \begin{bmatrix} -1 \\ -1 \end{bmatrix}$$

If the matrices are large one should do this carefully. For example, proceed carefully to multiply each column of the right matrix by the left matrix:

$$\begin{bmatrix} 6 & 1 & 2 & 1 \\ 4 & 0 & 1 & -1 \\ 2 & 1 & 3 & 0 \end{bmatrix}\begin{bmatrix} 4 & 2 & 1 \\ 1 & 1 & 0 \\ 2 & -3 & -1 \\ 6 & 1 & 1 \end{bmatrix} = \begin{bmatrix} 35 & 8 & 5 \\ 12 & 4 & 2 \\ 15 & -4 & -1 \end{bmatrix}$$

REMARK It is common practice in most elementary mathematics to write $f(x)$ rather than xf for the value of a function f at x. Following this practice, we write $T\bar{u}$ for the value of T at $\bar{u}$. This causes transformation products to be in reverse order from English speech; that is, ST means "do T then do S," which is opposite to the left to right order of "ST." It is also desirable to write the matrix on the same side of the vector as the transformation symbol and to have property **2** hold. In order not to have a separate definition of $A\bar{u}$ it therefore becomes necessary to express $\bar{u}$ as a column matrix and define A as the matrix whose columns are just the transformation applied to the standard basis.

EXAMPLE 7 Matrix products are not always defined, for the number of columns in A must be the same as the number of rows of B in order to define AB. The product

$$\begin{bmatrix} 2 & 1 \\ 1 & 2 \end{bmatrix}\begin{bmatrix} 1 & 2 \\ 3 & 1 \\ -1 & 4 \end{bmatrix}$$

is not defined, because the left matrix is the matrix of a linear transformation S from R^2 into R^2, and the right matrix is the matrix of a linear transformation T from R^2 into R^3. Thus, ST is not defined, for after applying T, we obtain a vector in R^3, and hence (since S is not defined for vectors in R^3) we cannot then apply S.

If the order in this case is reversed, the product is defined:

$$\begin{bmatrix} 1 & 2 \\ 3 & 1 \\ -1 & 4 \end{bmatrix}\begin{bmatrix} 2 & 1 \\ 1 & 2 \end{bmatrix} = \begin{bmatrix} 4 & 5 \\ 7 & 5 \\ 2 & 7 \end{bmatrix}$$

This shows that it may be possible to find BA but not AB. In general, even this may not be possible. For example, neither product

$$\begin{bmatrix} 2 & 1 \\ 1 & 2 \end{bmatrix} \begin{bmatrix} 1 & 2 & 1 \\ 3 & 1 & 0 \\ -1 & 4 & 0 \end{bmatrix} \quad \text{nor} \quad \begin{bmatrix} 1 & 2 & 1 \\ 3 & 1 & 0 \\ -1 & 4 & 0 \end{bmatrix} \begin{bmatrix} 2 & 1 \\ 1 & 2 \end{bmatrix}$$

is defined. Note also that

$$\begin{bmatrix} 1 & 1 \\ 0 & 1 \end{bmatrix} \begin{bmatrix} 1 & -3 \\ 0 & 2 \end{bmatrix} = \begin{bmatrix} 1 & -1 \\ 0 & 2 \end{bmatrix}$$

and

$$\begin{bmatrix} 1 & -3 \\ 0 & 2 \end{bmatrix} \begin{bmatrix} 1 & 1 \\ 0 & 1 \end{bmatrix} = \begin{bmatrix} 1 & -2 \\ 0 & 2 \end{bmatrix}$$

This shows that AB is not generally equal to BA, even if both are defined. (Example 3 should lead us to expect this.)

EXAMPLE 8 In the next section we shall establish the associative law $A(BC) = (AB)C$, from which it follows that $A(AA) = (AA)A$. One can therefore use exponent notation without ambiguity, writing A^2 for AA, A^3 for $(AA)A$, and so forth. The usual laws for exponents will then hold, such as $A^mA^n = A^{m+n}$ and $(A^m)^n = A^{mn}$. It is not true in general, however, that $A^2 = 0$ must imply that $A = 0$. The following examples illustrate some of these considerations.

$$\begin{bmatrix} 1 & 2 \\ -1 & 1 \end{bmatrix}^2 = \begin{bmatrix} 1 & 2 \\ -1 & 1 \end{bmatrix} \begin{bmatrix} 1 & 2 \\ -1 & 1 \end{bmatrix} = \begin{bmatrix} -1 & 4 \\ -2 & -1 \end{bmatrix}$$

$$\begin{bmatrix} 1 & 2 \\ -1 & 1 \end{bmatrix}^3 = \begin{bmatrix} -1 & 4 \\ -2 & -1 \end{bmatrix} \begin{bmatrix} 1 & 2 \\ -1 & 1 \end{bmatrix} = \begin{bmatrix} -5 & 2 \\ -1 & -5 \end{bmatrix}$$

$$\begin{bmatrix} 1 & 2 \\ -1 & 1 \end{bmatrix}^4 = \begin{bmatrix} -5 & 2 \\ -1 & -5 \end{bmatrix} \begin{bmatrix} 1 & 2 \\ -1 & 1 \end{bmatrix} = \begin{bmatrix} -7 & -8 \\ 4 & -7 \end{bmatrix}$$

Note that

$$\begin{bmatrix} 1 & 2 \\ -1 & 1 \end{bmatrix}^4 = \begin{bmatrix} -1 & 4 \\ -2 & -1 \end{bmatrix} \begin{bmatrix} -1 & 4 \\ -2 & -1 \end{bmatrix} = \begin{bmatrix} -7 & -8 \\ 4 & -7 \end{bmatrix}$$

so that, indeed, $A^4 = (A^2)^2$.

The following is an example of a nonzero matrix whose square is the zero matrix.

$$\begin{bmatrix} 1 & -1 \\ 1 & -1 \end{bmatrix}^2 = \begin{bmatrix} 1 & -1 \\ 1 & -1 \end{bmatrix} \begin{bmatrix} 1 & -1 \\ 1 & -1 \end{bmatrix} = \begin{bmatrix} 0 & 0 \\ 0 & 0 \end{bmatrix}$$

We have defined the matrix product in terms of the transformation product. The remaining two examples further illustrate this interconnection.

EXAMPLE 9 Formulas for the sine and cosine of the sum of two angles can be obtained by calculating the product of the matrices of two rotations.

Suppose S and T are counterclockwise rotations in R^2 through the angles θ and φ respectively. (See Example 2, page 159.) The respective matrices of S and T are

$$A = \begin{bmatrix} \cos\theta & -\sin\theta \\ \sin\theta & \cos\theta \end{bmatrix} \quad \text{and} \quad B = \begin{bmatrix} \cos\varphi & -\sin\varphi \\ \sin\varphi & \cos\varphi \end{bmatrix}$$

as shown in Example 3, page 145.

We have

$$AB = \begin{bmatrix} \cos\theta\cos\varphi - \sin\theta\sin\varphi & -\cos\theta\sin\varphi - \sin\theta\cos\varphi \\ \sin\theta\cos\varphi + \cos\theta\sin\varphi & -\sin\theta\sin\varphi + \cos\theta\cos\varphi \end{bmatrix}$$

Because of property 2, the matrix AB is the matrix of ST. We showed in Example 2 that ST is just counterclockwise rotation through the angle $\theta + \varphi$, so its matrix is

$$\begin{bmatrix} \cos(\theta+\varphi) & -\sin(\theta+\varphi) \\ \sin(\theta+\varphi) & \cos(\theta+\varphi) \end{bmatrix}$$

Since this must be equal to AB (for Theorem 10 tells us that the matrix of a transformation is unique) we have derived the familiar formulas

$$\cos(\theta+\varphi) = \cos\theta\cos\varphi - \sin\theta\sin\varphi$$
$$\sin(\theta+\varphi) = \cos\theta\sin\varphi + \sin\theta\cos\varphi$$

This derivation provides a convenient device for reconstructing these formulas.

EXAMPLE 10 Suppose S and T are defined by the formulas

$$T(x,y) = (2x - y,\ 3x + y) \quad \text{and} \quad S(x,y) = (3x + 2y,\ -x - 2y)$$

Then, rewriting these in column form,

$$T\begin{bmatrix} x \\ y \end{bmatrix} = \begin{bmatrix} 2 & -1 \\ 3 & 1 \end{bmatrix} \begin{bmatrix} x \\ y \end{bmatrix} \quad \text{and} \quad S\begin{bmatrix} x \\ y \end{bmatrix} = \begin{bmatrix} 3 & 2 \\ -1 & -2 \end{bmatrix} \begin{bmatrix} x \\ y \end{bmatrix}$$

Therefore

$$ST\begin{bmatrix} x \\ y \end{bmatrix} = \begin{bmatrix} 3 & 2 \\ -1 & -2 \end{bmatrix} \begin{bmatrix} 2 & -1 \\ 3 & 1 \end{bmatrix} \begin{bmatrix} x \\ y \end{bmatrix} = \begin{bmatrix} 12 & -1 \\ -8 & -1 \end{bmatrix} \begin{bmatrix} x \\ y \end{bmatrix}$$

which can be rewritten as

$$ST(x,y) = (12x - y, -8x - y)$$

This formula can also be computed directly from the given formulas for S and T:

$$\begin{aligned} ST(x,y) &= S[T(x,y)] \\ &= S(2x - y, 3x + y) \\ &= (3(2x - y) + 2(3x + y), -(2x - y) - 2(3x + y)) \\ &= (12x - y, -8x - y) \end{aligned}$$

EXERCISES

1 If S is counterclockwise rotation through the angle $\pi/4$, and T is counterclockwise rotation through the angle π, describe each of the following:
a S^2 **b** T^2 **c** S^4 **d** ST **e** $(ST)^2$ **f** S^4T

2 Suppose S is a counterclockwise rotation through $\pi/2$ and T is a reflection in the line $y = x$. Draw figures such as those in Figures 8 and 9 indicating $(ST)\bar{u}$ and $(TS)\bar{u}$. Do S and T commute?

3 **a** Suppose S is a reflection in the y-axis and T is a projection onto the line $y = x$. Draw figures illustrating $(ST)\bar{u}$ and $(TS)\bar{u}$. Do S and T commute?
b Suppose S is as given in part **a** and S_1 is reflection in the x-axis. What is S_1S? Do S_1 and S commute? (A figure may be helpful.)
c For S, S_1, and T as in parts **a** and **b** let T_1 denote S_1S. Do T_1 and T commute? (A figure may be helpful.)

4 Suppose $\bar{v}$ is a unit vector in R^2 and P is the projection onto $\bar{v}$, while Q is the projection orthogonal to $\bar{v}$. We showed that $P^2 = P$ in Example **4**. Show that
a $Q^2 = Q$ **b** $P + Q = I$ **c** $PQ = 0$ **d** $QP = 0$.
(The use of figures or the dot product formulas might be helpful.)

5 Suppose P is the projection onto the unit vector $\bar{v}$ in R^2 and T is the reflection in the line through $\bar{v}$.

a Show that $T^2 = I$. (A figure may be helpful.)

b Show that $T^2 = I$ by using the fact that $P^2 = P$ and $T = 2P - I$ and calculating $(2P - I)^2$.

□ **6** For $Df = f'$ and $Tf(x) = \int_0^x f(t)\,dt$ find

a Dx **b** Tx^2 **c** $D^3 \sin x$ **d** $T^2 (\sin 2x)$

e $TD (\cos x)$ **f** $DT (\cos x)$ **g** $(3D^2 + 2T)x^2$

Do D and T commute?

7 Calculate

a $\begin{bmatrix} 2 & 0 \\ 1 & 1 \end{bmatrix} \begin{bmatrix} -1 & 1 \\ 0 & 1 \end{bmatrix}$ **b** $\begin{bmatrix} 3 & 1 \\ -1 & 0 \end{bmatrix} \begin{bmatrix} 0 & 1 \\ 0 & 2 \end{bmatrix}$

c $\begin{bmatrix} 1 & 2 \\ 3 & 4 \end{bmatrix} \begin{bmatrix} 5 & 6 \\ 7 & 8 \end{bmatrix}$ **d** $\begin{bmatrix} 1 & 1 \\ 0 & 1 \end{bmatrix} \begin{bmatrix} 1 & -1 \\ 0 & 1 \end{bmatrix}$

8 Calculate

a $\begin{bmatrix} 3 & 1 & -4 \\ 1 & 1 & 0 \\ -1 & 1 & 2 \end{bmatrix} \begin{bmatrix} 1 & -2 & -1 \\ 6 & -1 & 0 \\ 1 & 0 & 0 \end{bmatrix}$

b $\begin{bmatrix} -1 & 1 & 1 \\ 1 & -1 & 1 \\ 1 & 1 & -1 \end{bmatrix} \begin{bmatrix} 1 & -1 & -1 \\ -1 & 1 & -1 \\ -1 & -1 & 1 \end{bmatrix}$

c $\begin{bmatrix} 1 & 2 & 3 \\ 4 & 5 & 6 \\ 7 & 8 & 9 \end{bmatrix} \begin{bmatrix} 1 & -2 & 3 \\ -4 & 5 & -6 \\ 7 & -8 & 9 \end{bmatrix}$ **d** $\begin{bmatrix} 0 & 0 & 1 \\ 0 & 1 & 0 \\ 1 & 0 & 0 \end{bmatrix} \begin{bmatrix} 1 & 2 & 1 \\ 2 & 1 & 1 \\ 1 & 1 & 2 \end{bmatrix}$

9 Calculate

a $\begin{bmatrix} 2 & 1 & 0 \\ 0 & 1 & 1 \end{bmatrix} \begin{bmatrix} 1 & -1 & 2 \\ 1 & 2 & 3 \\ 0 & 1 & 1 \end{bmatrix}$ **b** $\begin{bmatrix} 2 & 1 & 0 \end{bmatrix} \begin{bmatrix} 1 \\ -1 \\ 1 \end{bmatrix}$

c $\begin{bmatrix} 4 & 1 & 1 & 2 & -6 \\ 1 & 2 & 1 & 1 & 0 \\ 0 & 0 & -3 & -2 & -1 \\ 1 & 1 & 2 & -1 & 0 \\ 0 & 0 & 0 & 1 & 1 \end{bmatrix} \begin{bmatrix} 0 & 1 & -1 & 1 & 1 \\ 2 & 1 & 4 & 1 & -6 \\ 0 & 3 & 0 & 2 & 1 \\ 2 & 1 & 0 & -1 & 1 \\ 1 & 1 & 1 & 1 & 1 \end{bmatrix}$

d $\begin{bmatrix} 2 & 0 & 1 & 4 & 1 \\ 1 & 2 & 1 & 0 & 6 \\ 0 & 3 & 1 & 2 & 1 \end{bmatrix} \begin{bmatrix} -1 & 1 \\ -1 & 1 \\ -1 & 1 \\ -1 & 1 \\ 0 & 0 \end{bmatrix}$

10 For each of the following matrices A find A^2, A^3, A^4, and $(A^2)^2$.

a $A = \begin{bmatrix} 1 & 1 \\ 0 & 1 \end{bmatrix}$ **b** $A = \begin{bmatrix} 0 & 1 \\ 1 & 0 \end{bmatrix}$

c $A = \begin{bmatrix} 1 & 2 \\ 3 & 4 \end{bmatrix}$ **d** $A = \begin{bmatrix} 2 & 1 & 0 \\ 1 & -1 & 1 \\ 1 & 2 & 1 \end{bmatrix}$

11 For

$$A = \begin{bmatrix} 2 & 1 & 0 \\ -1 & 1 & 2 \end{bmatrix} \quad B = \begin{bmatrix} 3 & 1 \\ 2 & -1 \\ 0 & 1 \end{bmatrix} \quad C = \begin{bmatrix} 2 & 0 \\ -1 & 1 \end{bmatrix}$$

which of the following are defined? Calculate those that are defined.

a AB **b** $(AB)C$ **c** $A(BC)$
d $C(AB)$ **e** BA **f** AC
g A^2 **h** B^2 **i** $AB - C^2$
j $BA - C^2$

12 What relationship must hold between the number of columns of A and rows of B in order that AB is defined? How many rows and columns does AB have?

13 The entries of AB are just the dot products of the rows of A with the columns of B. Which rows and columns do you use to obtain the entry in the third row and fourth column of AB?

14 Let

$$A = \begin{bmatrix} A_{11} & A_{12} \\ A_{21} & A_{22} \end{bmatrix} \quad B = \begin{bmatrix} B_{11} & B_{12} \\ B_{21} & B_{22} \end{bmatrix}$$

a Show that the columns of AB are linear combinations of the columns of A.
b Show that the rows of AB are linear combinations of the rows of B.
c Deduce from **a** or **b** that rank AB is not larger than rank A or rank B.
d Show that the null space of AB is a subspace of the null space of B.

15 For $T(x,y) = (3x - y, 2x)$ and $S(x,y) = (2x + y, 0)$ calculate $TS(x,y)$ and $ST(x,y)$ directly from the definition of transformation multiplication. Find the matrix of S, T, ST, and TS.

16 Suppose $\bar{w} = \frac{1}{3}(1,2,2)$, P is projection onto $\bar{w}$, and T is reflection in the line through $\bar{w}$. Let P_0 and T_0 denote the matrices of P and T, respectively.

a Find P_0 and T_0.

b Verify that $P_0^2 = P_0$ and $T_0^2 = I$.

17 Suppose $\bar{v} = \dfrac{1}{\sqrt{5}}(1,2)$ and $\bar{w} = \dfrac{1}{\sqrt{5}}(2,-1)$. Let P_0 and Q_0 denote the matrices of projection onto $\bar{v}$ and $\bar{w}$, respectively.

a Find P_0 and Q_0.

b Verify that $P_0^2 = P_0$, $Q_0^2 = Q_0$, $P_0Q_0 = Q_0P_0$, and $P_0 + Q_0 = I$.

□ **18** The concepts and results of this section extend to complex vector spaces.

a Find $\begin{bmatrix} 1 & 1+i \\ 2 & -i \end{bmatrix} \begin{bmatrix} 1-i & i \\ 1 & i \end{bmatrix}$. **b** Find $\begin{bmatrix} 1 & 1+i \\ 2 & -i \end{bmatrix}^2$.

c Find the matrix of $S + T$, iT, and ST where

$$S(x,y) = (ix,0) \qquad \text{and} \qquad T(x,y) = [(1+i)y, ix]$$

SECTION 5 Rules and Special Products

While it is *not* true in general that $ST = TS$ for transformations S and T (or for matrices), some useful rules of transformation (and hence matrix) multiplication are true. A summary of these rules follows:

1
- **a** $R(S + T) = RS + RT$
- **b** $(R + S)T = RT + ST$
- **c** $R(ST) = (RS)T$
- **d** $a(ST) = (aS)T = S(aT)$

The first two are sometimes called *distributive laws* and the latter two, *associative laws*. These rules, of course, assume that the products and sums are defined. The same rules are true automatically for matrix products as well. We shall derive only the third rule, the derivations of the other rules being similar. Let us therefore proceed to show that

$$R(ST) = (RS)T$$

Assume that after applying T we obtain a vector for which S is defined, and that after then applying S we obtain a vector for which R is defined. To establish this law it is then necessary to show that

$$[R(ST)]\bar{u} = [(RS)T]\bar{u}$$

for each vector $\bar{u}$, for which T is defined. In order to do this one shows that both sides are equal to $R[S(T\bar{u})]$: Apply the definition of product (see page 158) to obtain

$$[R(ST)]\bar{u} = R[(ST)\bar{u}]$$

Apply the definition again to replace $(ST)\bar{u}$ by $S(T\bar{u})$. Hence we have the equality $[R(ST)]\bar{u} = R[S(T\bar{u})]$. Similar arguments show that $[(RS)T]\bar{u} = R[S(T\bar{u})]$, thereby completing the proof of rule **1c**.

This proof of the associative law is one case where it is easier to use transformations than to use matrices. Indeed, a direct proof for matrices that $(AB)C = A(BC)$ requires some fairly complicated notation. (See Exercise **5** in the Appendix.)

Special rules hold for products involving the zero transformation 0 and the identity transformation I, which were defined in Section 3. These are

2 **a** $S0 = 0 \qquad 0T = 0$
b $SI = S \qquad IT = T$

These rules assume that 0 and I denote the appropriate zero and identity transformations so that $S0$, $0T$, SI, and IT are defined.

For example, if 0 denotes the zero transformation from V into W and S is a linear transformation from W into U, then

$$(S0)\bar{u} = S(0\bar{u}) = S\bar{0} = \bar{0}$$

Therefore, $S0$ is the zero transformation from V into U. This establishes the first part of rule **2a**.

Suppose T is a linear transformation from V into W and that I is the identity transformation on W. Then for each $\bar{u}$ in V we have

$$(IT)\bar{u} = I(T\bar{u}) = T\bar{u}$$

since $T\bar{u}$ is in W and I leaves fixed any vector in W. This establishes the second part of rule **2b**. The proofs of the other parts of rule **2** are omitted.

EXAMPLE 1 Some care needs to be exercised in manipulating transformations and matrices because the commutative law is *not* in general true. For example, the identity

$$(a + b)(a - b) = a^2 - b^2$$

which holds for all real numbers a and b, uses the fact that $ab = ba$. For transformations S and T such that $S + T$, S^2, T^2, ST, and TS are defined, we can use rules **1** to obtain

$$\begin{aligned}(S + T)(S - T) &= S(S - T) + T(S - T) \\ &= SS - ST + TS - TT \\ &= S^2 - ST + TS - T^2\end{aligned}$$

Thus one can *only* conclude that

$$(S + T)(S - T) = S^2 - T^2$$

when one knows that $ST = TS$. In fact, if this result were true for a given S and T, we must have

$$S^2 - T^2 = S^2 - ST + TS - T^2$$

Therefore, after subtracting S^2 and adding T^2 to both sides we must have $0 = -ST + TS$, which is the same as $ST = TS$. In other words,

3 $\quad (S + T)(S - T) = S^2 - T^2$ if and only if $ST = TS$

EXAMPLE 2 □ The expression $y'' - y$ can be written in operator form as $(D^2 - I)y$, where $Dy = y'$ and I is the identity, for we have

$$(D^2 - I)y = D^2y - Iy = y'' - y$$

Since $DI = ID$ (for the identity commutes with any operator) we can apply result **3** to obtain

$$D^2 - I = (D + I)(D - I)$$

We shall find such factorization particularly useful in our discussion of linear differential equations in Chapter 6.

EXAMPLE 3 *Matrix and Operator Polynomials.*
Polynomial notation can be used to simplify discussion of expressions such as $A^2 - I$ or $A^3 - 3A^2 + 3A - I$, where A represents a matrix (or transformation), and I is the identity matrix (or the identity operator). If $p(\lambda)$ is the polynomial

$$p(\lambda) = a_k\lambda^k + a_{k-1}\lambda^{k-1} + \cdots + a_1\lambda + a_0$$

where $a_k, a_{k-1}, \ldots, a_1, a_0$ are scalars, then for a square matrix A, the symbol $p(A)$ will be used to denote the matrix

$$p(A) = a_kA^k + a_{k-1}A^{k-1} + \cdots + a_1A + a_0I$$

If T is a linear transformation from V into V, we use the symbol $p(T)$ to denote the transformation

$$p(T) = a_kT^k + a_{k-1}T^{k-1} + \cdots + a_1T + a_0I$$

For example, suppose $A = \begin{bmatrix} 2 & 1 \\ -1 & 0 \end{bmatrix}$ and $p(\lambda) = 3\lambda^3 - 2\lambda^2 + 4$. Then

$$\begin{aligned} p(A) &= 3A^3 - 2A^2 + 4I \\ &= 3\begin{bmatrix} 4 & 3 \\ -3 & -2 \end{bmatrix} - 2\begin{bmatrix} 3 & 2 \\ -2 & -1 \end{bmatrix} + 4\begin{bmatrix} 1 & 0 \\ 0 & 1 \end{bmatrix} \\ &= \begin{bmatrix} 10 & 5 \\ -5 & 0 \end{bmatrix} \end{aligned}$$

We note that if $p(\lambda)$ and $q(\lambda)$ are polynomials, the matrices $p(A)$ and $q(A)$ *commute* (as do $p(T)$ and $q(T)$ for transformations T). For example, if $p(\lambda) = \lambda^2 - 3\lambda + 1$ and $q(\lambda) = 5\lambda^2 - \lambda - 2$, then

$$p(A) = A^2 - 3A + I \text{ and } q(A) = 5A^2 - A - 2I$$

Therefore, $p(A)q(A) = (A^2 - 3A + I)(5A^2 - A - 2I)$. Using rules **1** and **2** we obtain $p(A)q(A) = 5A^4 - 16A^3 + 6A^2 + 5A - 2I$.

The same rules also yield

$$\begin{aligned} q(A)p(A) &= (5A^2 - A - 2I)(A^2 - 3A + I) \\ &= 5A^4 - 16A^3 + 6A^2 + 5A - 2I \end{aligned}$$

Therefore, we indeed have $p(A)q(A) = q(A)p(A)$.

In the discussion of linear differential operators in Chapter 6 we shall make use of the fact that polynomial factorization gives a corresponding matrix or operator factorization. For example, if $p(\lambda) = \lambda^2 - 1$, $q(\lambda) = \lambda - 1$, and $r(\lambda) = \lambda + 1$, then $p(\lambda) = q(\lambda)r(\lambda)$, and for any square matrix A we have $p(A) = q(A)r(A)$. Of course, this is exactly the relation used in Example **2**, above.

The remaining examples of this chapter exhibit the use of these rules for matrices, as well as giving some further special rules.

EXAMPLE 4 Multiplying by 0 or I is quite easy. For example,

a $\begin{bmatrix} 0 & 0 \\ 0 & 0 \end{bmatrix} \begin{bmatrix} 2 & 1 & 3 \\ 4 & -1 & -7 \end{bmatrix} = \begin{bmatrix} 0 & 0 & 0 \\ 0 & 0 & 0 \end{bmatrix}$

b $\begin{bmatrix} 2 & 1 & 3 \\ 4 & -1 & -7 \end{bmatrix} \begin{bmatrix} 0 & 0 & 0 & 0 \\ 0 & 0 & 0 & 0 \\ 0 & 0 & 0 & 0 \end{bmatrix} = \begin{bmatrix} 0 & 0 & 0 & 0 \\ 0 & 0 & 0 & 0 \end{bmatrix}$

c $\begin{bmatrix} 1 & 0 \\ 0 & 1 \end{bmatrix} \begin{bmatrix} 2 & 1 & 3 \\ 4 & -1 & -7 \end{bmatrix} = \begin{bmatrix} 2 & 1 & 3 \\ 4 & -1 & -7 \end{bmatrix}$

d $\begin{bmatrix} 2 & 1 & 3 \\ 4 & -1 & -7 \end{bmatrix} \begin{bmatrix} 1 & 0 & 0 \\ 0 & 1 & 0 \\ 0 & 0 & 1 \end{bmatrix} = \begin{bmatrix} 2 & 1 & 3 \\ 4 & -1 & -7 \end{bmatrix}$

Note that the zero and identity matrices have the appropriate sizes in order that the products are defined. For example,

$$\begin{bmatrix} 0 & 0 & 0 \\ 0 & 0 & 0 \end{bmatrix} \begin{bmatrix} 2 & 1 & 3 \\ 4 & -1 & -7 \end{bmatrix}$$

is *not* defined.

EXAMPLE 5 If A has a row of zeros, the same row of AB consists of zeros. For example,

a $\begin{bmatrix} -2 & -3 & 1 \\ 0 & 0 & 0 \\ 1 & -1 & 0 \end{bmatrix} \begin{bmatrix} 6 & -2 & 1 \\ 3 & 1 & 2 \\ -1 & 1 & 1 \end{bmatrix} = \begin{bmatrix} -22 & 2 & -7 \\ 0 & 0 & 0 \\ 3 & -3 & -1 \end{bmatrix}$

If B has a column of zeros, the same column of AB consists of zeros. For example,

b $\begin{bmatrix} 4 & 1 & 0 \\ -1 & 2 & 3 \\ 1 & 0 & 1 \end{bmatrix} \begin{bmatrix} 1 & 0 & 1 \\ -2 & 0 & 1 \\ 1 & 0 & 1 \end{bmatrix} = \begin{bmatrix} 2 & 0 & 5 \\ -2 & 0 & 4 \\ 2 & 0 & 2 \end{bmatrix}$

EXAMPLE 6 Suppose A is a **diagonal matrix**; that is, A has the same number of rows as columns and each entry *not* on the left-to-right downward diagonal is zero. Each row of AB is then just the product of the corresponding row of B with the corresponding diagonal entry of A. For example,

a $\begin{bmatrix} 2 & 0 & 0 \\ 0 & -1 & 0 \\ 0 & 0 & 3 \end{bmatrix} \begin{bmatrix} 4 & 2 & 1 \\ -1 & 0 & 6 \\ 2 & 1 & -3 \end{bmatrix} = \begin{bmatrix} 8 & 4 & 2 \\ 1 & 0 & -6 \\ 6 & 3 & -9 \end{bmatrix}$

b $\begin{bmatrix} 4 & 0 & 0 & 0 \\ 0 & 0 & 0 & 0 \\ 0 & 0 & 3 & 0 \\ 0 & 0 & 0 & -2 \end{bmatrix} \begin{bmatrix} 3 & 0 & -1 & 2 \\ -1 & 1 & 0 & 1 \\ 4 & -1 & -2 & 1 \\ 0 & 1 & 3 & -4 \end{bmatrix} = \begin{bmatrix} 12 & 0 & -4 & 8 \\ 0 & 0 & 0 & 0 \\ 12 & -3 & -6 & 3 \\ 0 & -2 & -6 & 8 \end{bmatrix}$

c $\begin{bmatrix} 2 & 0 & 0 \\ 0 & -1 & 0 \\ 0 & 0 & 3 \end{bmatrix}^2 = \begin{bmatrix} 2 & 0 & 0 \\ 0 & -1 & 0 \\ 0 & 0 & 3 \end{bmatrix} \begin{bmatrix} 2 & 0 & 0 \\ 0 & -1 & 0 \\ 0 & 0 & 3 \end{bmatrix} = \begin{bmatrix} 4 & 0 & 0 \\ 0 & 1 & 0 \\ 0 & 0 & 9 \end{bmatrix}$

Note in **c** how easy it is to take the powers of a diagonal matrix. Continuing with further powers we have

d $\begin{bmatrix} 2 & 0 & 0 \\ 0 & -1 & 0 \\ 0 & 0 & 3 \end{bmatrix}^3 = \begin{bmatrix} 2 & 0 & 0 \\ 0 & -1 & 0 \\ 0 & 0 & 3 \end{bmatrix} \begin{bmatrix} 2 & 0 & 0 \\ 0 & -1 & 0 \\ 0 & 0 & 3 \end{bmatrix}^2 = \begin{bmatrix} 8 & 0 & 0 \\ 0 & -1 & 0 \\ 0 & 0 & 27 \end{bmatrix}$

e $\begin{bmatrix} 2 & 0 & 0 \\ 0 & -1 & 0 \\ 0 & 0 & 3 \end{bmatrix}^{10} = \begin{bmatrix} 2^{10} & 0 & 0 \\ 0 & 1 & 0 \\ 0 & 0 & 3^{10} \end{bmatrix}$

EXAMPLE 7 If B is a diagonal matrix, each column of AB is just the product of the corresponding column of A with the corresponding diagonal entry of B.

a $\begin{bmatrix} 4 & 2 & 1 \\ -1 & 0 & 6 \\ 2 & 1 & -3 \end{bmatrix} \begin{bmatrix} 2 & 0 & 0 \\ 0 & -1 & 0 \\ 0 & 0 & 3 \end{bmatrix} = \begin{bmatrix} 8 & -2 & 3 \\ -2 & 0 & 18 \\ 4 & -1 & -9 \end{bmatrix}$

b $\begin{bmatrix} 3 & 0 & -1 & 2 \\ -1 & 1 & 0 & 1 \\ 4 & -1 & -2 & 1 \\ 0 & 1 & 3 & -4 \end{bmatrix} \begin{bmatrix} 4 & 0 & 0 & 0 \\ 0 & 0 & 0 & 0 \\ 0 & 0 & 3 & 0 \\ 0 & 0 & 0 & -2 \end{bmatrix} = \begin{bmatrix} 12 & 0 & -3 & -4 \\ -4 & 0 & 0 & -2 \\ 16 & 0 & -6 & -2 \\ 0 & 0 & 9 & 8 \end{bmatrix}$

EXAMPLE 8 If A and B are both diagonal matrices having n rows and n columns, they commute. In other words, $AB = BA$.

a $\begin{bmatrix} 2 & 0 & 0 \\ 0 & -1 & 0 \\ 0 & 0 & 3 \end{bmatrix} \begin{bmatrix} -2 & 0 & 0 \\ 0 & 4 & 0 \\ 0 & 0 & -6 \end{bmatrix} = \begin{bmatrix} -4 & 0 & 0 \\ 0 & -4 & 0 \\ 0 & 0 & -18 \end{bmatrix}$

b $\begin{bmatrix} -2 & 0 & 0 \\ 0 & 4 & 0 \\ 0 & 0 & -6 \end{bmatrix} \begin{bmatrix} 2 & 0 & 0 \\ 0 & -1 & 0 \\ 0 & 0 & 3 \end{bmatrix} = \begin{bmatrix} -4 & 0 & 0 \\ 0 & -4 & 0 \\ 0 & 0 & -18 \end{bmatrix}$

EXAMPLE 9 Much pathology can occur for matrix products. For example, each of the following can occur:

a We can have $AB \neq BA$.

$$\begin{bmatrix} 2 & 1 \\ -1 & 0 \end{bmatrix} \begin{bmatrix} 1 & 0 \\ 3 & 1 \end{bmatrix} = \begin{bmatrix} 5 & 1 \\ -1 & 0 \end{bmatrix}$$

$$\begin{bmatrix} 1 & 0 \\ 3 & 1 \end{bmatrix} \begin{bmatrix} 2 & 1 \\ -1 & 0 \end{bmatrix} = \begin{bmatrix} 2 & 1 \\ 5 & 3 \end{bmatrix}$$

b We can have $A \neq 0$, $B \neq 0$, and yet $AB = 0$.

$$\begin{bmatrix} 0 & 1 \\ 0 & 0 \end{bmatrix} \begin{bmatrix} 0 & 4 \\ 0 & 0 \end{bmatrix} = \begin{bmatrix} 0 & 0 \\ 0 & 0 \end{bmatrix}$$

$$\begin{bmatrix} 1 & 1 \\ 1 & 1 \end{bmatrix} \begin{bmatrix} 1 & 1 \\ -1 & -1 \end{bmatrix} = \begin{bmatrix} 0 & 0 \\ 0 & 0 \end{bmatrix}$$

c We can have $A \neq 0$ and $A^2 = 0$.

$$\begin{bmatrix} 0 & 1 \\ 0 & 0 \end{bmatrix}^2 = \begin{bmatrix} 0 & 0 \\ 0 & 0 \end{bmatrix}$$

$$\begin{bmatrix} 1 & -1 \\ 1 & -1 \end{bmatrix}^2 = \begin{bmatrix} 0 & 0 \\ 0 & 0 \end{bmatrix}$$

d We can have $A \neq 0$, $A^2 \neq 0$, $A^3 = 0$.

$$\begin{bmatrix} 0 & 1 & 1 \\ 0 & 0 & 1 \\ 0 & 0 & 0 \end{bmatrix}^2 = \begin{bmatrix} 0 & 1 & 1 \\ 0 & 0 & 1 \\ 0 & 0 & 0 \end{bmatrix} \begin{bmatrix} 0 & 1 & 1 \\ 0 & 0 & 1 \\ 0 & 0 & 0 \end{bmatrix} = \begin{bmatrix} 0 & 0 & 1 \\ 0 & 0 & 0 \\ 0 & 0 & 0 \end{bmatrix}$$

$$\begin{bmatrix} 0 & 1 & 1 \\ 0 & 0 & 1 \\ 0 & 0 & 0 \end{bmatrix}^3 = \begin{bmatrix} 0 & 1 & 1 \\ 0 & 0 & 1 \\ 0 & 0 & 0 \end{bmatrix} \begin{bmatrix} 0 & 0 & 1 \\ 0 & 0 & 0 \\ 0 & 0 & 0 \end{bmatrix} = \begin{bmatrix} 0 & 0 & 0 \\ 0 & 0 & 0 \\ 0 & 0 & 0 \end{bmatrix}$$

e We can have $A^2 = A$, with $A \neq 0$ and $A \neq I$.

$$\begin{bmatrix} \frac{1}{2} & \frac{1}{2} \\ \frac{1}{2} & \frac{1}{2} \end{bmatrix}^2 = \begin{bmatrix} \frac{1}{2} & \frac{1}{2} \\ \frac{1}{2} & \frac{1}{2} \end{bmatrix} \begin{bmatrix} \frac{1}{2} & \frac{1}{2} \\ \frac{1}{2} & \frac{1}{2} \end{bmatrix} = \begin{bmatrix} \frac{1}{2} & \frac{1}{2} \\ \frac{1}{2} & \frac{1}{2} \end{bmatrix}$$

This is the matrix of the projection onto the vector $\frac{1}{\sqrt{2}}(1,1)$.

f We can have $A^2 = I$ with $A \neq I$ and $A \neq -I$.

$$\begin{bmatrix} -1 & 0 \\ 0 & 1 \end{bmatrix}^2 = \begin{bmatrix} -1 & 0 \\ 0 & 1 \end{bmatrix} \begin{bmatrix} -1 & 0 \\ 0 & 1 \end{bmatrix} = \begin{bmatrix} 1 & 0 \\ 0 & 1 \end{bmatrix}$$

This is the matrix of reflection in the y-axis. Certainly reflecting twice returns us to our initial position.

EXAMPLE 10 The row operations introduced in Chapter 1, Section 1, can all be expressed as multiplication on the left by a suitable matrix.

Suppose E is obtained from I by interchanging two rows of I. Then EA simply interchanges the same two rows of A.

a $$\begin{bmatrix} 1 & 0 & 0 \\ 0 & 0 & 1 \\ 0 & 1 & 0 \end{bmatrix} \begin{bmatrix} 4 & 2 & 1 & -1 \\ 3 & 1 & 0 & 1 \\ -2 & 1 & 6 & -5 \end{bmatrix} = \begin{bmatrix} 4 & 2 & 1 & -1 \\ -2 & 1 & 6 & -5 \\ 3 & 1 & 0 & 1 \end{bmatrix}$$

Certainly $E^2 = I$; for example,

b $$\begin{bmatrix} 1 & 0 & 0 \\ 0 & 0 & 1 \\ 0 & 1 & 0 \end{bmatrix} \begin{bmatrix} 1 & 0 & 0 \\ 0 & 0 & 1 \\ 0 & 1 & 0 \end{bmatrix} = \begin{bmatrix} 1 & 0 & 0 \\ 0 & 1 & 0 \\ 0 & 0 & 1 \end{bmatrix}$$

Suppose F is obtained from I by multiplying one row of I by $a \neq 0$. Then FA just multiplies the same row of A by a.

c $$\begin{bmatrix} 1 & 0 & 0 \\ 0 & 1 & 0 \\ 0 & 0 & a \end{bmatrix} \begin{bmatrix} 4 & 2 & 1 & -1 \\ 3 & 1 & 0 & 1 \\ -2 & 1 & 6 & -5 \end{bmatrix} = \begin{bmatrix} 4 & 2 & 1 & -1 \\ 3 & 1 & 0 & 1 \\ -2a & a & 6a & -5a \end{bmatrix}$$

Given such an F, suppose F_1 is obtained from I by multiplying the same row by $1/a$. Then $F_1F = I$.

d $$\begin{bmatrix} 1 & 0 & 0 \\ 0 & 1 & 0 \\ 0 & 0 & 1/a \end{bmatrix} \begin{bmatrix} 1 & 0 & 0 \\ 0 & 1 & 0 \\ 0 & 0 & a \end{bmatrix} = \begin{bmatrix} 1 & 0 & 0 \\ 0 & 1 & 0 \\ 0 & 0 & 1 \end{bmatrix}$$

Suppose G is obtained from I by adding to one row, say row i, of I, b times another row, say row j, $j \neq i$. Then GA is found by adding to row i of A, b times row j of A.

$$\mathbf{e}\quad \begin{bmatrix} 1 & 0 & b \\ 0 & 1 & 0 \\ 0 & 0 & 1 \end{bmatrix} \begin{bmatrix} 4 & 2 & 1 & -1 \\ 3 & 1 & 0 & 1 \\ -2 & 1 & 6 & -5 \end{bmatrix} = \begin{bmatrix} 4-2b & 2+b & 1+6b & -1-5b \\ 3 & 1 & 0 & 1 \\ -2 & 1 & 6 & -5 \end{bmatrix}$$

Given such a G, suppose G_1 is obtained from I by adding to row i of I, $-b$ times row j of I. Then $G_1G = I$.

$$\mathbf{f}\quad \begin{bmatrix} 1 & 0 & -b \\ 0 & 1 & 0 \\ 0 & 0 & 1 \end{bmatrix} \begin{bmatrix} 1 & 0 & b \\ 0 & 1 & 0 \\ 0 & 0 & 1 \end{bmatrix} = \begin{bmatrix} 1 & 0 & 0 \\ 0 & 1 & 0 \\ 0 & 0 & 1 \end{bmatrix}$$

EXAMPLE 11 Example **10** shows that a row operation on a matrix A can be performed by multiplying on the left of A by a suitable matrix, such as the matrices E, F, or G given above. The effect of such an operation can be undone by a related row operation given by multiplying on the left by a suitable matrix, respectively E, F_1, and G_1.

This information can be used along with the associative law of multiplication (rule **1c**) to prove that the method of row reduction of Chapter 1 does in fact preserve the solutions of a system.

For example, suppose $A\bar{u} = \bar{0}$. Then for any matrix B for which the product BA is defined we have $(BA)\bar{u} = B(A\bar{u}) = B\bar{0} = \bar{0}$. We conclude that

4 Every vector in the null space of A is also in the null space of BA.

Suppose E, F, and G are matrices like the E, F, and G of Example **10**. Then from property **4** we have:

5 Every vector in the null space of A is also in the null space of each of the matrices EA, FA, and GA.

Select E, F_1, and G_1 as in Example **10**. We can use result **4** with A replaced respectively by EA, FA, and GA and B replaced by E, F_1, and G_1 respectively to conclude that each of the following statements is true.

6
- a Every vector in the null space of EA is also in the null space of $E(EA)$.
- b Every vector in the null space of FA is also in the null space of $F_1(FA)$.
- c Every vector in the null space of GA is also in the null space of $G_1(GA)$.

We showed in Example **10** that $EE = I$, $F_1F = I$, and $G_1G = I$. Thus the associative law gives

$$E(EA) = (EE)A = IA = A$$
$$F_1(FA) = (F_1F)A = IA = A$$
$$G_1(GA) = (G_1G)A = IA = A$$

We combine this information with that given in statements **5** and **6** to conclude that

7 Each of the matrices A, EA, FA, and GA has the same null space.

Since a row operation on A can be performed by multiplying on the left of A by a matrix such as E, F, or G, we have shown that row operations of the kind given in Chapter 1 *do not change* the solutions to $A\bar{u} = \bar{0}$.

A similar argument can be given to show that row operations applied to the augmented matrix do not change the solutions to $A\bar{u} = \bar{v}$.

EXERCISES

1 For the given matrices verify that $A(BC) = (AB)C$ and that $A(B + C) = AB + AC$.

$$A = \begin{bmatrix} 2 & 1 \\ 0 & 1 \end{bmatrix} \quad B = \begin{bmatrix} 3 & -1 \\ 1 & 1 \end{bmatrix} \quad C = \begin{bmatrix} -1 & 1 \\ 1 & 1 \end{bmatrix}$$

2 Solve for B where $2A - 3B = AC$.

3 Show that $A^2 - 4I = (A + 2I)(A - 2I)$.

4 Is it always true that $(A + B)^2 = A^2 + 2AB + B^2$?

5 Suppose R is a counterclockwise rotation through the angle θ, T is a counterclockwise rotation through the angle φ, and S is a counterclockwise rotation through the angle ψ. Describe $R(ST)$, $(RS)T$, $R(S^2T^3)$, and $(RS^2)T^3$.

6 Find each of the following products.

a $\begin{bmatrix} 0 & 0 \\ 0 & 0 \end{bmatrix} \begin{bmatrix} 1 & 2 \\ 1 & 1 \end{bmatrix}$ **b** $\begin{bmatrix} 0 & 0 \\ 0 & 0 \end{bmatrix} \begin{bmatrix} 1 & 2 & 1 \\ 1 & 1 & 1 \end{bmatrix}$

c $\begin{bmatrix} 1 & 0 \\ 0 & 1 \end{bmatrix} \begin{bmatrix} 1 & 2 \\ 1 & 1 \end{bmatrix}$ **d** $\begin{bmatrix} 1 & 0 \\ 0 & 1 \end{bmatrix} \begin{bmatrix} 1 & 2 & 1 \\ 1 & 1 & 1 \end{bmatrix}$

e $\begin{bmatrix} 0 & 0 & 0 & 0 \\ 0 & 0 & 0 & 0 \end{bmatrix} \begin{bmatrix} 3 & 1 \\ 2 & -1 \\ 0 & 1 \\ 1 & 1 \end{bmatrix}$ **f** $\begin{bmatrix} 3 & 1 \\ 2 & -1 \\ 0 & 1 \\ 1 & 1 \end{bmatrix} \begin{bmatrix} 1 & 0 \\ 0 & 1 \end{bmatrix}$

g $\begin{bmatrix} 2 & 1 & 0 \\ 1 & 1 & 1 \end{bmatrix} \begin{bmatrix} 1 & 0 & 0 \\ 0 & 1 & 0 \\ 0 & 0 & 1 \end{bmatrix} \begin{bmatrix} 0 & 0 \\ 0 & 0 \\ 0 & 0 \end{bmatrix} \begin{bmatrix} 2 & 1 \\ 1 & 1 \\ 1 & 1 \end{bmatrix}$

7 Find each of the following products.

a $\begin{bmatrix} 2 & 1 \\ 0 & 0 \end{bmatrix} \begin{bmatrix} 3 & 2 \\ 1 & 1 \end{bmatrix}$ **b** $\begin{bmatrix} 3 & 2 \\ 1 & 1 \end{bmatrix} \begin{bmatrix} -1 & 0 \\ 1 & 0 \end{bmatrix}$

c $\begin{bmatrix} 4 & 1 & 2 \\ 0 & 0 & 0 \\ 0 & 0 & 0 \end{bmatrix} \begin{bmatrix} 0 & 3 & 1 & 6 \\ 0 & 1 & 2 & 1 \\ 0 & 0 & 0 & 0 \end{bmatrix}$ **d** $\begin{bmatrix} 3 & 1 & 0 \\ 0 & 0 & 0 \\ 1 & 2 & 0 \end{bmatrix}^2$

8 Find each of the following products.

a $\begin{bmatrix} 3 & 0 \\ 0 & -1 \end{bmatrix} \begin{bmatrix} 4 & 2 & 6 & \frac{1}{2} \\ -2 & 1 & 0 & -\frac{1}{2} \end{bmatrix}$

b $\begin{bmatrix} -4 & 0 & 0 & 0 \\ 0 & \frac{1}{2} & 0 & 0 \\ 0 & 0 & 3 & 0 \\ 0 & 0 & 0 & -\frac{1}{2} \end{bmatrix} \begin{bmatrix} 6 & 6 & 1 & 6 \\ -1 & 0 & 1 & 1 \\ 0 & 1 & 3 & 0 \\ 0 & 2 & -1 & 1 \end{bmatrix}$

c $\begin{bmatrix} 6 & 6 & 1 & 6 \\ -1 & 0 & 1 & 1 \\ 0 & 1 & 3 & 0 \\ 0 & 2 & -1 & 1 \end{bmatrix} \begin{bmatrix} -4 & 0 & 0 & 0 \\ 0 & \frac{1}{2} & 0 & 0 \\ 0 & 0 & 3 & 0 \\ 0 & 0 & 0 & -\frac{1}{2} \end{bmatrix}$

d $\begin{bmatrix} 2 & 1 \\ 1 & 0 \\ 1 & 1 \\ 1 & -1 \end{bmatrix} \begin{bmatrix} \frac{1}{2} & 0 \\ 0 & \frac{3}{2} \end{bmatrix}$

9 Find each of the following products.

a $\begin{bmatrix} 1 & 0 & 0 \\ 0 & 2 & 0 \\ 0 & 0 & 3 \end{bmatrix}^3$ **b** $\begin{bmatrix} 3 & 0 & 0 \\ 0 & 7 & 0 \\ 0 & 0 & -1 \end{bmatrix}^2 \begin{bmatrix} 1 & 0 & 0 \\ 0 & 0 & 0 \\ 0 & 0 & -2 \end{bmatrix}^3$

c $\begin{bmatrix} 1 & 0 & 0 \\ 0 & -1 & 0 \\ 0 & 0 & 1 \end{bmatrix}^{27}$ **d** $\begin{bmatrix} 0 & 0 & 0 & 0 \\ 0 & 1 & 0 & 0 \\ 0 & 0 & 3 & 0 \\ 0 & 0 & 0 & -1 \end{bmatrix}^3$

10 For each of the following matrices A find $p(A)$, $q(A)$, and $r(A)$, where $p(\lambda) = \lambda^3 - \lambda^2 + \lambda$, $q(\lambda) = \lambda - 3$, and $r(\lambda) = 2\lambda^2 - 4$. Note the simplicity of these calculations when A is diagonal.

a $\begin{bmatrix} 1 & 2 \\ 0 & -1 \end{bmatrix}$ **b** $\begin{bmatrix} 3 & 1 & 0 \\ 2 & -1 & 1 \\ 1 & 1 & 2 \end{bmatrix}$ **c** $\begin{bmatrix} 4 & 0 & 0 \\ 0 & -1 & 0 \\ 0 & 0 & 2 \end{bmatrix}$

11 Suppose $A = \begin{bmatrix} a & 0 & 0 \\ 0 & b & 0 \\ 0 & 0 & c \end{bmatrix}$, and $p(\lambda)$ is a polynomial. What is $p(A)$? In particular what is $p(A)$ when $p(\lambda) = (\lambda - a)(\lambda - b)(\lambda - c)$?

12 Find a matrix B such that

$$B^3 = \begin{bmatrix} 8 & 0 & 0 \\ 0 & 27 & 0 \\ 0 & 0 & -8 \end{bmatrix}$$

13 Compute $A^3 - 3A^2 + A - 2I$ for $A = \begin{bmatrix} 1 & 0 & 0 \\ 0 & 2 & 0 \\ 0 & 0 & 3 \end{bmatrix}$ and for $A = \begin{bmatrix} 2 & 1 \\ 1 & 3 \end{bmatrix}$.

14 Suppose A has two rows and two columns, and $AB = BA$ for every diagonal matrix B with two rows and two columns. Show that A is a diagonal matrix.

15 Describe linear transformations S and T such that

a $ST \neq TS$

b $ST = TS$

c $S \neq 0$, $T \neq 0$, and $ST = 0$

d $T^2 = T$, $S^2 = S$, and $ST = 0$

16 a Find A such that $A^2 \neq A$ and $A^3 = A$.

b Find A such that $A \neq I$ and $A^3 = I$.

c Find A such that $A^3 \neq 0$ and $A^4 = 0$.

17 Suppose $P^2 = P$ and $Q^2 = Q$. Show that

a $(I - P)^2 = I - P$

b If $PQ = QP$, then $(P + Q - QP)^2 = P + Q - QP$.

c Find matrices P and Q such that $P^2 = P$, $Q^2 = Q$, and $PQ \neq QP$.

18 Suppose $AC = I$ and $BD = I$. Show that $(AB)(DC) = I$.

19 Show that $A^3 - I = (A^2 + A + I)(A - I)$.

20 Use the techniques of Example **10** to find, by noting the form of the left matrix,

a $\begin{bmatrix} 0 & 1 & 0 \\ 1 & 0 & 0 \\ 0 & 0 & 1 \end{bmatrix} \begin{bmatrix} 3 & 2 & 1 \\ 1 & -1 & 1 \\ 2 & 1 & 1 \end{bmatrix}$ **b** $\begin{bmatrix} 1 & 0 & 0 \\ 0 & \frac{1}{2} & 0 \\ 0 & 0 & 1 \end{bmatrix} \begin{bmatrix} 3 & 2 & 1 \\ 1 & -1 & 1 \\ 2 & 1 & 1 \end{bmatrix}$

c $\begin{bmatrix} 1 & 0 & 0 \\ 3 & 1 & 0 \\ 0 & 0 & 1 \end{bmatrix} \begin{bmatrix} 3 & 2 & 1 \\ 1 & -1 & 1 \\ 2 & 1 & 1 \end{bmatrix}$ **d** $\begin{bmatrix} 1 & 0 & 0 \\ -3 & 1 & 0 \\ 0 & 0 & 1 \end{bmatrix} \begin{bmatrix} 3 & 2 & 1 \\ 1 & -1 & 1 \\ 2 & 1 & 1 \end{bmatrix}$

21 For each of the following matrices A find a matrix A_1 such that $A_1A = I$ by using the techniques of Example **10**.

a $A = \begin{bmatrix} 0 & 1 & 0 \\ 1 & 0 & 0 \\ 0 & 0 & 1 \end{bmatrix}$ **b** $A = \begin{bmatrix} 1 & 0 & 0 \\ 0 & \frac{1}{2} & 0 \\ 0 & 0 & 1 \end{bmatrix}$

c $A = \begin{bmatrix} 1 & 0 & 0 \\ 3 & 1 & 0 \\ 0 & 0 & 1 \end{bmatrix}$ **d** $A = \begin{bmatrix} 1 & 0 & 0 \\ -3 & 1 & 0 \\ 0 & 0 & 1 \end{bmatrix}$

22 Find $EDCBA$ for

$$A = \begin{bmatrix} 2 & 1 \\ 1 & 1 \end{bmatrix} \quad B = \begin{bmatrix} 0 & 1 \\ 1 & 0 \end{bmatrix} \quad C = \begin{bmatrix} 1 & 0 \\ -2 & 1 \end{bmatrix}$$

$$D = \begin{bmatrix} 1 & 0 \\ 0 & -1 \end{bmatrix} \quad E = \begin{bmatrix} 1 & -1 \\ 0 & 1 \end{bmatrix}$$

Can you find a matrix A' such that $A'A = I$?

23 Find matrices A and B such that the null space of BA is *not* the same as the null space of A.

□ **24** Calculate, using the tricks of this section,

a $\begin{bmatrix} 0 & 0 \\ 0 & 0 \end{bmatrix} \begin{bmatrix} i & 2i \\ 1+i & 3-i \end{bmatrix}$ **b** $\begin{bmatrix} 1 & 0 \\ 0 & 1 \end{bmatrix} \begin{bmatrix} i & 2i \\ 1+i & 3-i \end{bmatrix}$

c $\begin{bmatrix} 2 & 1 \\ 0 & 0 \end{bmatrix} \begin{bmatrix} i & 2i \\ 1+i & 3-i \end{bmatrix}$ **d** $\begin{bmatrix} i & 2i \\ 1+i & 3-i \end{bmatrix} \begin{bmatrix} 0 & i \\ 0 & 1 \end{bmatrix}$

e $\begin{bmatrix} i & 0 \\ 0 & 2i \end{bmatrix} \begin{bmatrix} i & 2i \\ 1+i & 3-i \end{bmatrix}$ **f** $\begin{bmatrix} i & 2i \\ 1+i & 3-i \end{bmatrix} \begin{bmatrix} i & 0 \\ 0 & 2i \end{bmatrix}$

g $\begin{bmatrix} i & 0 \\ 0 & -i \end{bmatrix}^4$

Supplement on Linearity

Suppose T is a linear transformation of R^2 into R^2, or of R^3 into R^3. We can use our descriptions of lines, line segments, and planes (see pages 60ff and 124ff) and the linearity property

1 $$T(a\bar{u} + b\bar{v}) = aT\bar{u} + bT\bar{v}$$

to show how applying T affects these geometric objects.

LINEAR TRANSFORMATIONS MAP LINES INTO LINES AND PLANES INTO PLANES: Suppose T is a linear transformation from R^2 into R^2 and that

2 $$\bar{w} = t\bar{v} + \bar{u}_0$$

is the vector equation of the line through the terminal point of $\bar{u}_0$ which is parallel to $\bar{v}$. Then the linearity property **1** gives

3 $$T\bar{w} = tT\bar{v} + T\bar{u}_0$$

Thus, if $T\bar{v} = \bar{0}$, then T transforms the line **2** into the point $T\bar{u}_0$. If $T\bar{v} \neq \bar{0}$, then equation **3** shows that T transforms the line L of equation **2** into the line L' through the terminal point of $T\bar{u}_0$, which is parallel to $T\bar{v}$. Figures 10 and 11 illustrate two examples of this.

Suppose T is a linear transformation from R^3 into R^3. Just as in the above argument, T transforms the line

$$\bar{w} = t\bar{v} + \bar{u}_0$$

into $T\bar{u}_0$ (if $T\bar{v} = \bar{0}$) or onto the line $tT\bar{v}_0 + T\bar{u}_0$ (if $T\bar{v} \neq \bar{0}$).

Figure 10 *T is counterclockwise rotation through $\pi/4$.*

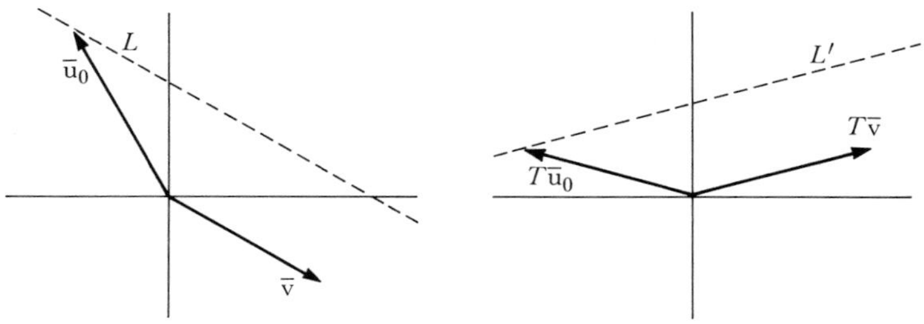

Figure 11 *T is projection onto* $\bar{v}_1$*, where* $\bar{v} \cdot \bar{v}_1 = \bar{0}$*, so that each vector from the origin to L is sent into* $T\bar{u}_0$*.*

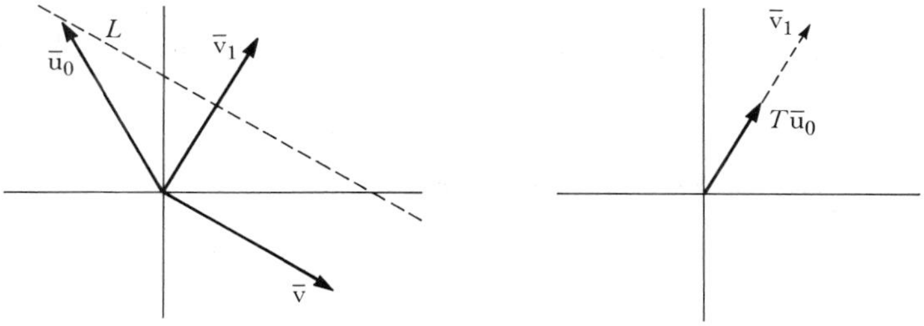

The plane

$$\bar{w} = s\bar{u} + t\bar{v} + \bar{w}_0$$

is transformed into

$$T\bar{w} = sT\bar{u} + tT\bar{v} + T\bar{w}_0$$

If $T\bar{u}$ and $T\bar{v}$ are independent, this again is a plane. Otherwise, this is a line (if $T\bar{u}$ and $T\bar{v}$ are dependent but not both zero) or the single point $T\bar{w}_0$ (if $T\bar{u} = T\bar{v} = \bar{0}$).

TRANSFORMATIONS OF LINE SEGMENTS, PARALLELOGRAMS, AND PARALLELOPIPEDS: If T is a linear transformation, then T transforms the line segment

$$t\bar{v} + (1 - t)\bar{u} \qquad 0 \leq t \leq 1$$

into the set

$$tT\bar{v} + (1 - t)T\bar{u} \qquad 0 \leq t \leq 1$$

If $T\bar{v} \neq T\bar{u}$, this is just the line segment from $T\bar{v}$ to $T\bar{u}$. If $T\bar{v} = T\bar{u}$, then

$$tT\bar{v} + (1 - t)T\bar{u} = [t + (1 - t)]T\bar{v} = T\bar{v}$$

Therefore, this set consists of the single point $T\bar{v}$.

The parallelogram

$$s\bar{u} + t\bar{v} \qquad 0 \leq s \leq 1 \qquad 0 \leq t \leq 1$$

is transformed into the set

$$sT\bar{u} + tT\bar{v} \qquad 0 \leq s \leq 1 \qquad 0 \leq t \leq 1 \tag{4}$$

which is (if $T\bar{u}$ and $T\bar{v}$ are independent) the parallelogram determined by $T\bar{u}$ and $T\bar{v}$. If $T\bar{u}$ and $T\bar{v}$ are dependent, the set described by relation **4** is a line segment or a point. We often call these figures *degenerate parallelograms*. Figures 12, 13, and 14 indicate some of the possibilities. The transformation of Figure 14 is sometimes called a *shear*.

Figure 12 *Rotation through $\pi/4$ sends P onto P'.*

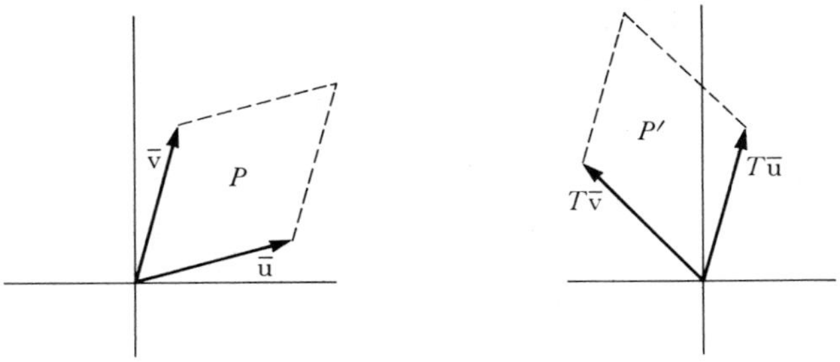

Figure 13 *Projection onto $\bar{w}$ sends P onto the line segment OA.*

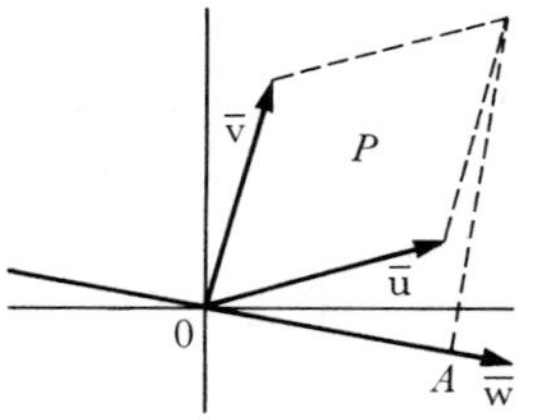

Figure 14 *If* $T\begin{bmatrix} x \\ y \end{bmatrix} = \begin{bmatrix} 1 & 1 \\ 0 & 1 \end{bmatrix} \begin{bmatrix} x \\ y \end{bmatrix}$, *then T transforms P onto P'.*

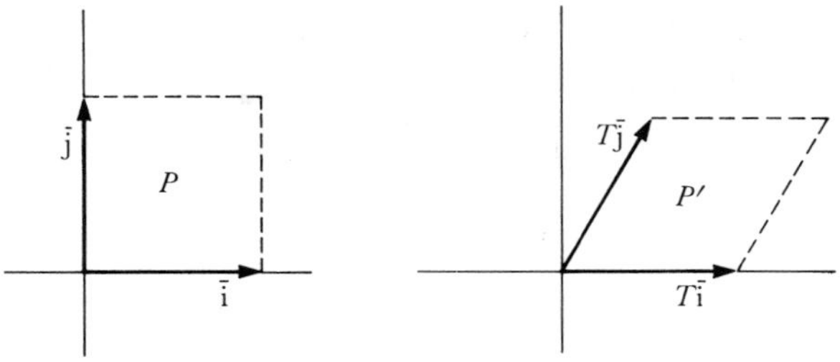

Analogous results show that if $\bar{u}$, $\bar{v}$, and $\bar{w}$ are independent vectors in R^3 and if T is linear, then T transforms the parallelopiped

5 $$r\bar{u} + s\bar{v} + t\bar{w} \qquad 0 \leq r \leq 1 \qquad 0 \leq s \leq 1 \qquad 0 \leq t \leq 1$$

into the set

$$rT\bar{u} + sT\bar{v} + tT\bar{w} \qquad 0 \leq r \leq 1 \qquad 0 \leq s \leq 1 \qquad 0 \leq t \leq 1$$

which is also a parallelopiped if $T\bar{u}$, $T\bar{v}$, and $T\bar{w}$ are independent.

EXERCISES

1 Suppose $T\bar{u} = \begin{bmatrix} 2 & 1 \\ 1 & 1 \end{bmatrix}\bar{u}$. What does T do to the set of points $t\bar{u} + s\bar{v}$, where $0 \leq t \leq 1$, $0 \leq s \leq 1$, and $\bar{u} = \begin{bmatrix} 1 \\ -2 \end{bmatrix}$, $\bar{v} = \begin{bmatrix} 3 \\ -1 \end{bmatrix}$?

2 Suppose $T\bar{u} = \begin{bmatrix} 1 & 1 \\ 0 & 1 \end{bmatrix}\bar{u}$. Graph the set of points $tT\bar{u} + T\bar{u}_0$ where

$$\bar{u} = \begin{bmatrix} 3 \\ 1 \end{bmatrix} \qquad \bar{u}_0 = \begin{bmatrix} -1 \\ 0 \end{bmatrix} \qquad 0 \leq t \leq 1$$

3 Suppose $T\bar{u} = \begin{bmatrix} 1 & 1 \\ 2 & 2 \end{bmatrix}\bar{u}$. What does T do to the square determined by the four points (0,0), (0,1), (1,0), (1,1)?

4 Let T be the projection onto the line through (0,0) and (1,2). Graph the set of points

$$s\bar{u} + t\bar{v} + \bar{w} \qquad 0 \leq s \leq 1 \qquad 0 \leq t \leq 1$$

where $\bar{u} = (1,0)$, $\bar{v} = (0,1)$, $\bar{w} = (1,1)$, and describe the set into which this is mapped by T.

5 Show that if T is a linear transformation of R^2 onto R^2, which maps one nondegenerate parallelogram onto a line segment, then T maps *every* nondegenerate parallelogram onto a line segment.

6 Suppose $T\bar{u} = \begin{bmatrix} 2 & 1 & -1 \\ 2 & 1 & 0 \\ 1 & 0 & 1 \end{bmatrix}\bar{u}$. Into what does T map the cube determined by $\bar{i}$, $\bar{j}$, $\bar{k}$?

7 If $T\bar{u}$, $T\bar{v}$, and $T\bar{w}$ are not independent, describe the possible figures into which T transforms the parallelopiped **5**.

8 Show that if T is linear and $T\bar{u}$, $T\bar{v}$ are independent, then T maps the triangle $\triangle$ determined by the endpoints of $\bar{u}$, $\bar{v}$, $\bar{u} - \bar{v}$ into a triangle $\triangle_1$. Does T map points inside $\triangle$ into points inside $\triangle_1$? (Hint: Describe points inside $\triangle$ as sums of certain kinds of multiples of $\bar{u}$ and $\bar{v}$.)

9 If T is linear and $T\bar{u}$, $T\bar{v}$, and $T\bar{w}$ are independent, show that T maps the tetrahedron with sides $\bar{u}$, $\bar{v}$, $\bar{w}$, $\bar{u} - \bar{v}$, $\bar{u} - \bar{w}$, $\bar{v} - \bar{w}$ into a tetrahedron.

CHAPTER 4

THE INVERSE

Three algebraic operations, the sum, scalar multiple and product, were defined for linear transformations and for their representations as matrices in the previous chapter. We can now manipulate matrices or transformations much as though they were real numbers, taking products and powers, distributing products over addition, subtracting one from another, and combining products in various ways. Some care needs to be exercised due to the fact that the commutative law of multiplication doesn't hold, and some results can occur which have no real number analog, such as the existence of two nonzero matrices whose product is zero.

In this chapter we wish to examine the question of division for matrices and transformations. Section 1 will discuss the inverse of a matrix, which is analogous to the reciprocal of a nonzero number. The method of row reduction can be used to determine if a matrix has an inverse and to calculate the inverse if it has one. Section 2 indicates some of the ways in which inverses can be manipulated algebraically. Section 3 discusses the inverse of a transformation. Section 4 relates the theory of invertibility to coordinate changes, while Section 5 introduces the theory of determinants.

SECTION 1 The Inverse of a Matrix

If a, b, and c are numbers, and $a \neq 0$ one can solve the equation $ab = c$ for b by multiplying by $a^{-1} = 1/a$ to obtain $b = a^{-1}c$. The number a^{-1} has the property $a^{-1}a = 1$.

This section will discuss a similar process for solving matrix equations by multiplying by the "inverse" of a matrix. The discussion will be confined to square matrices, that is, matrices with the same number of rows as columns.

The matrix B is called **an inverse** of A if

1 $$BA = I$$

The matrix A is said to be **invertible** if A has an inverse. We wish to give conditions that guarantee that A is invertible. It will also be shown that a matrix has at most one inverse. A method for calculating this inverse will be given.

The basic relation between invertibility and rank is expressed in the following theorem.

THEOREM 11 A is invertible if and only if the rank of A is the same as the number of columns of A; that is, $A\bar{u} = \bar{0}$ has a unique solution.

PROOF: To say that A is invertible is to say there is a B such that $BA = I$. Thus, if $A\bar{u} = \bar{0}$, we can multiply by B to obtain

$$B(A\bar{u}) = (BA)\bar{u} = \bar{u}$$

Since we also have $B(A\bar{u}) = B(\bar{0}) = \bar{0}$, we conclude that $\bar{u} = \bar{0}$.

Conversely, suppose $A\bar{u} = \bar{0}$ has a unique solution so that rank $A = n$, where n is the number of columns of A. Our problem is to construct B so that $BA = I$. We do this by using Theorem **9**. This theorem tells us (since A is a square matrix) that the rows of A are a basis for R^n. Let us simplify our discussion by assuming that $n = 2$. Put

$$A = \begin{pmatrix} A_{11} & A_{12} \\ A_{21} & A_{22} \end{pmatrix}$$

Since the rows of A are a basis for R^2, we can find numbers a_1, a_2 and b_1, b_2 such that

2
$$\begin{aligned} (1,0) &= a_1(A_{11},A_{12}) + a_2(A_{21},A_{22}) \\ (0,1) &= b_1(A_{11},A_{12}) + b_2(A_{21},A_{22}) \end{aligned}$$

Now just rewrite this to obtain

$$\begin{pmatrix} 1 & 0 \\ 0 & 1 \end{pmatrix} = \begin{pmatrix} a_1 & a_2 \\ b_1 & b_2 \end{pmatrix} \begin{pmatrix} A_{11} & A_{12} \\ A_{21} & A_{22} \end{pmatrix}$$

so that, if

$$B = \begin{pmatrix} a_1 & a_2 \\ b_1 & b_2 \end{pmatrix}$$

then $BA = I$. This completes our proof of Theorem **11**.

The numbers a_1, a_2, b_1, b_2 in equations **2** are uniquely determined by the rows of A so

3 If A has an inverse, it has only one inverse.

We now prove the following theorem.

THEOREM 12 If A is square and $BA = I$, then $AB = I$.

PROOF: If $BA = I$, then Theorem **11** implies that rank A = number of columns of A. Hence Theorem **3** implies that the columns of A are independent. To simplify our discussion, we assume that A has two columns, and let

$$A = \begin{bmatrix} A_{11} & A_{12} \\ A_{21} & A_{22} \end{bmatrix}$$

The columns of A form a basis for R^2 so we can find numbers c_1, c_2, d_1, d_2 such that

$$\begin{bmatrix} 1 \\ 0 \end{bmatrix} = c_1 \begin{bmatrix} A_{11} \\ A_{21} \end{bmatrix} + c_2 \begin{bmatrix} A_{12} \\ A_{22} \end{bmatrix} \qquad \begin{bmatrix} 0 \\ 1 \end{bmatrix} = d_1 \begin{bmatrix} A_{11} \\ A_{21} \end{bmatrix} + d_2 \begin{bmatrix} A_{12} \\ A_{22} \end{bmatrix}$$

and hence

$$\begin{bmatrix} 1 & 0 \\ 0 & 1 \end{bmatrix} = \begin{bmatrix} A_{11} & A_{12} \\ A_{21} & A_{22} \end{bmatrix} \begin{bmatrix} c_1 & d_1 \\ c_2 & d_2 \end{bmatrix}$$

Thus if we put

$$C = \begin{bmatrix} c_1 & d_1 \\ c_2 & d_2 \end{bmatrix}$$

then $AC = I$. A simple trick will now show us that C must be B. We have

$$B = BI = B(AC) = (BA)C = IC = C$$

This completes the proof of Theorem **12**.

If A is invertible, we usually write A^{-1} (read "A inverse") for the unique inverse of A.

Theorem **11** gives a condition for A to be invertible. Other conditions will be discussed in Section 3. We shall now show how to construct the inverse of an invertible matrix A. For simplicity, suppose that A has two rows and two columns. Suppose

$$A^{-1} = \begin{bmatrix} B_{11} & B_{12} \\ B_{21} & B_{22} \end{bmatrix}$$

so that $A^{-1} \begin{bmatrix} 1 \\ 0 \end{bmatrix} = \begin{bmatrix} B_{11} \\ B_{21} \end{bmatrix}$ and $A^{-1} \begin{bmatrix} 0 \\ 1 \end{bmatrix} = \begin{bmatrix} B_{12} \\ B_{22} \end{bmatrix}$

Multiplying each of these by A and using the fact that $AA^{-1} = I$, we have

$$\begin{bmatrix} 1 \\ 0 \end{bmatrix} = A \begin{bmatrix} B_{11} \\ B_{21} \end{bmatrix} \quad \text{and} \quad \begin{bmatrix} 0 \\ 1 \end{bmatrix} = A \begin{bmatrix} B_{12} \\ B_{22} \end{bmatrix}$$

Therefore,

4 The columns of A^{-1} are the solutions to $A\bar{u} = \begin{bmatrix} 1 \\ 0 \end{bmatrix}$ and $A\bar{u} = \begin{bmatrix} 0 \\ 1 \end{bmatrix}$.

We can solve each of these equations by forming the augmented matrix of each and reducing. Since the same operations will be performed on each matrix, we can combine the procedures and proceed as follows: If

$$A = \begin{bmatrix} A_{11} & A_{12} \\ A_{21} & A_{22} \end{bmatrix}$$

form the matrix

5
$$A_1 = \left[\begin{array}{cc:cc} A_{11} & A_{12} & 1 & 0 \\ A_{21} & A_{22} & 0 & 1 \end{array}\right]$$

Reduce this to

6
$$B_1 = \left[\begin{array}{cc:cc} 1 & 0 & B_{11} & B_{12} \\ 0 & 1 & B_{21} & B_{22} \end{array}\right]$$

Then

$$A^{-1} = \begin{bmatrix} B_{11} & B_{12} \\ B_{21} & B_{22} \end{bmatrix}$$

This is just a device for solving the two equations

$$A\bar{u} = \begin{bmatrix} 1 \\ 0 \end{bmatrix} \quad \text{and} \quad A\bar{u} = \begin{bmatrix} 0 \\ 1 \end{bmatrix}$$

at the same time.

This method clearly generalizes to give a procedure for finding A^{-1}.

To find A^{-1}, we form $A_1 = [A \vdots I]$ and reduce to $[I \vdots B]$. Then $A^{-1} = B$.

Some examples of the use of this method are given below. It is important that the student learn to find inverses quickly and correctly.

EXAMPLE 1 The matrix

$$A = \begin{bmatrix} 2 & 1 \\ -4 & -2 \end{bmatrix} \quad \text{reduces to} \quad \begin{bmatrix} 1 & \frac{1}{2} \\ 0 & 0 \end{bmatrix}$$

Therefore $A\bar{u} = \bar{0}$ has solutions $\bar{u} \neq \bar{0}$. Theorem **11** tells us that A is not invertible.

EXAMPLE 2 Since the matrix

$$A = \begin{bmatrix} 3 & 1 \\ -1 & 6 \end{bmatrix} \quad \text{reduces to} \quad \begin{bmatrix} 1 & 0 \\ 0 & 1 \end{bmatrix}$$

A is invertible. To find A^{-1} form

$$\left[\begin{array}{cc:cc} 3 & 1 & 1 & 0 \\ -1 & 6 & 0 & 1 \end{array}\right] \quad \text{and reduce it to} \quad \left[\begin{array}{cc:cc} 1 & 0 & \frac{6}{19} & -\frac{1}{19} \\ 0 & 1 & \frac{1}{19} & \frac{3}{19} \end{array}\right]$$

We conclude that

$$A^{-1} = \begin{bmatrix} \frac{6}{19} & -\frac{1}{19} \\ \frac{1}{19} & \frac{3}{19} \end{bmatrix}$$

The student should check to see that indeed $A^{-1}A = I$ and $AA^{-1} = I$.

EXAMPLE 3 Since the matrix

$$\begin{bmatrix} 1 & -7 & -14 \\ 2 & 1 & -1 \\ 1 & 3 & 4 \end{bmatrix} \quad \text{reduces to} \quad \begin{bmatrix} 1 & 0 & -\frac{7}{5} \\ 0 & 1 & \frac{9}{5} \\ 0 & 0 & 0 \end{bmatrix}$$

it does *not* have an inverse.

The matrix

$$A = \begin{bmatrix} 3 & 1 & 0 \\ 1 & -1 & 2 \\ 1 & 1 & 1 \end{bmatrix} \quad \text{reduces to} \quad \begin{bmatrix} 1 & 0 & 0 \\ 0 & 1 & 0 \\ 0 & 0 & 1 \end{bmatrix}$$

Therefore A is invertible. Forming

$$\left[\begin{array}{ccc:ccc} 3 & 1 & 0 & 1 & 0 & 0 \\ 1 & -1 & 2 & 0 & 1 & 0 \\ 1 & 1 & 1 & 0 & 0 & 1 \end{array}\right]$$

and reducing it to

$$\left[\begin{array}{ccc:ccc} 1 & 0 & 0 & \frac{3}{8} & \frac{1}{8} & -\frac{2}{8} \\ 0 & 1 & 0 & -\frac{1}{8} & -\frac{3}{8} & \frac{6}{8} \\ 0 & 0 & 1 & -\frac{2}{8} & \frac{2}{8} & \frac{4}{8} \end{array}\right] \quad \text{we have} \quad A^{-1} = \begin{bmatrix} \frac{3}{8} & \frac{1}{8} & -\frac{2}{8} \\ -\frac{1}{8} & -\frac{3}{8} & \frac{6}{8} \\ -\frac{2}{8} & \frac{2}{8} & \frac{4}{8} \end{bmatrix}$$

The student should of course carry out these calculations and check his results by showing that $A^{-1} A = I$.

EXAMPLE 4 The matrix

$$A = \begin{bmatrix} 1 & 1 & 1 & 1 \\ 0 & 1 & 1 & 1 \\ 0 & 0 & 1 & 1 \\ 0 & 0 & 0 & 1 \end{bmatrix} \text{clearly reduces to } I_4$$

To find its inverse reduce

$$\left[\begin{array}{cccc:cccc} 1 & 1 & 1 & 1 & 1 & 0 & 0 & 0 \\ 0 & 1 & 1 & 1 & 0 & 1 & 0 & 0 \\ 0 & 0 & 1 & 1 & 0 & 0 & 1 & 0 \\ 0 & 0 & 0 & 1 & 0 & 0 & 0 & 1 \end{array}\right]$$

to

$$\left[\begin{array}{cccc:cccc} 1 & 0 & 0 & 0 & 1 & -1 & 0 & 0 \\ 0 & 1 & 0 & 0 & 0 & 1 & -1 & 0 \\ 0 & 0 & 1 & 0 & 0 & 0 & 1 & -1 \\ 0 & 0 & 0 & 1 & 0 & 0 & 0 & 1 \end{array}\right]$$

so that

$$A^{-1} = \begin{bmatrix} 1 & -1 & 0 & 0 \\ 0 & 1 & -1 & 0 \\ 0 & 0 & 1 & -1 \\ 0 & 0 & 0 & 1 \end{bmatrix}$$

EXAMPLE 5 Determining invertibility and finding inverses for diagonal matrices is particularly simple, for if one of the diagonal entries is zero, the matrix is not invertible. If all the diagonal entries are *not* zero, the inverse is then obtained by inverting these diagonal entries. For example,

$$\begin{bmatrix} 1 & 0 & 0 \\ 0 & 0 & 0 \\ 0 & 0 & 3 \end{bmatrix}$$

is not invertible, while

$$\begin{bmatrix} 1 & 0 & 0 \\ 0 & 2 & 0 \\ 0 & 0 & 3 \end{bmatrix}^{-1} = \begin{bmatrix} 1 & 0 & 0 \\ 0 & \frac{1}{2} & 0 \\ 0 & 0 & \frac{1}{3} \end{bmatrix}$$

and

$$\begin{bmatrix} 3 & 0 & 0 & 0 & 0 \\ 0 & -2 & 0 & 0 & 0 \\ 0 & 0 & \frac{1}{4} & 0 & 0 \\ 0 & 0 & 0 & 6 & 0 \\ 0 & 0 & 0 & 0 & -1 \end{bmatrix}^{-1} = \begin{bmatrix} \frac{1}{3} & 0 & 0 & 0 & 0 \\ 0 & -\frac{1}{2} & 0 & 0 & 0 \\ 0 & 0 & 4 & 0 & 0 \\ 0 & 0 & 0 & \frac{1}{6} & 0 \\ 0 & 0 & 0 & 0 & -1 \end{bmatrix}$$

EXAMPLE 6 Suppose A is **upper triangular**, that is, the entries below the diagonal are zeros. If one of the diagonal entries of A is zero, A is not invertible. For example,

$$A = \begin{bmatrix} 2 & 1 & 1 \\ 0 & 0 & 3 \\ 0 & 0 & 1 \end{bmatrix}$$

is not invertible.

If all the diagonal entries of A are *nonzero*, there is a method for calculating A^{-1} which is quicker than the method of forming matrix **5** and reducing to matrix **6**. For example, if

$$A = \begin{bmatrix} 1 & 2 & -1 \\ 0 & 3 & 1 \\ 0 & 0 & -4 \end{bmatrix}$$

A^{-1} must then be of the form

$$A^{-1} = \begin{bmatrix} 1 & a & b \\ 0 & \frac{1}{3} & c \\ 0 & 0 & -\frac{1}{4} \end{bmatrix}$$

and we can determine a, b, and c from the relation $AA^{-1} = I$. Since we have

$$AA^{-1} = \begin{bmatrix} 1 & a + \frac{2}{3} & b + 2c + \frac{1}{4} \\ 0 & 1 & 3c - \frac{1}{4} \\ 0 & 0 & 1 \end{bmatrix}$$

we must have $a + \frac{2}{3} = 0$, $b + 2c + \frac{1}{4} = 0$, and $3c - \frac{1}{4} = 0$. These are easily solved to give $a = -\frac{2}{3}$, $c = \frac{1}{12}$, and $b = -\frac{5}{12}$ so that

$$A^{-1} = \begin{bmatrix} 1 & -\frac{2}{3} & -\frac{5}{12} \\ 0 & \frac{1}{3} & \frac{1}{12} \\ 0 & 0 & -\frac{1}{4} \end{bmatrix}$$

In other words, for upper triangular matrices the diagonal entries of A^{-1} are the reciprocals of the diagonal entries of A, and the entries below the diagonal of A^{-1} are all zero. The entries above the diagonal can then be simply calculated from the relation $AA^{-1} = I$.

EXERCISES

1 Use Theorem **11** to show that each of the following matrices is not invertible.

a $\begin{bmatrix} 1 & 1 \\ 2 & 2 \end{bmatrix}$ **b** $\begin{bmatrix} 5 & 2 & 6 \\ 1 & 2 & 2 \\ -1 & 2 & 0 \end{bmatrix}$ **c** $\begin{bmatrix} 1 & 2 & 0 & 3 \\ -1 & 1 & 0 & 4 \\ 4 & 1 & 0 & 2 \\ 3 & 2 & 0 & 1 \end{bmatrix}$

2 Use the method of forming matrix **5** and reducing to matrix **6** to find the inverse of each of the following matrices. Check by showing that $A^{-1}A = AA^{-1} = I$.

a $\begin{bmatrix} 6 & 1 \\ 2 & 4 \end{bmatrix}$ **b** $\begin{bmatrix} -1 & -1 \\ 1 & -1 \end{bmatrix}$ **c** $\begin{bmatrix} 1 & 2 \\ 3 & 4 \end{bmatrix}$

d $\begin{bmatrix} 1 & 0 \\ 2 & 1 \end{bmatrix}$ **e** $\begin{bmatrix} 2 & 1 & 1 \\ 1 & 3 & 1 \\ -1 & 4 & 0 \end{bmatrix}$ **f** $\begin{bmatrix} 1 & 1 & 1 \\ 0 & 1 & 1 \\ 0 & 0 & 1 \end{bmatrix}$

g $\begin{bmatrix} 3 & 1 & 2 \\ 4 & 1 & -6 \\ 1 & 0 & 1 \end{bmatrix}$ **h** $\begin{bmatrix} 1 & 2 & 1 & 1 \\ 0 & 2 & 0 & 1 \\ 0 & 0 & -\frac{1}{2} & 1 \\ 0 & 0 & 0 & -1 \end{bmatrix}$ **i** $\begin{bmatrix} 1 & 1 & 1 & 1 \\ 1 & 1 & 1 & -1 \\ 1 & 1 & -1 & 1 \\ 1 & -1 & 1 & 1 \end{bmatrix}$

3 Find the inverse of each of the following.

a $\begin{bmatrix} 1 & 1 \\ 0 & 1 \end{bmatrix}$ b $\begin{bmatrix} 1 & 1 & 1 \\ 0 & 1 & 1 \\ 0 & 0 & 1 \end{bmatrix}$ c $\begin{bmatrix} 1 & 1 & 1 & 1 & 1 \\ 0 & 1 & 1 & 1 & 1 \\ 0 & 0 & 1 & 1 & 1 \\ 0 & 0 & 0 & 1 & 1 \\ 0 & 0 & 0 & 0 & 1 \end{bmatrix}$

4 Find the inverse of each of the following.

a $\begin{bmatrix} 2 & 0 \\ 0 & -2 \end{bmatrix}$ b $\begin{bmatrix} \frac{1}{4} & 0 & 0 \\ 0 & \frac{1}{2} & 0 \\ 0 & 0 & \frac{1}{3} \end{bmatrix}$ c $\begin{bmatrix} 1 & 0 & 0 & 0 & 0 \\ 0 & 4 & 0 & 0 & 0 \\ 0 & 0 & 2 & 0 & 0 \\ 0 & 0 & 0 & -1 & 0 \\ 0 & 0 & 0 & 0 & 3 \end{bmatrix}$

5 Which of the following matrices are invertible? Use the method of Example **6** to find the inverse if the matrix is invertible.

a $\begin{bmatrix} 2 & 0 & -1 \\ 0 & 1 & 3 \\ 0 & 0 & 0 \end{bmatrix}$ b $\begin{bmatrix} 3 & 2 & 1 \\ 0 & 2 & 2 \\ 0 & 0 & -1 \end{bmatrix}$ c $\begin{bmatrix} 3 & 1 & 1 & 2 \\ 0 & 1 & 1 & 3 \\ 0 & 0 & 2 & 1 \\ 0 & 0 & 0 & -1 \end{bmatrix}$

d $\begin{bmatrix} 2 & -1 & 1 & 2 \\ 0 & 0 & 1 & 3 \\ 0 & 0 & 2 & 1 \\ 0 & 0 & 0 & -3 \end{bmatrix}$ e $\begin{bmatrix} 2 & 0 & 0 & 0 \\ 1 & 3 & 0 & 0 \\ -1 & 1 & 2 & 0 \\ 1 & 2 & 1 & 1 \end{bmatrix}$

6 If $ad - bc \neq 0$, show that

$$\begin{bmatrix} a & b \\ c & d \end{bmatrix}^{-1} = \frac{1}{ad - bc}\begin{bmatrix} d & -b \\ -c & a \end{bmatrix}$$

7 Suppose

$$A = \begin{bmatrix} 1 & 0 & 0 \\ 0 & 1 & 0 \end{bmatrix} \quad \text{and} \quad B = \begin{bmatrix} 1 & 0 \\ 0 & 1 \\ 0 & 0 \end{bmatrix}$$

Find AB and BA. Why does this not contradict Theorem **12**?

8 Show that if A has m rows and n columns, and B has n rows and m columns, and if $m > n$, then AB is not invertible. (Hint: Apply Theorem **1** to B to find $\bar{u} \neq \bar{0}$ such that $B\bar{u} = \bar{0}$. Then apply property **2** to AB.)

□ **9** The method of forming the matrix **5** and reducing extends to complex matrices. Find the inverse of

a $\begin{bmatrix} i & 1 \\ 2 & -i \end{bmatrix}$ **b** $\begin{bmatrix} 1+i & 0 \\ 0 & 1-i \end{bmatrix}$

SECTION 2 The Algebra of Inverses

In the previous section we said that a matrix B was an inverse of the matrix A if $BA = I$. Such a matrix cannot always be found, but if A has an inverse it has only one inverse, which is usually denoted by A^{-1}.

In this section we exhibit some of the algebraic manipulations which are possible with inverses and show how to obtain inverses of various combinations of matrices. These are presented as a series of examples.

EXAMPLE 1 *Division.*

If A and C are given matrices and one seeks to find a matrix B such that

$$AB = C$$

one can proceed as follows: First express this as a system of equations with the entries of B as unknowns, then use reduction methods to solve this system. If it is known that A is invertible, a simple trick enables us to express B in terms of A^{-1} and C. Multiplying on the left of each side of the equation $AB = C$ by A^{-1}, we have

$$A^{-1}(AB) = A^{-1}C$$

Now regroup in the expression $A^{-1}(AB)$ and use the fact that $A^{-1}A = I$ to obtain

$$A^{-1}(AB) = (A^{-1}A)B = IB = B$$

In other words, if A is invertible and $AB = C$, then

1 $$B = A^{-1}C$$

Unless it is known that A^{-1} and C commute, the order in expression **1** is critical, for $A^{-1}C$ is *not* the same as CA^{-1}.

EXAMPLE 2 *Cancellation.*

Just as one can cancel the number a from the equation $ba = ca$, if $a \neq 0$, one can also cancel A from the matrix equation

$$BA = CA$$

if one knows that A is invertible. This cancellation is done by multiplying through by A^{-1} to obtain

$$(BA)A^{-1} = (CA)A^{-1}$$

then regrouping and using $AA^{-1} = I$ to obtain $B = C$. Again, the order is important; for example, even if P is invertible it cannot be canceled out of the equation

$$PQ = RP$$

for multiplying on the left by P^{-1} would give

$$P^{-1}(PQ) = P^{-1}RP$$

that is, $Q = P^{-1}RP$, while multiplying on the right would give

$$(PQ)P^{-1} = (RP)P^{-1}$$

so that $PQP^{-1} = R$. In neither of these cases can it be concluded that Q and R are equal, unless it is known that P commutes with R or Q.

EXAMPLE 3 *The Inverse of a Product.*

Suppose A and B are invertible. Let us show that AB is invertible and that the inverse of AB is the product of B^{-1} and A^{-1}; that is,

2 $$(AB)^{-1} = B^{-1}A^{-1}$$

To establish this we must show that

3 $$(B^{-1}A^{-1})(AB) = I$$

for this is what it means to say that $B^{-1}A^{-1}$ is the inverse of AB. Statement 3 is a simple consequence of the associativity of matrix products:

$$\begin{aligned}(B^{-1}A^{-1})(AB) &= B^{-1}(A^{-1}A)B \\ &= B^{-1}IB \\ &= B^{-1}B = I\end{aligned}$$

It is in fact true that if A and B are square and AB is invertible then both A and B must be invertible. A proof of this is sketched in Exercise **15** on page 201.

A word of warning: Note the order reversal in statement **2**: The inverse of AB is $B^{-1}A^{-1}$, *not* $A^{-1}B^{-1}$. This order reversal holds for products of more than two matrices; for example,

$$(ABC)^{-1} = C^{-1}B^{-1}A^{-1}$$

EXAMPLE 4 *Inverses of Polynomial Expressions.*

The inverse of a matrix can often be found from the definition and the fact that the inverse is unique; that is, B is *the* unique inverse of A if $BA = I$.

For example, the inverse of the inverse A^{-1} must be A, for A is a matrix such that $AA^{-1} = I$. If A is invertible, the inverse of A^2 must be $(A^{-1})^2$, for $(A^{-1})^2$ is a matrix such that $(A^{-1})^2A^2 = I$.

By analogy with the algebra of numbers we usually write A^{-2} instead of $(A^{-1})^2$, and, in general,

$$A^{-k} = (A^k)^{-1} = (A^{-1})^k$$

This method of finding inverses is also useful when A satisfies a polynomial relation. For example, if $p(A) = 0$, where $p(\lambda) = \lambda^2 - \lambda + 2$, then

$$A^2 - A + 2I = 0$$

which can be rewritten as $A - A^2 = 2I$, and finally as

$$\tfrac{1}{2}(I - A)A = I$$

We therefore must have

$$A^{-1} = \tfrac{1}{2}(I - A)$$

EXAMPLE 5 *The Transpose.*

The **transpose** of A, denoted by A^t, is the matrix obtained from A by interchanging the rows and columns of A. For example, if

$$A = \begin{bmatrix} 3 & 1 & 2 \\ 1 & 4 & -1 \\ 6 & 7 & 0 \end{bmatrix} \quad \text{then} \quad A^t = \begin{bmatrix} 3 & 1 & 6 \\ 1 & 4 & 7 \\ 2 & -1 & 0 \end{bmatrix}$$

Let us show that

4 A^t is invertible if A is invertible and $(A^t)^{-1} = (A^{-1})^t$; that is, the inverse of the transpose is the transpose of the inverse.

This is a rather simple consequence of two properties of the transpose:

5 **a** $I^t = I$
b $(AB)^t = B^tA^t$

Part **5a** follows from the fact that the diagonal entries of a matrix and its transpose are always the same; hence I^t has ones on its diagonal and zeros off its diagonal, so it must be I. The proof of part **b** requires a calculation which we give in the two-row, two-column case. Suppose

$$A = \begin{bmatrix} A_{11} & A_{12} \\ A_{21} & A_{22} \end{bmatrix} \qquad B = \begin{bmatrix} B_{11} & B_{12} \\ B_{21} & B_{22} \end{bmatrix}$$

so that

$$AB = \begin{bmatrix} A_{11}B_{11} + A_{12}B_{21} & A_{11}B_{12} + A_{12}B_{22} \\ A_{21}B_{11} + A_{22}B_{21} & A_{21}B_{12} + A_{22}B_{22} \end{bmatrix}$$

and

$$A^t = \begin{bmatrix} A_{11} & A_{21} \\ A_{12} & A_{22} \end{bmatrix} \qquad B^t = \begin{bmatrix} B_{11} & B_{21} \\ B_{12} & B_{22} \end{bmatrix}$$

Thus

$$B^tA^t = \begin{bmatrix} B_{11}A_{11} + B_{21}A_{12} & B_{11}A_{21} + B_{21}A_{22} \\ B_{12}A_{11} + B_{22}A_{12} & B_{12}A_{21} + B_{22}A_{22} \end{bmatrix}$$

which is just the transpose of AB.

Now we are ready to prove that $(A^t)^{-1}$ is $(A^{-1})^t$. First take the transpose of each side of $AA^{-1} = I$ to obtain

$$(AA^{-1})^t = I^t$$

Since the transpose of a product is the product of transposes in reverse order (property **5b**) we have

$$(AA^{-1})^t = (A^{-1})^tA^t$$

Hence

$$(A^{-1})^tA^t = I^t = I$$

so that $(A^{-1})^t$ is indeed the inverse of A^t, that is

$$(A^{-1})^t = (A^t)^{-1}$$

which is what we wished to prove.

EXAMPLE 6 *Orthogonal Matrices.*

A matrix A is said to be **orthogonal** if A is invertible and its inverse is equal to its transpose, that is, if

$$A^tA = I$$

For example, the inverse of

$$A = \begin{pmatrix} 1\sqrt{2} & -1\sqrt{2} \\ 1\sqrt{2} & 1\sqrt{2} \end{pmatrix} \quad \text{is} \quad A^{-1} = \begin{pmatrix} 1\sqrt{2} & 1\sqrt{2} \\ -1\sqrt{2} & 1\sqrt{2} \end{pmatrix}$$

and A^{-1} is just A^t. The terminology comes from the fact that the equation

$$A^tA = I$$

gives conditions on the entries of A which are just the condition that the columns of A be *orthonormal*. For example, if

$$A = \begin{bmatrix} A_{11} & A_{21} \\ A_{12} & A_{22} \end{bmatrix} \qquad \text{then} \qquad A^t = \begin{bmatrix} A_{11} & A_{21} \\ A_{12} & A_{22} \end{bmatrix}$$

Thus

$$A^tA = \begin{bmatrix} A_{11}^2 + A_{21}^2 & A_{11}A_{12} + A_{21}A_{22} \\ A_{12}A_{11} + A_{22}A_{21} & A_{12}^2 + A_{22}^2 \end{bmatrix}$$

and it follows that $A^tA = I$ if and only if

6
$$\begin{matrix} A_{11}^2 + A_{21}^2 = 1 & A_{11}A_{12} + A_{21}A_{22} = 0 \\ A_{12}A_{11} + A_{22}A_{21} = 0 & A_{12}^2 + A_{22}^2 = 1 \end{matrix}$$

These are precisely the conditions that the columns of A be orthonormal.

A final note of caution: It is *not* enough that the columns of A be orthogonal for A to be called an orthogonal matrix. The terminology is unfortunate but deeply embedded in mathematics: a matrix A is *orthogonal*, that is $A^tA = I$, if and only if its columns are *orthonormal*. (See Exercise **13** below.)

EXERCISES

1 **a** Assume A is invertible and solve $BA = C$ for B.
b Assume A and B are invertible and $ACB = ADB$. Show that $C = D$.
c Solve $B = P^{-1}AP$ for A.

2 State assumptions that allow one to solve for B; then solve for B.
a $AB + aB = D$ **b** $AB + 2C = DB$ **c** $AB = C + 3B$
d $BA + BC = C$ **e** $BA = CA$.

3 **a** Assume A, B, C, and D are invertible and find $(ABCD)^{-1}$.
b Find the inverse of $ABA^{-1}B^{-1}$.

4 Assume A is invertible and find the inverse of each of the following.
a A^3 **b** $3A$ **c** $-A^4$ **d** $(A^{-1})^2$

5 Find the inverse of A if
a $A^2 - 2A - 3I = 0$ **b** $A^6 - 5A^4 + 3A^2 + 2I = 0$

6 **a** Suppose $B \neq 0$ and $AB = 0$. Can A be invertible?
b Suppose $A^2 = 0$. Can A be invertible?
c Suppose $P^2 = P$ and $P \neq I$. Can P be invertible?

7 **a** What is $(A^t)^t$?
b What is $(A + B)^t$?
c What is $(A^2)^t$?

8 Given $A = \begin{bmatrix} 2 & 1 \\ 3 & 0 \end{bmatrix}$ and $B = \begin{bmatrix} -1 & 2 \\ 1 & 1 \end{bmatrix}$.

a Find A^{-1} and B^{-1}.
b Verify that $(AB)^{-1} = B^{-1}A^{-1}$.
c Verify that $(AB)^t = B^tA^t$.
d Verify that $(A^t)^{-1} = (A^{-1})^t$.

9 Show that the product

$$\begin{bmatrix} 3 & 1 \\ 2 & 1 \\ 1 & 1 \end{bmatrix} \begin{bmatrix} 2 & 1 & 0 \\ 1 & -1 & 1 \end{bmatrix}$$

is not invertible, then interchange the order of the factors and show that the resulting product is invertible.

10 **a** Show that if A is an orthogonal matrix then its *rows* are orthonormal. (Hint: Consider $AA^t = I$.)
b Show that if A and B are orthogonal matrices then AB is orthogonal.

11 **a** Show that $(A\bar{u}) \cdot \bar{v} = \bar{u} \cdot (A^t\bar{v})$ for

$$A = \begin{bmatrix} A_{11} & A_{12} \\ A_{21} & A_{22} \end{bmatrix} \qquad \bar{u} = \begin{pmatrix} x_1 \\ y_1 \end{pmatrix} \qquad \bar{v} = \begin{pmatrix} x_2 \\ y_2 \end{pmatrix}$$

b Show that if A is an orthogonal matrix then

$$(A\bar{u}) \cdot (A\bar{v}) = \bar{u} \cdot \bar{v} \qquad \text{for all } \bar{u}, \bar{v}$$

12 Show that

$$\begin{bmatrix} 1/\sqrt{3} & 1/\sqrt{3} & 1/\sqrt{3} \\ 1/\sqrt{2} & -1/\sqrt{2} & 0 \\ 1/\sqrt{6} & 1/\sqrt{6} & -2/\sqrt{6} \end{bmatrix}$$

is an orthogonal matrix.

13 Show that A^tA is a diagonal matrix if and only if the columns of A are orthogonal. Find a matrix whose columns are orthogonal and whose rows are *not* orthogonal.

14 Solve for B:

$$\begin{bmatrix} 1 & 1 \\ 1 & -1 \end{bmatrix} B = \begin{bmatrix} 2 & 1 & 4 \\ 1 & 3 & 1 \end{bmatrix}$$

15 Suppose that A and B are square and AB is invertible.
a Show that B is invertible.
b Show that A is invertible. (Hint: Take transposes and use part **a**.)

16 **a** Show that if B is invertible then rank AB = rank A. (Hint: Apply Exercise **14c**, page 168, to the pair A, B, and then to the pair AB, B^{-1}.)

b Show that if A is invertible then rank AB = rank B.

c Apply the results of **a** and **b** to prove the two parts of Exercise **15**.

□ **17** A complex matrix A is *unitary* if A^{-1} is the transpose conjugate of A (see Exercise **11**, page 282). Show that A is unitary if and only if its columns are orthonormal with respect to the complex dot product. (See Exercise **17**, page 116.)

SECTION 3 The Inverse of a Linear Transformation

Now let us turn to the definition and properties of the inverse of a linear transformation. We could define this in terms of the inverse of the matrix of the transformation but prefer instead to give the following definition which is analogous to our definition of inverse of a matrix.

A linear transformation T from V into V is **invertible** if there is a linear transformation S from V into V such that $ST = I$. The transformation S is an **inverse** of T.

If V is finite-dimensional, then the condition $ST = I$ implies that $TS = I$. This is the transformation analog of Theorem 12 and its proof will be found in the Appendix. The trick used at the end of the proof of Theorem 12 can then be used to show that

1 If V is finite-dimensional and $ST = I$, $S_1T = I$, then $S = S_1$; that is, if T has an inverse, it has only one inverse.

In this case we call S *the* inverse of T, usually denoting it by T^{-1}.

If T is a linear transformation from R^n into R^n let A be its matrix, so that

$$T\bar{u} = A\bar{u} \qquad \text{for all } \bar{u} \text{ in } R^n$$

If S is another linear transformation with matrix B then to say that $ST = I$, the identity transformation, is the same as saying that $BA = I$, the identity matrix. In summary,

2 If T is a linear transformation from R^n into R^n then T is invertible if and only if the matrix A of T is invertible. Furthermore the matrix of T^{-1} is A^{-1}.

All the algebraic constructions of Section 2 carry over immediately to transformations on finite-dimensional spaces. For example if S and T are invertible, so is ST and $(ST)^{-1} = T^{-1}S^{-1}$.

The following examples describe some invertible and noninvertible transformations, after which we shall discuss some conditions for invertibility.

EXAMPLE 1 *Rotations.*

Let us describe the inverse of T where T is counterclockwise rotation in R^2 through the angle θ. Denote by S the clockwise rotation through θ, that is, the counterclockwise rotation through $-\theta$. Clearly

$$S T \bar{u} = \bar{u} \qquad \text{for all } \bar{u} \text{ in } R^2$$

(See Figure 1.)

Figure 1

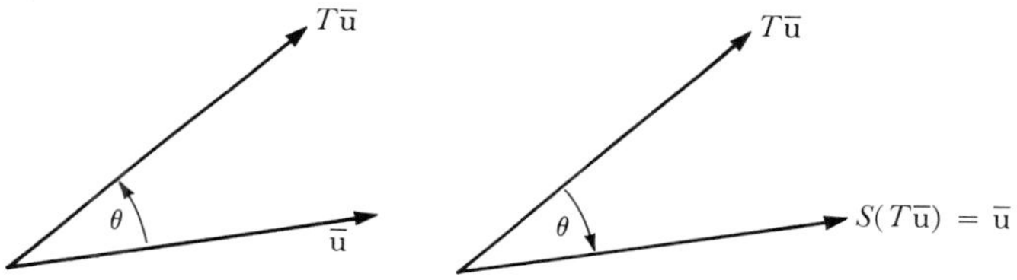

The matrices of T and S are, respectively,

3 $$A = \begin{bmatrix} \cos\theta & -\sin\theta \\ \sin\theta & \cos\theta \end{bmatrix} \qquad B = \begin{bmatrix} \cos(-\theta) & -\sin(-\theta) \\ \sin(-\theta) & \cos(-\theta) \end{bmatrix}$$

Since $ST = I$, we must have $BA = I$. A simple calculation shows that the following is also true

$$\begin{bmatrix} \cos\theta & \sin\theta \\ -\sin\theta & \cos\theta \end{bmatrix} \begin{bmatrix} \cos\theta & -\sin\theta \\ \sin\theta & \cos\theta \end{bmatrix} = \begin{bmatrix} 1 & 0 \\ 0 & 1 \end{bmatrix}$$

The inverse of A is unique, however, and hence we must have

$$B = \begin{bmatrix} \cos\theta & \sin\theta \\ -\sin\theta & \cos\theta \end{bmatrix}$$

Comparing this with the expression 3 yields the conclusion that

$$\cos\theta = \cos(-\theta)$$
$$\sin\theta = -\sin(-\theta)$$

EXAMPLE 2 *Reflection and Projection.*

The inverse of a reflection is also easy to describe. Suppose T is reflection in the line through $\bar{v}$, where $\bar{v}$ is some nonzero vector in R^2. Apply T to a vector $\bar{u}$, obtaining $T\bar{u}$. To get $\bar{u}$ back again, merely reflect again; that is

$$T^2\bar{u} = \bar{u} \qquad \text{for all } \bar{u} \text{ in } R^2$$

In other words, reflecting twice always returns us to our original position, so that $T^2 = I$ and hence $T^{-1} = T$ (much as the reciprocal of -1 is -1).

The situation with a projection is quite different. For example, if P is the projection along $\bar{v}$, then P is *not* invertible. There are several ways to see this. Let us show here that there *cannot* be a linear transformation Q such that $QP = I$. If there were such a Q, then we could multiply by P on the right of each side of $QP = I$ to obtain

$$(QP)P = IP$$

Since $P^2 = P$ we would have

$$(QP)P = QP^2 = QP$$

and hence

$$QP = IP = P$$

Since P is not the identity the two equations $QP = P$ and $QP = I$ cannot both be true; yet if $QP = I$ we have just shown that $QP = P$. This confirms that there *cannot* be any Q such that $QP = I$. (See also Example 4 below.)

EXAMPLE 3 If T happens to be given as a matrix multiplication then its inverse can be determined by finding the inverse of its matrix. For example if

$$T\bar{u} = \begin{bmatrix} 2 & 1 \\ -1 & 3 \end{bmatrix}\bar{u} \qquad \bar{u} \text{ in } R^2$$

then the reduction method of Section 1 gives

$$\begin{bmatrix} 2 & 1 \\ -1 & 3 \end{bmatrix}^{-1} = \begin{bmatrix} \frac{3}{7} & -\frac{1}{7} \\ \frac{1}{7} & \frac{2}{7} \end{bmatrix}$$

so the inverse T^{-1} is given by

$$T^{-1}\bar{u} = \begin{bmatrix} \frac{3}{7} & -\frac{1}{7} \\ \frac{1}{7} & \frac{2}{7} \end{bmatrix}\bar{u}$$

DISCUSSION We now give some alternative conditions for a linear transformation (or matrix) to be invertible. These will be stated for transformations from R^n into R^n although they hold for arbitrary finite-dimensional spaces.

First note the following transformation version of Theorem **11** (page 188):

THEOREM 11 A linear transformation T from R^n into R^n is invertible if and only if $T\bar{u} \neq \bar{0}$ whenever $\bar{u} \neq \bar{0}$.

The statement that $T\bar{u} \neq \bar{0}$ whenever $\bar{u} \neq \bar{0}$ is really another way of saying that $T\bar{u} = \bar{0}$ has a unique solution.

An important property of invertible transformations is that they map independent sets into independent sets; that is:

THEOREM 13 If T is an invertible linear transformation and $\bar{u}_1, \bar{u}_2, \ldots, \bar{u}_n$ are independent then $T\bar{u}_1, T\bar{u}_2, \ldots, T\bar{u}_n$ are independent.

PROOF: To simplify the proof of this suppose $n = 2$; that is, $\bar{u}_1$ and $\bar{u}_2$ are independent. Our task is to show that if T is invertible then $T\bar{u}_1$ and $T\bar{u}_2$ are also independent. In other words we must show that if

4 $$c_1 T\bar{u}_1 + c_2 T\bar{u}_2 = \bar{0}$$

then c_1 and c_2 must be zero. Use the linearity of T to rewrite equation **4** as

$$T(c_1\bar{u}_1 + c_2\bar{u}_2) = \bar{0}$$

By assumption T is invertible, so Theorem **11** above implies that

$$c_1\bar{u}_1 + c_2\bar{u}_2 = \bar{0}$$

The vectors $\bar{u}_1$ and $\bar{u}_2$ are independent so we must have $c_1 = c_2 = 0$. This shows that if T is invertible and equation **4** holds then c_1 and c_2 must be zero; in other words $T\bar{u}_1$ and $T\bar{u}_2$ must be independent. This completes the proof of Theorem **13** in the special case $n = 2$.

A converse of Theorem **13** is discussed in the Appendix.

EXAMPLE 4 It is sometimes easy to establish that a linear transformation is invertible or not invertible by using Theorem **11**. For example, if P is the projection along $\bar{v}$ and $\bar{u}$ is perpendicular to $\bar{v}$, then $P\bar{u} = \bar{0}$ so P cannot be invertible. (Another proof that such projections are not invertible is sketched in Example **2** above.)

As another example, suppose T is a rotation followed by a reflection. If $\bar{u}$ is not zero, then rotating will not give the zero vector, and following this by a reflection will not give the zero vector, that is, $T\bar{u} \neq 0$ if $\bar{u} \neq 0$, so that T must be invertible. The description of T^{-1} is left to the exercises.

EXAMPLE 5 Other conditions for invertibility can frequently be established as consequences of Theorem **11** or **13**. For example let us show that

5 The linear transformation T from R^n into R^n is invertible if and only if $T\bar{u}_1, T\bar{u}_2, \ldots, T\bar{u}_k$ spans R^n for some set $\bar{u}_1, \bar{u}_2, \ldots, \bar{u}_k$.

Suppose $T\bar{u}_1, T\bar{u}_2, \ldots, T\bar{u}_k$ spans R^n. Let A be the matrix of T, so that $A\bar{u}_1, A\bar{u}_2, \ldots, A\bar{u}_k$ also spans R^n. The column space of A must therefore be R^n, so rank $A = n$. Since T is defined for vectors in R^n, the matrix A is square so A must be invertible (Theorem **11**), and therefore result **2** implies that T is invertible.

EXAMPLE 6 □ In infinite-dimensional spaces the condition that $ST = I$ *does not* imply that $TS = I$. For example, suppose D is differentiation and T is integration, defined by

$$Df = f' \qquad \text{and} \qquad Tf(x) = \int_0^x f(t)\, dt$$

As shown in calculus, we have, for continuous functions f,

$$\frac{d}{dx}\int_0^x f(t)\, dt = f(x)$$

and for functions f such that f' is continuous,

$$\int_0^x f'(t)\, dt = f(x) - f(0)$$

These can be written in operator form as

6 $$(DT)f = f \qquad \text{and} \qquad (TD)f = f - f(0)$$

Thus, if V is the collection of all functions f such that f and all its derivatives are defined and continuous for $0 \leq x \leq 1$, then D and T are linear transformations from V into V. Formula **6** tells us that $DT = I$ and $TD \neq I$, where I is the identity operator on V. For example, $TD\,(\cos x) = T\,(-\sin x) = \cos x - 1 \neq \cos x$.

Let D_1 be the operator defined by $D_1 f = f' + f(0)$. Then

$$D_1 T(f) = D_1\left(\int_0^x f(t)\, dt\right) = f - \int_0^0 f(t)\, dt = f$$

This shows that it is possible in the infinite-dimensional case to have more than one operator S such that $ST = I$.

EXERCISES

1 Suppose T is a counterclockwise rotation through the angle θ.
 a Describe T^{-1} if $\theta = \pi/4, \ -3\pi/2, \pi$.
 b Describe the inverse of T^2.
 c For what values of θ does $T^{-1} = T$?
 d For what values of θ does $T^{-1} = T^3$?

2 Suppose T is rotation followed by reflection. Describe T^{-1}.

3 Suppose T is defined by $T\bar{u} = \begin{bmatrix} 1 & 2 & 1 \\ -1 & 1 & 3 \\ 0 & 1 & 1 \end{bmatrix} \bar{u}$. Find the matrix of T^{-1}.

4 Suppose $T(x,y) = (3x - y, x + 2y)$. Show that T is invertible and find a formula for T^{-1}. (Hint: Find the matrix of T.)

5 Suppose $\bar{u}_1$ and $\bar{u}_2$ are a basis for R^2 and T is a linear transformation which satisfies $T\bar{u}_1 = 3\bar{u}_1$ and $T\bar{u}_2 = 2\bar{u}_1$. Show that T is *not* invertible. (Hint: Find $\bar{u} \neq \bar{0}$ such that $T\bar{u} = \bar{0}$.)

6 Let T be the reflection in R^2 in the line $y = x$.
 a Describe T^{-1}.
 b Find the matrix A of T and verify that A^{-1} is the matrix of T^{-1}.

7 Let P be the projection along $\bar{v} = (1,1)$. Find the matrix of P and verify that it is not invertible.

8 Show that if T^2 is invertible then T is invertible.

9 a If T is invertible and $\bar{u}_1$ and $\bar{u}_2$ are independent show that $T\bar{u}_1$ and $T\bar{u}_2$ are independent.
 b If T is an invertible linear transformation from V into V where V has dimension 2 and if $T\bar{u}_1$ and $T\bar{u}_2$ are independent for some independent set $\bar{u}_1, \bar{u}_2$, show that T is invertible. (Hint: Show that $T\bar{u} = \bar{0}$ has only one solution by expressing $\bar{u}$ in terms of $\bar{u}_1$ and $\bar{u}_2$.)

10 Let A_1 be a matrix with two rows and two columns. Which of the following imply that A is invertible?
 a Rank $A = 2$.
 b The columns of A are independent.
 c $A^2 = 0$.
 d The rows of A span R^2.
 e There is a $\bar{u} = 0$ such that $A^t\bar{u} = \bar{0}$.
 f There is a $\bar{u} \neq 0$ such that $A\bar{u} \neq \bar{0}$.
 g The reduced matrix of A is P and $P^2 = P$.

11 Suppose the third, fourth, and sixth columns of A are an independent set. Show that these same columns of BA are independent if B is invertible.

12 Suppose S and T are transformations of V into V such that $ST = TS = I$, the identity operator on V.
 a Show that $S\bar{v} = \bar{u}$ if and only if $T\bar{u} = \bar{v}$.

b Show that if T is linear S must also be linear. [Hint: Use part **a** to show that $S(\bar{v}_1 + \bar{v}_2) = S\bar{v}_1 + S\bar{v}_2$ by setting $\bar{u}_1 = S\bar{v}_1$, $\bar{u}_2 = S\bar{v}_2$, and finding $T(\bar{u}_1 + \bar{u}_2)$.]

☐ **13** Suppose V is the vector space $C[0,1]$ of functions that are defined and continuous for $0 \le x \le 1$. Suppose $Tf = gf$, where g is in $C[0,1]$ and $g(x) \ne 0$, $0 \le x \le 1$. What is T^{-1}? In particular, what is T^{-1} if $g = e^{ax}$?

☐ **14** **a** Solve for B:

$$\begin{bmatrix} i & 1-i \\ 0 & 2i \end{bmatrix} B = \begin{bmatrix} 3-i & i \\ 1 & 2i \end{bmatrix}$$

b Proceed as in Example **3** to find a formula for the coordinates of (x,y) with respect to $(i, 1 + 2i)$ and $(1 - i, 0)$.

SECTION 4 Changing Coordinates

A given geometric object, such as a plane, can be coordinatized in many ways. In the next chapter methods will be discussed for choosing coordinate systems in such a way as to simplify the description of a given linear transformation. In this section we make use of the inverse to obtain formulas for changing from one coordinate system to another.

To simplify our results, let us first show how to change coordinates in the two-dimensional space R^2. Denote the standard basis for R^2 by $\bar{\imath}$ and $\bar{\jmath}$; that is,

1 $$\bar{\imath} = (1,0) \qquad \text{and} \qquad \bar{\jmath} = (0,1)$$

Let $\bar{u}_1$ and $\bar{u}_2$ be another basis for R^2. If $\bar{u} = (x,y)$ is some arbitrary vector in R^2, then

2 $$\bar{u} = x\bar{\imath} + y\bar{\jmath}$$

Furthermore, $\bar{u}$ must be a linear combination of $\bar{u}_1$ and $\bar{u}_2$, and hence there are numbers c_1 and c_2 such that

3 $$\bar{u} = c_1\bar{u}_1 + c_2\bar{u}_2$$

This pair (c_1,c_2) was called the coordinates of $\bar{u}$ with respect to the basis $\bar{u}_1$, $\bar{u}_2$. An expression for c_1 and c_2 in terms of x and y can be obtained as follows. If

4 $$\bar{u}_1 = (a,b) \qquad \bar{u}_2 = (c,d)$$

then equate the two expressions **2** and **3** for $\bar{u}$

$$x\bar{\imath} + y\bar{\jmath} = c_1\bar{u}_1 + c_2\bar{u}_2$$

and replace $\bar{u}_1$ and $\bar{u}_2$ and $\bar{i}$ and $\bar{j}$ with their respective expressions **4** and **1** to obtain

$$x(1,0) + y(0,1) = c_1(a,b) + c_2(c,d)$$

That is,

$$(x,y) = (c_1a + c_2c,\ c_1b + c_2d)$$

In column form this becomes

$$\begin{bmatrix} x \\ y \end{bmatrix} = \begin{bmatrix} c_1a + c_2c \\ c_1b + c_2d \end{bmatrix}$$

The right-hand matrix is just the product

$$\begin{bmatrix} a & c \\ b & d \end{bmatrix} \begin{bmatrix} c_1 \\ c_2 \end{bmatrix}$$

Combining these gives the formula

$$\begin{bmatrix} x \\ y \end{bmatrix} = \begin{bmatrix} a & c \\ b & d \end{bmatrix} \begin{bmatrix} c_1 \\ c_2 \end{bmatrix}$$

To express c_1 and c_2 in terms of x and y merely invert the matrix

$$P = \begin{bmatrix} a & c \\ b & d \end{bmatrix}$$

to obtain the formula

$$\begin{bmatrix} c_1 \\ c_2 \end{bmatrix} = P^{-1} \begin{bmatrix} x \\ y \end{bmatrix}$$

The general procedure in R^n is now easy to describe.

To express the coordinates of an arbitrary vector $\bar{u} = (x_1, x_2, \ldots, x_n)$ with respect to a basis $\bar{u}_1, \bar{u}_2, \ldots, \bar{u}_n$, form the matrix P whose *columns* are the vectors $\bar{u}_1, \bar{u}_2, \ldots, \bar{u}_n$. Take the inverse P^{-1} of P. The coordinates $(c_1, c_2, \ldots, c_n)$ of $\bar{u}$ with respect to $\bar{u}_1, \bar{u}_2, \ldots, \bar{u}_n$ are then given by

5 $$\begin{bmatrix} c_1 \\ c_2 \\ \vdots \\ c_n \end{bmatrix} = P^{-1} \begin{bmatrix} x_1 \\ x_2 \\ \vdots \\ x_n \end{bmatrix}$$

This formula enables us to change the description of a vector from the standard coordinate system description $(x_1, x_2, \ldots, x_n)$ to the new coordinates $(c_1, c_2, \ldots, c_n)$ with respect to the new axes given by $\bar{u}_1$, $\bar{u}_2, \ldots, \bar{u}_n$. After looking at an example we shall examine the situation when there is a third coordinate system.

EXAMPLE 1 In this example we illustrate the basic formula **5** and also give a geometric description of coordinates with respect to a basis. Let $\bar{i} = (1,0)$ and $\bar{j} = (0,1)$ be the standard basis for R^2 and let

$$\bar{u}_1 = (3,1) \qquad \bar{u}_2 = (-2,2)$$

Let P be the matrix with these as columns:

$$P = \begin{pmatrix} 3 & -2 \\ 1 & 2 \end{pmatrix}$$

Since P has rank 2, the vectors $\bar{u}_1$ and $\bar{u}_2$ are a basis for R^2. Applying the procedure used to obtain formula **5** we first calculate to obtain

$$P^{-1} = \begin{pmatrix} \frac{2}{8} & \frac{2}{8} \\ -\frac{1}{8} & \frac{3}{8} \end{pmatrix}$$

Hence, if $\bar{u} = (x,y)$, then

$$\bar{u} = c_1\bar{u}_1 + c_2\bar{u}_2$$

where

6 $$\begin{bmatrix} c_1 \\ c_2 \end{bmatrix} = P^{-1}\begin{bmatrix} x \\ y \end{bmatrix} = \begin{bmatrix} \frac{2}{8}x + \frac{2}{8}y \\ -\frac{1}{8}x + \frac{3}{8}y \end{bmatrix}$$

Let us see what this means geometrically. As usual, represent $\bar{i}$ and $\bar{j}$ as unit vectors along the horizontal and vertical axes, respectively, and $\bar{u}$ as the arrow from the origin 0 to (x,y). The coordinate x is determined as follows (see Figure 2):

Figure 2

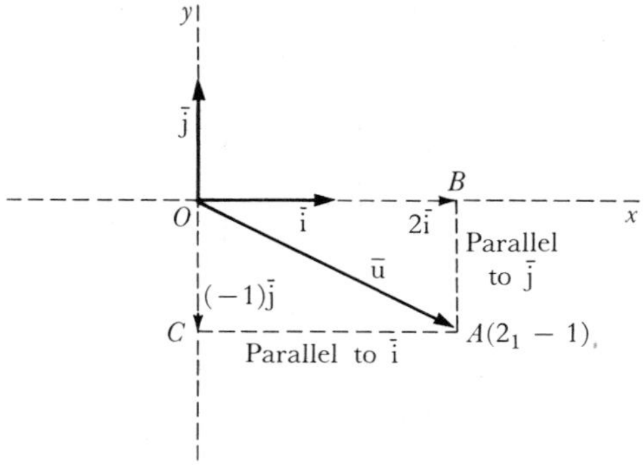

Figure 3

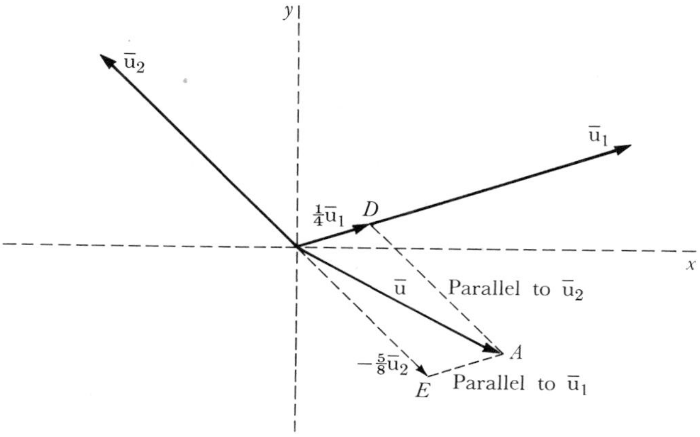

7a Let A be the terminal point of $\bar{u}$ and let B be the terminal point of $x\bar{i}$. Then x is such that AB is parallel to $\bar{j}$.

A similar description holds for y. Now refer to Figure 3, where $\bar{u}_1$, $\bar{u}_2$, and $\bar{u}$ have been sketched. The coordinate c_1 is determined as follows:

7b c_1 is that multiple of $\bar{u}_1$ so that the terminal point D of $c_1\bar{u}_1$ and the terminal point A of $\bar{u}$ lie on a line parallel to $\bar{u}_2$.

A similar description holds for c_2. Note that for the $\bar{u}$ pictured in Figure 2,

$$x = 2 \qquad y = -1$$

Formula **6** gives

$$c_1 = \tfrac{1}{4} \qquad c_2 = -\tfrac{5}{8}$$

which is certainly in agreement with Figure 3. (For example, OE is $\frac{5}{8}$ as long as $\bar{u}_2$ and points in the opposite direction, so that c_2 should be $-\frac{5}{8}$.)

DISCUSSION Some simple algebra applied to formula **5** will enable us to interrelate the coordinates of a given vector $\bar{u}$ with respect to several different bases. For example, suppose $\bar{u} = (x,y)$ and:

c_1 and c_2 are the coordinates of $\bar{u}$ with respect to $\bar{u}_1$ and $\bar{u}_2$,

d_1 and d_2 are the coordinates of $\bar{u}$ with respect to $\bar{v}_1$ and $\bar{v}_2$.

Let P be the matrix whose columns are $\bar{u}_1$ and $\bar{u}_2$, while Q is the matrix whose columns are $\bar{v}_1$ and $\bar{v}_2$. Formula **5** gives

$$\begin{bmatrix} c_1 \\ c_2 \end{bmatrix} = P^{-1}\begin{bmatrix} x \\ y \end{bmatrix} \quad \text{and} \quad \begin{bmatrix} d_1 \\ d_2 \end{bmatrix} = Q^{-1}\begin{bmatrix} x \\ y \end{bmatrix}$$

To express c_1, c_2 in terms of d_1, d_2, rewrite these as

$$P\begin{bmatrix} c_1 \\ c_2 \end{bmatrix} = \begin{bmatrix} x \\ y \end{bmatrix} \quad \text{and} \quad Q\begin{bmatrix} d_1 \\ d_2 \end{bmatrix} = \begin{bmatrix} x \\ y \end{bmatrix}$$

then equate to obtain

$$P\begin{bmatrix} c_1 \\ c_2 \end{bmatrix} = Q\begin{bmatrix} d_1 \\ d_2 \end{bmatrix}$$

Multiply through by P^{-1} to obtain the formula

$$\begin{bmatrix} c_1 \\ c_2 \end{bmatrix} = P^{-1}Q\begin{bmatrix} d_1 \\ d_2 \end{bmatrix}$$

which expresses c_1 and c_2 in terms of d_1 and d_2. Inverting $P^{-1}Q$ gives $Q^{-1}P$, so

$$\begin{bmatrix} d_1 \\ d_2 \end{bmatrix} = Q^{-1}P\begin{bmatrix} c_1 \\ c_2 \end{bmatrix}$$

expresses d_1 and d_2 in terms of c_1 and c_2.

The general case is a straightforward extension of these formulas. If the columns of P are the vectors $\bar{u}_1, \bar{u}_2, \ldots, \bar{u}_n$ and the columns of Q are the vectors $\bar{v}_1, \bar{v}_2, \ldots, \bar{v}_n$ and if each of these sets is a basis for R^n, then the coordinates $c_1, c_2, \ldots, c_n$ of $\bar{u}$ relative to $\bar{u}_1, \bar{u}_2, \ldots, \bar{u}_n$ and the coordinates $d_1, d_2, \ldots, d_n$ of $\bar{u}$ relative to $\bar{v}_1, \bar{v}_2, \ldots, \bar{v}_n$ are related by the two equations

8
$$\begin{bmatrix} c_1 \\ c_2 \\ \vdots \\ c_n \end{bmatrix} = P^{-1}Q\begin{bmatrix} d_1 \\ d_2 \\ \vdots \\ d_n \end{bmatrix} \qquad \begin{bmatrix} d_1 \\ d_2 \\ \vdots \\ d_n \end{bmatrix} = Q^{-1}P\begin{bmatrix} c_1 \\ c_2 \\ \vdots \\ c_n \end{bmatrix}$$

EXAMPLE 2 Continuing the discussion of Example **1**, we introduce another basis and examine the formulas **8**. Let $\bar{u}_1$, $\bar{u}_2$ be as in Example **1**, that is $\bar{u}_1 = (3,1)$

and $\bar{u}_2 = (-2,2)$, and suppose

$$\bar{v}_1 = (1,1) \qquad \bar{v}_2 = (-1,-2)$$

Let P be the matrix with $\bar{u}_1$ and $\bar{u}_2$ as columns, and Q the matrix with $\bar{v}_1$ and $\bar{v}_2$ as columns:

$$P = \begin{bmatrix} 3 & -2 \\ 1 & 2 \end{bmatrix} \qquad Q = \begin{bmatrix} 1 & -1 \\ 1 & -2 \end{bmatrix}$$

Now calculate the inverses to obtain

$$P^{-1} = \begin{bmatrix} \frac{2}{8} & \frac{2}{8} \\ -\frac{1}{8} & \frac{3}{8} \end{bmatrix} \qquad Q^{-1} = \begin{bmatrix} 2 & -1 \\ 1 & -1 \end{bmatrix}$$

Suppose $\bar{u} = (2,-1)$. The coordinates c_1, c_2 of $\bar{u}$ with respect to $\bar{u}_1$, $\bar{u}_2$ were found to be

$$\begin{bmatrix} c_1 \\ c_2 \end{bmatrix} = P^{-1}\begin{bmatrix} 2 \\ -1 \end{bmatrix} = \begin{bmatrix} \frac{2}{8} & \frac{2}{8} \\ -\frac{1}{8} & \frac{3}{8} \end{bmatrix}\begin{bmatrix} 2 \\ -1 \end{bmatrix} = \begin{bmatrix} \frac{1}{4} \\ -\frac{5}{8} \end{bmatrix}$$

The coordinates d_1, d_2 of $\bar{u}$ with respect to $\bar{v}_1$, $\bar{v}_2$ are given by

$$\begin{bmatrix} d_1 \\ d_2 \end{bmatrix} = Q^{-1}\begin{bmatrix} 2 \\ -1 \end{bmatrix} = \begin{bmatrix} 2 & -1 \\ 1 & -1 \end{bmatrix}\begin{bmatrix} 2 \\ -1 \end{bmatrix} = \begin{bmatrix} 5 \\ 3 \end{bmatrix}$$

Since

$$P^{-1}Q = \begin{bmatrix} \frac{4}{8} & -\frac{6}{8} \\ \frac{2}{8} & -\frac{5}{8} \end{bmatrix} \qquad \text{and} \qquad Q^{-1}P = \begin{bmatrix} 5 & -6 \\ 2 & -4 \end{bmatrix}$$

a straightforward calculation shows that

$$P^{-1}Q\begin{bmatrix} 5 \\ 3 \end{bmatrix} = \begin{bmatrix} \frac{4}{8} & -\frac{6}{8} \\ \frac{2}{8} & -\frac{5}{8} \end{bmatrix}\begin{bmatrix} 5 \\ 3 \end{bmatrix} = \begin{bmatrix} \frac{1}{4} \\ -\frac{5}{8} \end{bmatrix}$$

and

$$Q^{-1}P\begin{bmatrix} \frac{1}{4} \\ -\frac{5}{8} \end{bmatrix} = \begin{bmatrix} 5 & -6 \\ 2 & -4 \end{bmatrix}\begin{bmatrix} \frac{1}{4} \\ -\frac{5}{8} \end{bmatrix} = \begin{bmatrix} 5 \\ 3 \end{bmatrix}$$

as we expect from formulas **8**.

EXAMPLE 3 *Changing Coordinates in Other Spaces.*
In Example **5**, page 83, and Example **6**, page 90, we discussed coordinates relative to bases in a vector space of polynomials. Let us summarize

those results here and then show how to calculate a change in coordinates in this setting. Let V be the set of all functions f defined by expressions of the form

9 $$f(x) = a_0 + a_1x + a_2x^2$$

where a_0, a_1, and a_2 are real numbers. The three polynomials

$$1,\ x,\ x^2$$

are a basis for this space V, which is, of course, just another way of saying that each f in V has a unique expression in the form **9**. The coordinates of f relative to the basis $1, x, x^2$ are just the corresponding coefficients a_0, a_1, a_2 in the expression for f. The three polynomial functions f_1, f_2, f_3,

$$f_1(x) = \frac{(x-1)(x-2)}{2} \qquad f_2(x) = \frac{x(x-2)}{-1} \qquad f_2(x) = \frac{x(x-1)}{2}$$

are also a basis for V (see page 90) and any given f in V can be expressed in the form

$$c_1f_1(x) + c_2f_2(x) + c_3f_3(x)$$

where the coordinates c_1, c_2, and c_3 relative to f_1, f_2, and f_3 are given by

10 $$c_1 = f(0) \qquad c_2 = f(1) \qquad c_3 = f(2)$$

Comparing the relations **9** and **10** we see that if the coordinates of f relative to $1, x, x^2$ are a_0, a_1, a_2, then

$$\begin{aligned} c_1 &= f(0) = a_0 + a_1 \cdot 0 + a_2 \cdot 0^2 = a_0 \\ c_2 &= f(1) = a_0 + a_1 \cdot 1 + a_2 \cdot 1^2 = a_0 + a_1 + a_2 \\ c_3 &= f(2) = a_0 + a_1 \cdot 2 + a_2 \cdot 2^2 = a_0 + 2a_1 + 4a_2 \end{aligned}$$

This set of equations can be rewritten in the form

$$\begin{bmatrix} c_1 \\ c_2 \\ c_3 \end{bmatrix} = \begin{bmatrix} 1 & 0 & 0 \\ 1 & 1 & 1 \\ 1 & 2 & 4 \end{bmatrix} \begin{bmatrix} a_0 \\ a_1 \\ a_2 \end{bmatrix}$$

This shows how to change from coordinates with respect to $1, x, x^2$ into coordinates with respect to f_1, f_2, f_3. To reverse this calculation put

$$P = \begin{bmatrix} 1 & 0 & 0 \\ 1 & 1 & 1 \\ 1 & 2 & 4 \end{bmatrix} \quad \text{and calculate} \quad P^{-1} = \begin{bmatrix} 1 & 0 & 0 \\ -\frac{3}{2} & 2 & -\frac{1}{2} \\ \frac{1}{2} & -1 & \frac{1}{2} \end{bmatrix}$$

Hence

$$\begin{bmatrix} a_0 \\ a_1 \\ a_2 \end{bmatrix} = P^{-1} \begin{bmatrix} c_1 \\ c_2 \\ c_3 \end{bmatrix} = \begin{bmatrix} 1 & 0 & 0 \\ -\frac{3}{2} & 2 & -\frac{1}{2} \\ \frac{1}{2} & -1 & \frac{1}{2} \end{bmatrix} \begin{bmatrix} c_1 \\ c_2 \\ c_3 \end{bmatrix}$$

This shows how to calculate the coefficients of a polynomial of degree not exceeding 2 from knowledge of its values at 0, 1, and 2:

$$\begin{aligned} a_0 &= c_1 &&= f(0) \\ a_1 &= -\tfrac{3}{2}c_1 + 2c_2 - \tfrac{1}{2}c_3 &&= -\tfrac{3}{2}f(0) + 2f(1) - \tfrac{1}{2}f(2) \\ a_2 &= \tfrac{1}{2}c_1 - c_2 + \tfrac{1}{2}c_3 &&= \tfrac{1}{2}f(0) - f(1) + \tfrac{1}{2}f(2) \end{aligned}$$

EXERCISES

1 Let P be the matrix of formula **5**. How do we know that P has an inverse?

2 **a** Calculate the coordinates c_1 and c_2 of $\bar{u} = (x,y)$ with respect to the basis $\bar{u}_1 = (-1,-1)$ and $\bar{u}_2 = (2,-1)$.

b Draw a careful picture like Figure 3, page 211, for the case when $\bar{u} = (0,3)$ and use it to estimate the values of c_1 and c_2 in this case. Compare your estimate with the true values obtained from part **a**.

3 **a** Show that

$$\bar{u}_1 = (3,1,1) \qquad \bar{u}_2 = (1,-1,1) \qquad \bar{u}_3 = (0,2,1)$$

is a basis for R^3.

b Let P be the matrix whose columns are $\bar{u}_1$, $\bar{u}_2$, and $\bar{u}_3$. Find P^{-1} and use this to express the coordinates $\bar{c}_1, \bar{c}_2, \bar{c}_3$ with respect to $\bar{u}_1, \bar{u}_2, \bar{u}_3$ of a vector $\bar{u} = (x,y,z)$.

c Use the formula of part **b** to find the coordinates of (0,1,4) with respect to $\bar{u}_1, \bar{u}_2, \bar{u}_3$.

4 Let $\bar{u}_1, \bar{u}_2$ be the basis of Exercise **2** and let $\bar{v}_1 = (3,1)$, $\bar{v}_2 = (2,2)$. Express coordinates with respect to $\bar{u}_1$ and $\bar{u}_2$ in terms of coordinates with respect to $\bar{v}_1, \bar{v}_2$ as in formula **8**. Then express coordinates with respect to $\bar{v}_1, \bar{v}_2$ in terms of coordinates with respect to $\bar{u}_1, \bar{u}_2$.

5 **a** Show that

$$\bar{u}_1 = \left(\frac{1}{\sqrt{3}}, \frac{1}{\sqrt{3}}, \frac{1}{\sqrt{3}}\right) \qquad \bar{u}_2 = \left(\frac{1}{\sqrt{2}}, -\frac{1}{\sqrt{2}}, 0\right)$$

$$\bar{u}_3 = \left(\frac{1}{\sqrt{6}}, \frac{1}{\sqrt{6}}, -\frac{2}{\sqrt{6}}\right)$$

is an orthonormal basis for R^3.

b Let P be the matrix whose columns are $\bar{u}_1, \bar{u}_2, \bar{u}_3$. Calculate P^{-1} without using row reduction. (See Example **6**, page 199.)

c Calculate the coordinates of $\bar{u} = (x,y,z)$ with respect to $\bar{u}_1, \bar{u}_2, \bar{u}_3$ by applying formula **5**.

d Calculate the coordinates of $\bar{u} = (x,y,z)$ with respect to $\bar{u}_1, \bar{u}_2, \bar{u}_3$ by projection methods (see statement **2**, page 120) and compare your result with that of part **c**.

6 Let V be the vector space of polynomial functions of degree not exceeding 2 discussed in Example **3**. Let g_1, g_2, g_3 be the polynomial functions defined by

$$g_1(x) = \frac{x(x-1)}{2}$$

$$g_2(x) = \frac{(x+1)(x-1)}{-1}$$

$$g_3(x) = \frac{x(x+1)}{2}$$

a Show that g_1, g_2, g_3 are a basis for V.
b Show that any f in V can be expressed as

$$f(x) = f(-1)g_1(x) + f(0)g_2(x) + f(1)g_3(x)$$

c Express the coefficients of f in terms of its values $f(-1), f(0), f(1)$. (See Example **3**.)
d Express the values $f(-1), f(0), f(1)$ in terms of the values $f(0), f(1), f(2)$. (Hint: Show how coordinates with respect to g_1, g_2, g_3 can be expressed in terms of coordinates with respect to f_1, f_2, f_3.)

7 Let P have the columns $\bar{u}_1, \bar{u}_2$ where these are a basis for R^2. Show that the first column of P^{-1} consists of the coordinates of $(1,0)$ with respect to the basis $\bar{u}_1, \bar{u}_2$, and the second column of P^{-1} consists of the coordinates of $(0,1)$ with respect to the basis $\bar{u}_1, \bar{u}_2$.

8 Let P, Q be as in Example **2**. Show that the first column of $P^{-1}Q$ consists of the coordinates of $\bar{v}_1$ with respect to $\bar{u}_1, \bar{u}_2$ while the second column consists of the coordinates of $\bar{v}_2$ with respect to $\bar{u}_1, \bar{u}_2$.

9 Suppose each of the sets of vectors

$$\{\bar{u}_1,\bar{u}_2\} \qquad \{\bar{v}_1,\bar{v}_2\} \qquad \{\bar{w}_1,\bar{w}_2\}$$

is a basis for R^2. Let P, Q, R be the matrices with these as columns, respectively. Give formulas to express coordinates with respect to each of these bases in terms of coordinates with respect to each of the other bases.

10 Suppose P has the columns $\bar{u}_1, \bar{u}_2$ where these are a basis for R^2. Suppose the basis $\bar{v}_1, \bar{v}_2$ is not known, but it is known that the coordinates d_1, d_2 of an arbitrary vector $\bar{u}$, are known with respect to the basis in terms of coordinates c_1, c_2 with respect to $\bar{u}_1, \bar{u}_2$; that is

$$\begin{bmatrix} d_1 \\ d_2 \end{bmatrix} = A \begin{bmatrix} c_1 \\ c_2 \end{bmatrix}$$

Show how to find $\bar{v}_1$ and $\bar{v}_2$.

□ **11** Find the coordinates of (x,y) with respect to the basis $(i,1)$, $(1,i)$ for C^2. Express your result in the form **5**.

SECTION 5 Determinants

A useful condition for a matrix to be invertible can be given using the determinant. This section sketches some of the relevant properties of the determinant. A complete discussion of determinants can be found elsewhere (see Bibliography).

The second-order determinant is defined by

1
$$\det\begin{bmatrix} a & b \\ c & d \end{bmatrix} = ad - bc$$

Higher-order determinants are defined as follows: Suppose A is a matrix with n rows and n columns. Form the matrix obtained from A by omitting the first row and jth column of A. This matrix has $n - 1$ rows and $n - 1$ columns. Denote its determinant by a_j and denote the entry in the first row and jth column of A by A_{1j}. Then the determinant of A is defined by

2
$$\det A = A_{11}a_1 - A_{12}a_2 + A_{13}a_3 - \cdots - (-1)^{n+1}A_{1n}a_n$$

For example, the third-order determinant is defined by

3
$$\det\begin{bmatrix} A_{11} & A_{12} & A_{13} \\ A_{21} & A_{22} & A_{23} \\ A_{31} & A_{32} & A_{33} \end{bmatrix} = A_{11}\det\begin{bmatrix} A_{22} & A_{23} \\ A_{32} & A_{33} \end{bmatrix} - A_{12}\det\begin{bmatrix} A_{21} & A_{23} \\ A_{31} & A_{33} \end{bmatrix} + A_{13}\det\begin{bmatrix} A_{21} & A_{22} \\ A_{31} & A_{32} \end{bmatrix}$$

The fourth-order determinant is the sum of multiples of third-order determinants, and so forth.

EXAMPLE 1

a $\det\begin{bmatrix} 3 & 1 \\ 2 & -4 \end{bmatrix} = 3 \cdot (-4) - 1 \cdot 2 = -14$

b $$\det\begin{bmatrix} 3 & 1 & 2 \\ 4 & 1 & -6 \\ 1 & 0 & 1 \end{bmatrix} = 3\det\begin{bmatrix} 1 & -6 \\ 0 & 1 \end{bmatrix} - 1\det\begin{bmatrix} 4 & -6 \\ 1 & 1 \end{bmatrix} + 2\det\begin{bmatrix} 4 & 1 \\ 1 & 0 \end{bmatrix}$$

$$= 3[1 \cdot 1 - 0 \cdot (-6)] - 1[4 \cdot 1 - 1 \cdot (-6)] + 2[4 \cdot 0 - 1 \cdot 1] = -9$$

c

$$\det\begin{bmatrix}4 & 1 & 2 & -1\\ 3 & -1 & 2 & 1\\ 6 & -1 & 0 & -1\\ 2 & -1 & 1 & 0\end{bmatrix} = 4\det\begin{bmatrix}-1 & 2 & 1\\ -1 & 0 & -1\\ -1 & 1 & 0\end{bmatrix} - 1\det\begin{bmatrix}3 & 2 & 1\\ 6 & 0 & -1\\ 2 & 1 & 0\end{bmatrix}$$

$$+ 2\det\begin{bmatrix}3 & -1 & 1\\ 6 & -1 & -1\\ 2 & -1 & 0\end{bmatrix} - (-1)\det\begin{bmatrix}3 & -1 & 2\\ 6 & -1 & 0\\ 2 & -1 & 1\end{bmatrix}$$

These four third-order determinants can each be reduced to a sum of multiples of three second-order determinants by using formula **3**. The result is a sum of multiples of twelve second-order determinants, each of which can be evaluated using formula **1**.

DISCUSSION Fortunately, there is an easier way to evaluate high-order determinants by judicious use of row reduction. The basic properties of the determinant used in this process are listed in statement **4**, below. These properties are easily established by direct calculation in the second-order case, and their general proofs are omitted.

4

- **a** If we interchange two rows of A, the determinant changes by a factor of (-1).
- **b** If we add to one row of A a multiple of another row, the determinant is *not* changed.
- **c** If all entries below the diagonal of A are zero, the determinant of A is the product of the diagonal entries of A.

Thus, to find the determinant of A, we apply row operations of the two types given above until we obtain a matrix B such that every entry below the diagonal of B is zero. Then

5 $$\det A = \pm \det B$$

where the sign is positive if we interchanged an even number of rows and negative otherwise. Furthermore,

6 $\det B =$ the product of the diagonal entries of B

EXAMPLE 2 To find $\det A$ where

$$A = \begin{bmatrix}3 & 2 & 0\\ 4 & 4 & 1\\ 1 & -1 & 3\end{bmatrix}$$

apply the following row operations, and use rules **4a** and **b**:

1. Interchange rows one and three.
2. Add to row two -4 times row one, and add to row three -3 times row one.
3. Add to row three $-\frac{5}{8}$ times row two.

We now have the matrix

$$B = \begin{bmatrix} 1 & -1 & 3 \\ 0 & 8 & -11 \\ 0 & 0 & -\frac{17}{8} \end{bmatrix}$$

Using rule **4c** we have

$$\det B = 1 \cdot 8 \cdot (-17/8) = -17$$

Operation **1** multiplies the determinant by -1, while operations **2** and **3** do not change the value. So we have

$$\det A = -\det B = 17$$

DISCUSSION Suppose A has two identical rows. One can then interchange those two rows and the matrix will be unchanged, while, from rule **4a**, its determinant changes by a factor of -1. This can happen only if $\det A = 0$.

If A has a row of zeros, we can add another row to that row. This will not affect the value of the determinant (rule **4b**), and the new matrix will have two rows that are identical. We conclude from the above argument that $\det A$ must be zero. In summary,

7 If two rows of A are identical, or if A has a row of zeros, $\det A = 0$.

It follows that if one row of A is a linear combination of the other rows of A, repeated application of rule **4b** to that row will result in a row of zeros. In other words,

8 If the rows of A are dependent, then $\det A = 0$.

The condition that the rows be dependent is (Theorem **11**, page 188, and Theorem **3**, page 71) the same as the condition that A does not have an inverse, and thus, property **8** can be rephrased as:

9 If A is *not* invertible, then $\det A = 0$.

The converse of this result can be established by using a row-reduction argument. It is also a simple consequence of the following result:

10 $\det AB = \det A \det B$

This is a consequence of rules **4**. We omit its proof. We shall use it to prove that

11 If A is invertible, then $\det A \neq 0$.

To prove this, suppose that A is invertible. Then $A^{-1}A = I$. Hence from equation **10** we have

$$\det I = \det A^{-1}A = \det A^{-1} \det A$$

Since $\det I = 1$ (from rule **4c**), we cannot have $\det A = 0$. This proves statement **11**. In summary, we have the following useful result:

THEOREM 14 A has an inverse if and only if $\det A \neq 0$.

EXAMPLE 3 The above results often make it easy to calculate determinants and to determine invertibility, as indicated by the following examples.

a $$\det\begin{bmatrix} 2 & 1 & 1 \\ 0 & 0 & 0 \\ 4 & 3 & 1 \end{bmatrix} = 0 \qquad \text{(from property 7)}$$

b $$\det\begin{bmatrix} 3 & 0 & 0 \\ 2 & 1 & 0 \\ 1 & 2 & 0 \end{bmatrix} = 0$$

This follows from the fact that the columns of this matrix are dependent (since a set of vectors which includes the zero vector must be dependent); the matrix, therefore, is not invertible.

c $$\det\begin{bmatrix} 4 & 1 & 2 \\ 8 & 2 & 4 \\ 3 & 1 & 0 \end{bmatrix} = 0$$ since the second row is a multiple of the first row

d $$\det\begin{bmatrix} 2 & 1 \\ 3 & 5 \end{bmatrix} = 10 - 3 = 7 \neq 0$$

Therefore, this matrix is invertible.

e To find the determinant of

$$A = \begin{bmatrix} 3 & 1 & 2 \\ 2 & 6 & 4 \\ -2 & 1 & 3 \end{bmatrix}$$

we need only eliminate in the first column and then use formula **3**. Adding suitable multiples of the first row to the other two rows gives the matrix

$$\begin{bmatrix} 3 & 1 & 2 \\ 0 & \frac{16}{3} & \frac{8}{3} \\ 0 & \frac{5}{3} & \frac{13}{3} \end{bmatrix}$$

whose determinant is (using formula **3**):

$$3 \det \begin{bmatrix} \frac{16}{3} & \frac{8}{3} \\ \frac{5}{3} & \frac{13}{3} \end{bmatrix} - 1 \det \begin{bmatrix} 0 & \frac{8}{3} \\ 0 & \frac{13}{3} \end{bmatrix} + 2 \det \begin{bmatrix} 0 & \frac{16}{3} \\ 0 & \frac{5}{3} \end{bmatrix}$$

Since the last two determinants are zero (from Theorem **14**), it follows that

$$\det A = 3 \det \begin{bmatrix} \frac{16}{3} & \frac{8}{3} \\ \frac{5}{3} & \frac{13}{3} \end{bmatrix} = 3(\tfrac{16}{3} \cdot \tfrac{13}{3} - \tfrac{8}{3} \cdot \tfrac{5}{3}) = 56$$

We conclude, in particular, that A is invertible.

f To find

$$\det \begin{bmatrix} 2 & 4 & 2 & 2 \\ 3 & 1 & 0 & 1 \\ 2 & 1 & 0 & -1 \\ 4 & 2 & 1 & 1 \end{bmatrix}$$

eliminate in the first column to obtain

$$\begin{bmatrix} 2 & 4 & 2 & 2 \\ 0 & -5 & -3 & -2 \\ 0 & -3 & -2 & -3 \\ 0 & -6 & -3 & -3 \end{bmatrix}$$

An argument similar to that given in **e** shows that our desired determinant is

$$2 \det \begin{bmatrix} -5 & -3 & -2 \\ -3 & -2 & -3 \\ -6 & -3 & -3 \end{bmatrix}$$

We can now proceed to eliminate in the first column of this matrix.

EXAMPLE 4 We shall later wish to find det $(\lambda I - A)$. For example, if

$$A = \begin{bmatrix} a & b \\ c & d \end{bmatrix} \quad \text{then} \quad \lambda I - A = \begin{bmatrix} \lambda - a & -b \\ -c & \lambda - d \end{bmatrix}$$

so that

$$\det(\lambda I - A) = (\lambda - a)(\lambda - d) - (-b)(-c) = \lambda^2 - (a + d)\lambda + ad - bc$$

Notice that this is a polynomial of degree 2. It can be shown, using induction and formula **2**, that

12 det $(\lambda I - A)$ is a polynomial of degree n if A has n rows and n columns.

For example,

$$\det\left\{\lambda\begin{bmatrix} 1 & 0 & 0 \\ 0 & 1 & 0 \\ 0 & 0 & 1 \end{bmatrix} - \begin{bmatrix} A_{11} & A_{12} & A_{13} \\ A_{21} & A_{22} & A_{23} \\ A_{31} & A_{32} & A_{33} \end{bmatrix}\right\} = \det\begin{bmatrix} \lambda - A_{11} & -A_{12} & -A_{13} \\ -A_{21} & \lambda - A_{22} & -A_{23} \\ -A_{31} & -A_{32} & \lambda - A_{33} \end{bmatrix}$$

$$= (\lambda - A_{11}) \det\begin{bmatrix} \lambda - A_{22} & -A_{23} \\ -A_{32} & \lambda - A_{33} \end{bmatrix} - (-A_{12}) \det\begin{bmatrix} -A_{21} & -A_{23} \\ -A_{31} & \lambda - A_{33} \end{bmatrix}$$

$$+ (-A_{13}) \det\begin{bmatrix} -A_{21} & \lambda - A_{22} \\ -A_{31} & -A_{32} \end{bmatrix}$$

Since the first term has degree 3, while the degree of the last two terms cannot exceed degree 1, we conclude that det $(\lambda I - A)$ has degree 3.

EXAMPLE 5 □ *The Wronskian.*
The *Wronskian* of two functions at $x = x_0$ is defined by

$$w(f,g,x_0) = \det\begin{bmatrix} f(x_0) & g(x_0) \\ f'(x_0) & g'(x_0) \end{bmatrix}$$

If f and g are dependent, there are numbers c_1 and c_2 not both zero such that

$$c_1 f + c_2 g = 0$$

Differentiating this gives

$$c_1 f' + c_2 g' = 0$$

so that, in particular,

$$c_1\begin{bmatrix} f(x_0) \\ f'(x_0) \end{bmatrix} + c_2\begin{bmatrix} g(x_0) \\ g'(x_0) \end{bmatrix} = \begin{bmatrix} 0 \\ 0 \end{bmatrix}$$

Since c_1 and c_2 are not both zero, Theorem **14** gives the following result:

13 If f and g are dependent, then for each x_0, $w(f,g,x_0) = 0$.

In general, the converse of this is *not* true.

The Wronskian for more than two functions is defined analogously. For example,

$$w(f,g,h,x_0) = \det\begin{bmatrix} f(x_0) & g(x_0) & h(x_0) \\ f'(x_0) & g'(x_0) & h'(x_0) \\ f''(x_0) & g''(x_0) & h''(x_0) \end{bmatrix}$$

EXAMPLE 6 ☐ *Area and Volume.*

The second-order determinant has a natural interpretation in terms of area, while the third-order determinant is related to volume. For example, suppose $\bar{u} = a\bar{i} + b\bar{j}$ and $\bar{v} = c\bar{i} + d\bar{j}$. Then $\bar{u}$ and $\bar{v}$ determine a parallelogram, as shown in Figure 4.

As Figure 5 shows, the vector $\bar{w}$, which is the projection of $\bar{v}$ orthogonal

Figure 4

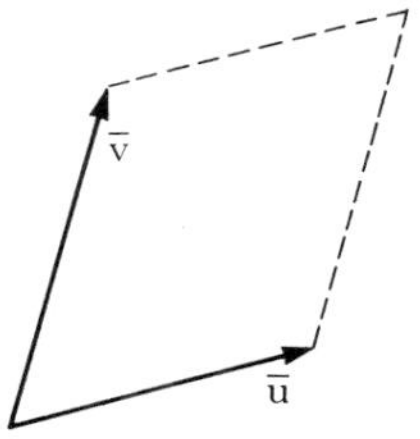

Figure 5

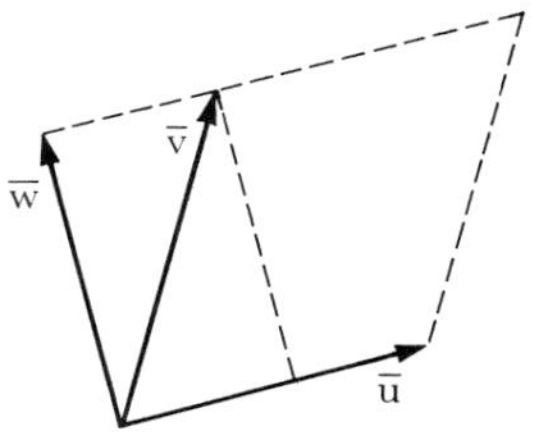

to $\bar{u}$, has length equal to the altitude of this parallelogram. Thus the area of the parallelogram is just $|\bar{u}|\,|\bar{w}|$. We recall that

$$\bar{w} = \bar{v} - \frac{(\bar{v} \cdot \bar{u})\bar{u}}{\bar{u} \cdot \bar{u}}$$

as established in formula **7**, page 109.

Since $\bar{w}$ is orthogonal to $\bar{u}$, we have

$$\begin{aligned}\bar{w} \cdot \bar{w} &= \left(\bar{v} - \frac{\bar{v} \cdot \bar{u}}{\bar{u} \cdot \bar{u}}\bar{u}\right) \cdot \left(\bar{v} - \frac{\bar{v} \cdot \bar{u}}{\bar{u} \cdot \bar{u}}\bar{u}\right) \\ &= \bar{v} \cdot \bar{v} - \frac{\bar{v} \cdot \bar{u}}{\bar{u} \cdot \bar{u}}(\bar{u} \cdot \bar{v}) \\ &= \frac{(\bar{v} \cdot \bar{v})(\bar{u} \cdot \bar{u}) - (\bar{v} \cdot \bar{u})^2}{\bar{u} \cdot \bar{u}}\end{aligned}$$

so that

$$|\bar{w}|^2|\bar{u}|^2 = |\bar{v}|^2|\bar{u}|^2 - (\bar{v} \cdot \bar{u})^2$$

is the square of the area of the parallelogram of Figure 4. With $\bar{u} = a\bar{i} + b\bar{j}$ and $\bar{v} = c\bar{i} + d\bar{j}$ we have

$$\begin{aligned}|\bar{w}|^2|\bar{u}|^2 &= (c^2 + d^2)(a^2 + b^2) - (ac + bd)^2 \\ &= a^2d^2 + b^2c^2 - 2acbd \\ &= (ad - bc)^2\end{aligned}$$

We conclude that

> The area of the parallelogram determined by $\bar{u} = a\bar{i} + b\bar{j}$ and $\bar{v} = c\bar{i} + d\bar{j}$ is the absolute value of
>
> $$\det\begin{bmatrix} a & b \\ c & d \end{bmatrix}.$$

A similar argument shows that

> The volume of the parallelopiped determined by $a_1\bar{i} + b_1\bar{j} + c_1\bar{k}$, $a_2\bar{i} + b_2\bar{j} + c_2\bar{k}$, and $a_3\bar{i} + b_3\bar{j} + c_3\bar{k}$ is
>
> $$\left|\det\begin{bmatrix} a_1 & b_1 & c_1 \\ a_2 & b_2 & c_2 \\ a_3 & b_3 & c_3 \end{bmatrix}\right|.$$

EXAMPLE 7 ☐ *The Determinant of the Transpose.*

Rules **4a** and **b** are also valid for column operations. This fact can be established directly for second-order determinants and then proved for higher-order determinants using formulas **2** and **3**. Thus properties **7** and **8** also hold for columns. We can use this information to prove directly that

14 $$\det A = \det A^t$$

Let us give an alternative proof of this which uses Theorem **14** and the results of Example **10**, page 176. Suppose A^t is not invertible, so that A is also not invertible. Thus Theorem **14** implies that both sides of equation **14** are zero. To complete the proof, we need only establish equality when A^t is invertible. Suppose A^t is invertible. Then we can apply a sequence of row operations to reduce A^t to I. As shown in Example **10**, each such row operation can be performed by multiplying on the left of A^t by a suitable matrix. Thus, we can find a sequence of matrices $E_1, E_2, \ldots, E_k$ of the types given in Example **10** such that

15 $$E_1E_2 \cdots E_kA^t = I.$$

Applying formula **10** we conclude that

$$\det A^t = \frac{1}{\det E_1 \det E_2 \cdots \det E_k}$$

Suppose E_i is obtained from I by interchanging two rows of I. In this case $E_i^t = E_i$. For example, if

$$E_i = \begin{bmatrix} 1 & 0 & 0 \\ 0 & 0 & 1 \\ 0 & 1 & 0 \end{bmatrix} \quad \text{then} \quad E_i^t = \begin{bmatrix} 1 & 0 & 0 \\ 0 & 0 & 1 \\ 0 & 1 & 0 \end{bmatrix}$$

Thus, in this case, $\det E_i^t = \det E_i$.

If E_i is obtained from I by multiplying one row of I by $a \neq 0$, we again have $E_i^t = E_i$. For example, if

$$E_i = \begin{bmatrix} 1 & 0 & 0 \\ 0 & a & 0 \\ 0 & 0 & 1 \end{bmatrix} \quad \text{then} \quad E_i^t = \begin{bmatrix} 1 & 0 & 0 \\ 0 & a & 0 \\ 0 & 0 & 1 \end{bmatrix}$$

Thus, again we have $\det E_i^t = \det E_i$.

Suppose E_i is obtained from I by adding to row s of I, b times row r. Then E_i^t is obtained from I by adding to row r of I, b times row s. For example, if

$$E_i = \begin{bmatrix} 1 & 0 & b \\ 0 & 1 & 0 \\ 0 & 0 & 1 \end{bmatrix} \quad \text{then} \quad E_i^t = \begin{bmatrix} 1 & 0 & 0 \\ 0 & 1 & 0 \\ b & 0 & 1 \end{bmatrix}$$

In this case since $\det E_i = 1$ and $\det E_i^t = 1$ (from rule **4b**) we again have $\det E_i = \det E_i^t$.

In summary, we have

17 $$\det E_1 \det E_2 \cdots \det E_k = \det E_1^t \det E_2^t \cdots \det E_k^t$$

Taking transposes of equation **15** and using formulas **1**, **2**, and **3** of the previous section, we conclude that

$$AE_k^t \cdots E_2^t E_1^t = I$$

Formula **10** therefore gives

$$\det A = \frac{1}{\det E_k^t \cdots \det E_2^t \det E_1^t}$$

Comparing this with equations **16** and **17** we conclude that equation **14** is true when A^t is invertible.

EXAMPLE 8 □ *The Classical Adjoint.*

Suppose B_{ij} is the determinant of the matrix obtained from A by omitting the ith row and jth column of A. Denote the entry in the ith row and jth column of A by A_{ij}. It can then be shown that, for each j, $1 \le j \le n$

18 $$\det A = (-1)^{1+j}A_{1j}B_{1j} + (-1)^{2+j}A_{2j}B_{2j} + \cdots + (-1)^{n+j}A_{nj}B_{nj}$$

Put

$$C = \begin{bmatrix} B_{11} & -B_{21} & B_{31} & \cdots & (-1)^{n+1}B_{n1} \\ -B_{12} & B_{22} & -B_{32} & \cdots & (-1)^{n+2}B_{n2} \\ \vdots & \vdots & \vdots & \cdots & \vdots \\ (-1)^{n+1}B_{1n} & (-1)^{n+2}B_{2n} & \vdots & \cdots & B_{nn} \end{bmatrix}$$

The use of formula **18** for $j = 1, 2, \ldots, n$ thus gives the formula

19 $$CA = (\det A)I$$

The numbers B_{ij} are called the *minors* of A, while the numbers $(-1)^{i+j}B_{ij}$ are called *cofactors*. The matrix C is called the *classical adjoint* of A, while formula **18** is called the *Laplace expansion of the determinant of A along the* jth *column*. There is also a Laplace expansion along the ith row, namely

$$(-1)^{i+1}A_{i1}B_{i1} + (-1)^{i+2}A_{i2}B_{i2} + \cdots + (-1)^{i+n}A_{in}B_{in}$$

which can be shown to be equal to det A.

In the second-order case

$$C = \begin{bmatrix} d & -b \\ -c & a \end{bmatrix}$$

so equation **19** gives the formula (easily verified directly):

$$\begin{bmatrix} d & -b \\ -c & a \end{bmatrix} \begin{bmatrix} a & b \\ c & d \end{bmatrix} = [ad - bc] \begin{bmatrix} 1 & 0 \\ 0 & 1 \end{bmatrix}$$

Suppose

$$A = \begin{bmatrix} 3 & 1 & 2 \\ 0 & 1 & 1 \\ -1 & 1 & 0 \end{bmatrix}$$

Since

$$B_{11} = \det \begin{bmatrix} 1 & 1 \\ 1 & 0 \end{bmatrix} = -1, B_{12} = \det \begin{bmatrix} 0 & 1 \\ -1 & 0 \end{bmatrix} = 1, B_{13} = \det \begin{bmatrix} 0 & 1 \\ -1 & 1 \end{bmatrix} = 1$$

$$B_{21} = \det \begin{bmatrix} 1 & 2 \\ 1 & 0 \end{bmatrix} = -2, B_{22} = \det \begin{bmatrix} 3 & 2 \\ -1 & 0 \end{bmatrix} = 2, B_{23} = \det \begin{bmatrix} 3 & 1 \\ -1 & 1 \end{bmatrix} = 4$$

$$B_{31} = \det \begin{bmatrix} 1 & 2 \\ 1 & 1 \end{bmatrix} = -1, B_{32} = \det \begin{bmatrix} 3 & 2 \\ 0 & 1 \end{bmatrix} = 3, B_{33} = \det \begin{bmatrix} 3 & 1 \\ 0 & 1 \end{bmatrix} = 3$$

the classical adjoint is

$$C = \begin{bmatrix} -1 & 2 & -1 \\ -1 & 2 & -3 \\ 1 & -4 & 3 \end{bmatrix}$$

The student can check that $CA = (\det A)I$.

EXAMPLE 9 □ *Cramer's Rule.*

Formula **19** gives us a way to calculate A^{-1}, for if $\det A \neq 0$, it follows that $(1/\det A)CA = I$ where C is the classical adjoint. Therefore

$$A^{-1} = \frac{1}{\det A} C$$

In this case, $A\bar{u} = \bar{v}$ has the unique solution

$$\bar{u} = A^{-1}\bar{v} = \frac{1}{\det A} C\bar{v}$$

Suppose $\bar{v} = (y_1, y_2, \ldots, y_n)$. The formula for C then gives

$$20 \quad C\bar{v} = \begin{bmatrix} B_{11}y_1 - B_{21}y_2 + B_{31}y_3 - \cdots + (-1)^{n+1}B_{n1}y_n \\ -B_{12}y_1 + B_{22}y_2 - B_{32}y_3 - \cdots + (-1)^{n+2}B_{n2}y_n \\ \cdots\cdots\cdots\cdots\cdots\cdots\cdots\cdots\cdots \\ (-1)^{n+1}B_{1n}y_1 + (-1)^{n+2}B_{2n}y_n \cdots + \cdots + B_{nn}y_n \end{bmatrix}$$

The first row is just equal to (from formula **18**)

$$\det \begin{bmatrix} y_1 & A_{12} & A_{13} & \cdots & A_{1n} \\ y_2 & A_{22} & A_{23} & \cdots & A_{2n} \\ \cdots & \cdots & \cdots & \cdots & \cdots \\ y_n & A_{n2} & A_{n3} & \cdots & A_{nn} \end{bmatrix}$$

In general, each row of matrix **20** is just the determinant of the matrix that is obtained by replacing the corresponding column of A by $\bar{v}$. This gives the following procedure when $\det A \neq 0$, known as *Cramer's rule*, for solving $A\bar{u} = \bar{v}$:

> Each coordinate of $\bar{u}$ is $(1/\det A)$ times the determinant of the matrix obtained by replacing the corresponding column of A by $\bar{v}$.

This method of solution is quite useful in the second-order case, but requires more calculation than the elimination method in high-order cases.

Consider the equation

$$\begin{bmatrix} 3 & 1 \\ 2 & 4 \end{bmatrix} \begin{bmatrix} x \\ y \end{bmatrix} = \begin{bmatrix} -1 \\ 6 \end{bmatrix}$$

Cramer's rule gives

$$x = \frac{\det \begin{bmatrix} -1 & 1 \\ 6 & 4 \end{bmatrix}}{\det \begin{bmatrix} 3 & 1 \\ 2 & 4 \end{bmatrix}} = \frac{-10}{10} = -1 \qquad y = \frac{\det \begin{bmatrix} 3 & -1 \\ 2 & 6 \end{bmatrix}}{\det \begin{bmatrix} 3 & 1 \\ 2 & 4 \end{bmatrix}} = \frac{20}{10} = 2$$

EXERCISES

1 Use formula **2** to find the determinant of

$$\begin{bmatrix} 1 & 2 & 3 & 1 \\ 1 & -1 & 1 & 4 \\ 2 & 1 & 0 & 1 \\ 1 & 1 & 2 & 4 \end{bmatrix}$$

2 Find the determinant of the matrix of Exercise **1** by using rules **4** as in Example **2**.

3 Find the determinants of each of the following matrices. (Hint: First see whether property **7**, **8**, or **9** is true.)

a $\begin{bmatrix} 2 & 1 \\ 1 & 2 \end{bmatrix}$ **b** $\begin{bmatrix} 2 & 1 \\ 4 & 2 \end{bmatrix}$ **c** $\begin{bmatrix} 1 & 0 \\ 3 & 0 \end{bmatrix}$

d $\begin{bmatrix} 3 & 1 & 2 \\ 1 & 0 & 6 \\ -1 & 1 & 1 \end{bmatrix}$ **e** $\begin{bmatrix} -1 & 1 & 3 \\ 2 & 1 & 1 \\ 4 & 2 & 2 \end{bmatrix}$ **f** $\begin{bmatrix} 1 & 1 & 2 \\ 4 & 3 & 6 \\ 1 & -1 & -2 \end{bmatrix}$

g $\begin{bmatrix} 1 & 1 & 2 & 1 \\ -1 & 1 & 0 & 1 \\ 2 & 1 & 1 & 0 \\ 1 & 3 & 1 & 0 \end{bmatrix}$ **h** $\begin{bmatrix} 4 & 1 & 2 & 6 \\ 0 & -1 & -2 & 1 \\ 1 & 2 & 4 & 1 \\ 0 & 6 & 12 & 0 \end{bmatrix}$ **i** $\begin{bmatrix} 3 & 1 & 0 & 6 & 1 \\ 1 & 2 & 1 & 3 & 1 \\ -1 & 1 & 1 & 1 & 1 \\ 1 & -1 & 1 & 1 & 1 \\ 0 & 0 & 1 & 1 & 1 \end{bmatrix}$

4 Which of the matrices of Exercise **3** are invertible? (Hint: Use Theorem **14**.)

5 For $A = \begin{bmatrix} 1 & 2 \\ 4 & 1 \end{bmatrix}$, verify that

$$\det \begin{bmatrix} 4 & 1 \\ 1 & 2 \end{bmatrix} = -\det A \quad \text{and that} \quad \det \begin{bmatrix} 1 + 3 \cdot 4 & 2 + 3 \cdot 1 \\ 4 & 1 \end{bmatrix} = \det A$$

6 For $A = \begin{bmatrix} 1 & 2 \\ 4 & 1 \end{bmatrix}$ and $B = \begin{bmatrix} -1 & 3 \\ 1 & 0 \end{bmatrix}$, find AB, det A, det B, det AB and verify equation **10**.

7 Show that if A is invertible, then det $A^{-1} = 1/(\det A)$.

8 Show that det $(PAP^{-1}) = \det A$.

9 For each of the following matrices A find $\det(\lambda I - A)$.

a $\begin{bmatrix} 3 & 1 \\ -1 & 2 \end{bmatrix}$ **b** $\begin{bmatrix} 2 & 1 & 0 \\ 1 & -1 & 1 \\ 1 & 2 & 1 \end{bmatrix}$

10 Derive the following formulas:

a If $A = \begin{bmatrix} a & b \\ c & d \end{bmatrix}$ then $\det(\lambda I - A) = \lambda^2 - (a + d)\lambda + ad - bc$.

b If $A = \begin{bmatrix} A_{11} & A_{12} & A_{13} \\ A_{21} & A_{22} & A_{23} \\ A_{31} & A_{32} & A_{33} \end{bmatrix}$ then

$$\det(\lambda I - A) = \lambda^3 - (A_{11} + A_{22} + A_{33})\lambda^2 + \left\{\det\begin{bmatrix} A_{11} & A_{12} \\ A_{21} & A_{22} \end{bmatrix} + \det\begin{bmatrix} A_{11} & A_{13} \\ A_{31} & A_{33} \end{bmatrix} + \det\begin{bmatrix} A_{22} & A_{23} \\ A_{32} & A_{33} \end{bmatrix}\right\}\lambda - \det A$$

Notice that the coefficient of λ^2 is the negative of the sum of the determinants obtained by successively omitting two rows and the corresponding two columns, while the coefficient of λ is the sum of the determinants obtained by successively omitting one row and its corresponding column.

11 Use the information of Exercise **10b** to make a guess about a formula for $\det(\lambda I - A)$, where A has four rows. Prove your conjecture.

□ **12** Find $\det A$ where $A = \begin{bmatrix} i & 1 \\ 2 + i & -i \end{bmatrix}$. Is A invertible?

□ **13** Find the Wronskian of e^x, e^{2x}, and e^{3x} at $x = 0$.

□ **14** **a** Find the Wronskian of 1, $\cos 2x$, and $\sin^2 x$ at $x = 0$. Are these functions dependent?

b Find the Wronskian of $\sin x$ and $\sin 2x$ at $x = 0$. Are these functions dependent?

□ **15** For $A = \begin{bmatrix} a & b \\ c & d \end{bmatrix}$ verify that $\det A = \det A^t$.

□ **16** For $A = \begin{bmatrix} 2 & 1 \\ 1 & 1 \end{bmatrix}$, $E_4 = \begin{bmatrix} 0 & 1 \\ 1 & 0 \end{bmatrix}$, $E_3 = \begin{bmatrix} 1 & 0 \\ -2 & 1 \end{bmatrix}$, $E_2 = \begin{bmatrix} 1 & 0 \\ 0 & -1 \end{bmatrix}$,

$E_1 = \begin{bmatrix} 1 & -1 \\ 0 & 1 \end{bmatrix}$ verify that $E_1E_2E_3E_4A = I$, and that therefore the row operations given by these matrices reduce A to I. Show that for each i, $\det E_i = \det E_i^t$. Show that $A^tE_4^tE_3^tE_2^tE_1^t = I$. Show also that

$$\det A = \frac{1}{\det E_1 \det E_2 \det E_3 \det E_4} = \det A^t$$

□ **17** Find the classical adjoint of $\begin{bmatrix} 1 & 1 & 2 \\ 4 & 3 & 6 \\ 1 & -1 & -2 \end{bmatrix}$ and verify equation **19** in this case.

□ **18** Solve each of the following by using Cramer's rule.

a $\begin{bmatrix} 2 & 1 \\ 1 & 2 \end{bmatrix} \begin{bmatrix} x \\ y \end{bmatrix} = \begin{bmatrix} 1 \\ -3 \end{bmatrix}$ **b** $\begin{bmatrix} -3 & 0 \\ 1 & 4 \end{bmatrix} \begin{bmatrix} x \\ y \end{bmatrix} = \begin{bmatrix} 6 \\ 0 \end{bmatrix}$

c $\begin{bmatrix} 1 & 1 & 2 \\ 4 & 3 & 6 \\ 1 & -1 & 2 \end{bmatrix} \begin{bmatrix} x \\ y \\ z \end{bmatrix} = \begin{bmatrix} -1 \\ 2 \\ -1 \end{bmatrix}$ **d** $\begin{aligned} 3x - y + z &= 1 \\ x - 2y + z &= 2 \\ 2x + y + 3z &= 0 \end{aligned}$

Supplement on Transformations and the Determinant

ORTHOGONAL TRANSFORMATIONS

A linear transformation T from R^n into R^n is called an *orthogonal transformation* if its matrix A is an orthogonal matrix, that is, if $A^{-1} = A^t$. (See Example **6**, page 199.) If we let T^* denote the transformation whose matrix is A^t, then we can write $T\bar{u} = A\bar{u}$ and $T^*\bar{u} = A^t\bar{u}$ for all $\bar{u}$ in R^n. The condition that $A^t = A^{-1}$ is then the same as the condition that $T^* = T^{-1}$.

Suppose T is a counterclockwise rotation in R^2 through the angle θ. The matrices of T and T^* are thus respectively given by (see Example **1**, page 203).

$$A = \begin{bmatrix} \cos\theta & -\sin\theta \\ \sin\theta & \cos\theta \end{bmatrix} \quad \text{and} \quad A^t = \begin{bmatrix} \cos\theta & \sin\theta \\ -\sin\theta & \cos\theta \end{bmatrix}$$

Since $\cos\theta = \cos(-\theta)$ and $\sin\theta = -\sin(-\theta)$, we see that A^t is the matrix of a counterclockwise rotation through the angle $-\theta$. That is, A^t is the matrix of T^{-1}. We therefore know that

1 A rotation in R^2 is an orthogonal transformation.

Suppose T is the linear transformation defined for $\bar{u}$ in R^2 by $T\bar{u} = A\bar{u}$ where

$$A = \begin{bmatrix} 0 & 1 \\ 1 & 0 \end{bmatrix}$$

Note that T is a reflection in the line $y = x$.

Since $A^t = A$ and $A^2 = I$ we see that A is an orthogonal matrix, that is,

> Reflection in the line $y = x$ is an orthogonal transformation.

We have

$$\det\begin{bmatrix} 0 & 1 \\ 1 & 0 \end{bmatrix} = -1 \qquad \det\begin{bmatrix} \cos\theta & -\sin\theta \\ \sin\theta & \cos\theta \end{bmatrix} = \cos^2\theta + \sin^2\theta = 1$$

so that

2 A rotation has determinant 1 and reflection in the line $y = x$ has determinant -1.

Let us show that

3 The determinant of an orthogonal matrix is ± 1,

from which we shall be able to deduce a partial converse to the result **1**. Suppose $A^t = A^{-1}$ so that $A^tA = I$. We know that $\det I = 1$, $\det A = \det A^t$ (see Example **7**, page 224) and $\det(A^tA) = \det A^t \det A$ (see equation **10**, page 219). Thus if $A^tA = I$, we must have $(\det A)^2 = 1$, and statement **3** is therefore true.

We now wish to prove that

4 An orthogonal transformation of R^2 of determinant 1 is a rotation.

Suppose T is an orthogonal transformation of determinant 1. Put

$$\bar{u}_1 = T\begin{bmatrix} 1 \\ 0 \end{bmatrix} = T\bar{i}, \qquad \bar{u}_2 = T\begin{bmatrix} 0 \\ 1 \end{bmatrix} = T\bar{j}$$

These are the columns of the matrix of T; and hence, since this matrix is orthogonal:

$$\bar{u}_1 \cdot \bar{u}_1 = \bar{u}_2 \cdot \bar{u}_2 = 1 \qquad \bar{u}_1 \cdot \bar{u}_2 = 0$$

Suppose $\bar{u}_1$ is as shown in Figure 6. There are only the two possibilities for $\bar{u}_2$ shown in Figure 6, for $\bar{u}_2$ must be orthogonal to $\bar{u}_1$ and have the same length as $\bar{u}_1$.

Figure 6

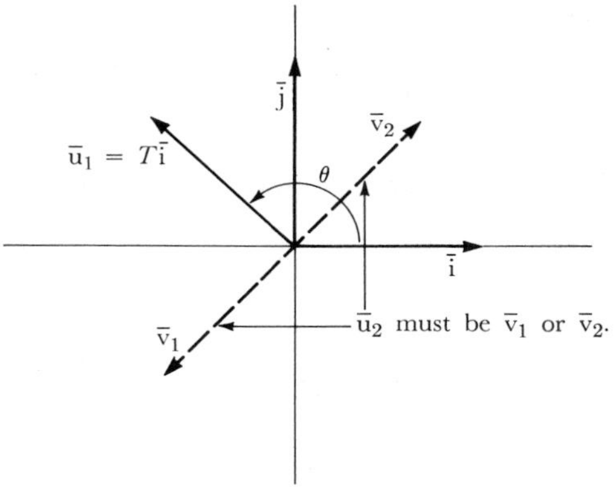

Suppose θ is the angle from $\bar{i}$ to $\bar{u}_1 = T\bar{i}$ so that (in column form)

$$\bar{u}_1 = \begin{bmatrix} \cos\theta \\ \sin\theta \end{bmatrix}$$

The vectors $\bar{v}_1$ and $\bar{v}_2$ are then given by

$$\bar{v}_1 = \begin{bmatrix} -\sin\theta \\ \cos\theta \end{bmatrix} \qquad \bar{v}_2 = \begin{bmatrix} \sin\theta \\ -\cos\theta \end{bmatrix}$$

(as seen from Figure 6).

Since

$$\det\begin{bmatrix} \cos\theta & \sin\theta \\ \sin\theta & -\cos\theta \end{bmatrix} = -(\cos^2\theta + \sin^2\theta) = -1$$

we cannot have $\bar{u}_2 = \bar{v}_2$. Thus we must have $\bar{u}_2 = \bar{v}_1$, which shows that the matrix of T must be

$$\begin{bmatrix} \cos\theta & -\sin\theta \\ \sin\theta & \cos\theta \end{bmatrix}$$

that is, T is a counterclockwise rotation through the angle θ. Statement **4** is therefore proved. A characterization of orthogonal transformations of R^2 of determinant -1 is given in the exercises. There are references to characterizations of orthogonal transformations in R^3 in the bibliography.

AREA AND VOLUME

Suppose $\bar{u} = (x_1,y_1)$, $\bar{v} = (x_2,y_2)$, and

$$B = \begin{bmatrix} x_1 & x_2 \\ y_1 & y_2 \end{bmatrix}$$

The area of the parallelogram determined by $\bar{u}$ and $\bar{v}$ is $|\det B|$. (See Example **6**, page 223.)

Suppose T has the matrix

$$A = \begin{bmatrix} a & b \\ c & d \end{bmatrix}$$

so that

$$T\bar{u} = \begin{bmatrix} ax_1 + by_1 \\ cx_1 + dy_1 \end{bmatrix} \qquad T\bar{v} = \begin{bmatrix} ax_2 + by_2 \\ cx_2 + dy_2 \end{bmatrix}$$

Since

$$\begin{bmatrix} ax_1 + by_1 & ax_2 + by_2 \\ cx_1 + dy_1 & cx_2 + dy_2 \end{bmatrix} = AB$$

we see that the area of the parallelogram determined by $T\bar{u}$ and $T\bar{v}$ is $|\det AB|$.

Formula **10**, page 219, gives

$$\det AB = \det A \det B$$

so we know that

5 The ratio of the area of the parallelogram determined by $T\bar{u}$ and $T\bar{v}$ to the area of the parallelogram determined by $\bar{u}$ and $\bar{v}$ is $\det A$, where A is the matrix of T.

A corresponding result holds for ratios of the volumes of the parallelopipeds determined by $\bar{u}$, $\bar{v}$, and $\bar{w}$ and by $T\bar{u}$, $T\bar{v}$, and $T\bar{w}$.

EXERCISES

1 Suppose $\bar{u}$ and $\bar{v}$ are orthonormal vectors in R^2. Show that there is an orthogonal linear transformation T from R^2 into R^2 such that $T\bar{i} = \bar{u}$ and $T\bar{j} = \bar{v}$.

2 Suppose $\bar{u}_1$, $\bar{u}_2$ and $\bar{v}_1$, $\bar{v}_2$ are each orthonormal sets in R^2. Let A be the matrix with columns $\bar{u}_1$ and $\bar{u}_2$ and B the matrix with columns $\bar{v}_1$ and $\bar{v}_2$. Suppose $\det A = \det B$. Show that there is a rotation T such that $T\bar{u}_1 = \bar{v}_1$ and $T\bar{u}_2 = \bar{v}_2$. (Hint: Use Exercise **1** to find orthogonal transformations T_1 and T_2 such that $T_1\bar{i} = \bar{u}_1$, $T_1\bar{j} = \bar{u}_2$, $T_2\bar{i} = \bar{v}_1$, and $T_2\bar{j} = \bar{v}_2$. Then show that $T = T_2T_1^{-1}$ is a rotation.)

3 Suppose B is an orthogonal matrix of determinant -1 and that $A = \begin{bmatrix} 0 & 1 \\ 1 & 0 \end{bmatrix}$. Put $C = AB$.

a What is $\det C$?

b How is C related to B?

c Show that $B = AC$.

d Use these facts to prove that an orthogonal transformation of R^2 of determinant -1 must be a rotation followed by a reflection in the line $y = x$.

e Replace A by $A = \begin{bmatrix} -1 & 0 \\ 0 & 1 \end{bmatrix}$ and use similar arguments to show that an orthogonal transformation of R^2 of determinant -1 must be a rotation followed by a reflection in the y-axis.

f Suppose the matrix of T is

$$\begin{bmatrix} \sqrt{3}/2 & 1/2 \\ 1/2 & -\sqrt{3}/2 \end{bmatrix}$$

Express T as a rotation followed by a reflection in the line $y = x$. Also express T as a rotation followed by a reflection in the y-axis.

4 Suppose T is a reflection in R^2 in the line through $\bar{w}$. Show that T is an orthogonal transformation of determinant -1.

5 Show that an orthogonal transformation in R^2 preserves the area of parallelograms.

6 Show that a rotation preserves the lengths of vectors and the angles between vectors. (Hint: Use the fact that $A\bar{u} \cdot A\bar{v} = \bar{u} \cdot A^tA\bar{v}$ and that $A^t = A^{-1}$ if A is the matrix of a rotation.)

7 Suppose $\det A = 1$. Show that if $T\bar{u} = A\bar{u}$, then T preserves areas of parallelograms.

8 Show that rotations preserve area. What about reflections? Projections?

9 Find the area of the parallelogram determined by the four vertices (1,0,0), (0,1,0), (0,0,1), (−2,1,1). Also, for

$$T\bar{u} = \begin{bmatrix} 1 & 2 & 1 \\ 0 & 1 & 1 \\ 0 & 1 & 3 \end{bmatrix} \bar{u}$$

find the area of the transform of this parallelogram by T.

10 Suppose T is an invertible linear transformation from R^2 into R^2. Show that if $\bar{u}$ and $\bar{v}$ are independent, then $T\bar{u}$ and $T\bar{v}$ are also independent, by using result **5**. Use this same result to show that if T is not invertible and $\bar{u}$ and $\bar{v}$ are independent, then $T\bar{u}$ and $T\bar{v}$ must be dependent. (Hint: The parallelogram determined by $T\bar{u}$ and $T\bar{v}$ is not degenerate if and only if these vectors are independent.)

CHAPTER 5

REPRESENTATIONS OF LINEAR TRANSFORMATIONS

In Chapter 3 we showed that a linear transformation from R^n into R^m can be represented as a multiplication by a matrix. This we called the matrix of the transformation. This was merely the observation that from knowledge of the transformation's effect on the standard basis one can calculate the effect on any vectors and that this formula can be stated in the form of a matrix product. This principle can be generalized into the observation that from knowledge of a linear transformation's effect on *any* basis one can calculate its effect on any vector, and the resulting formula can be put into the form of a matrix product. The first section of this chapter shows precisely how to construct the matrix of a linear transformation relative to any arbitrary basis, and then shows how changing the basis affects the matrix. Section 2 shows how easy it is to carry out calculations if a basis can be found for which the matrix has a simple form. Sections 3, 4, and 5 discuss the problem of finding a diagonal matrix representation. An application of these results is discussed in Section 6.

SECTION 1 The Matrix of a Linear Transformation with Respect to a Basis

Our purpose is to show that from knowledge of the effect of a linear transformation on a basis one can calculate the effect on an arbitrary vector. The formula for this will be stated in terms of multiplication by a matrix. For simplicity we begin our discussion with R^2.

Let T be a linear transformation from R^2 into R^2, that is, for each $\bar{u}$ in R^2, $T\bar{u}$ is in R^2 and for any pair $\bar{u}$, $\bar{v}$ in R^2 and any scalar a

1 **a** $T(\bar{u} + \bar{v}) = T\bar{u} + T\bar{v}$
b $T(a\bar{u}) = aT\bar{u}$

Let us review our construction of the matrix of T given in Section 2, Chapter 3. Apply T to the two standard basis vectors

$$\bar{i} = \begin{bmatrix} 1 \\ 0 \end{bmatrix} \quad \text{and} \quad \bar{j} = \begin{bmatrix} 0 \\ 1 \end{bmatrix}$$

This will give two vectors, say

2 $$T\begin{bmatrix} 1 \\ 0 \end{bmatrix} = \begin{bmatrix} A_{11} \\ A_{21} \end{bmatrix} \qquad T\begin{bmatrix} 0 \\ 1 \end{bmatrix} = \begin{bmatrix} A_{12} \\ A_{22} \end{bmatrix}$$

If x and y are arbitrary numbers, the basic property **1b** gives

$$T\begin{bmatrix} x \\ 0 \end{bmatrix} = \begin{bmatrix} xA_{11} \\ xA_{21} \end{bmatrix} \quad \text{and} \quad T\begin{bmatrix} 0 \\ y \end{bmatrix} = \begin{bmatrix} yA_{12} \\ yA_{22} \end{bmatrix}$$

and then property **1a** gives

$$T\begin{bmatrix} x \\ y \end{bmatrix} = \begin{bmatrix} xA_{11} \\ xA_{21} \end{bmatrix} + \begin{bmatrix} yA_{12} \\ yA_{22} \end{bmatrix} = \begin{bmatrix} xA_{11} + yA_{12} \\ xA_{21} + yA_{22} \end{bmatrix}$$

which can be rewritten as

3 $$T\begin{bmatrix} x \\ y \end{bmatrix} = A\begin{bmatrix} x \\ y \end{bmatrix} \qquad \text{where } A = \begin{bmatrix} A_{11} & A_{12} \\ A_{21} & A_{22} \end{bmatrix}$$

The matrix A was called the matrix of T; its columns are the two vectors $T\bar{\mathrm{i}}$ and $T\bar{\mathrm{j}}$ (statement **2**), and the effect $T\bar{\mathrm{u}}$ on an arbitrary vector is obtained by calculating $A\bar{\mathrm{u}}$ (statement **3**).

Now suppose $\bar{\mathrm{u}}_1$, $\bar{\mathrm{u}}_2$ is some other basis for R^2. Each of the vectors $T\bar{\mathrm{u}}_1$ and $T\bar{\mathrm{u}}_2$ can be expressed as linear combinations of $\bar{\mathrm{u}}_1$ and $\bar{\mathrm{u}}_2$, say

4 $$\begin{aligned} T\bar{\mathrm{u}}_1 &= B_{11}\bar{\mathrm{u}}_1 + B_{21}\bar{\mathrm{u}}_2 \\ T\bar{\mathrm{u}}_2 &= B_{12}\bar{\mathrm{u}}_1 + B_{22}\bar{\mathrm{u}}_2 \end{aligned}$$

This is the general form of statement **2**. Let us show how to use these to calculate the effect of T on an arbitrary vector. First note that for any c_1 and c_2, property **1b** gives

$$\begin{aligned} T(c_1\bar{\mathrm{u}}_1) &= c_1 T\bar{\mathrm{u}}_1 \\ T(c_2\bar{\mathrm{u}}_2) &= c_2 T\bar{\mathrm{u}}_2 \end{aligned}$$

Substitute relations **4** into these to obtain

$$\begin{aligned} T(c_1\bar{\mathrm{u}}_1) &= c_1(B_{11}\bar{\mathrm{u}}_1 + B_{21}\bar{\mathrm{u}}_2) \\ T(c_2\bar{\mathrm{u}}_2) &= c_2(B_{12}\bar{\mathrm{u}}_1 + B_{22}\bar{\mathrm{u}}_2) \end{aligned}$$

Now add together and collect coefficients of $\bar{\mathrm{u}}_1$ and $\bar{\mathrm{u}}_2$:

$$T(c_1\bar{\mathrm{u}}_1) + T(c_2\bar{\mathrm{u}}_2) = (c_1B_{11} + c_2B_{12})\bar{\mathrm{u}}_1 + (c_1B_{21} + c_2B_{22})\bar{\mathrm{u}}_2$$

Now use the basic property **1a** to replace the left-hand side by $T(c_1\bar{u}_1 + c_2\bar{u}_2)$. Thus for any scalars c_1 and c_2 we have shown that

$$T(c_1\bar{u}_1 + c_2\bar{u}_2) = (c_1B_{11} + c_2B_{12})\bar{u}_1 + (c_1B_{21} + c_2B_{22})\bar{u}_2$$

This tells as that relative to the basis $\bar{u}_1$, $\bar{u}_2$, the coordinates c_1, c_2 are transformed into the coordinates

$$c_1B_{11} + c_2B_{12}, \qquad c_1B_{21} + c_2B_{22}$$

If these are written as column matrices, then relative to the basis $\bar{u}_1$, $\bar{u}_2$ the coordinates

$$\begin{bmatrix} c_1 \\ c_2 \end{bmatrix} \quad \text{become} \quad \begin{bmatrix} c_1B_{11} + c_2B_{12} \\ c_1B_{21} + c_2B_{22} \end{bmatrix}$$

The latter matrix can be rewritten as the product

$$\begin{bmatrix} B_{11} & B_{12} \\ B_{21} & B_{22} \end{bmatrix} \begin{bmatrix} c_1 \\ c_2 \end{bmatrix}$$

Let us summarize: Relative to a given basis $\bar{u}_1$, $\bar{u}_2$, the coordinates of $T\bar{u}$ are the two entries of the product

5 $$B\begin{bmatrix} c_1 \\ c_2 \end{bmatrix}, \qquad B = \begin{bmatrix} B_{11} & B_{12} \\ B_{21} & B_{22} \end{bmatrix}$$

where c_1, c_2 are the coordinates of $\bar{u}$ and the columns of B are the coordinates of $T\bar{u}_1$ and $T\bar{u}_2$ (all with respect to $\bar{u}_1$, $\bar{u}_2$).

The general form of this is as follows: Relative to a given basis $\bar{u}_1$, $\bar{u}_2$, . . ., $\bar{u}_n$, the coordinates of $T\bar{u}$ are the entries of the product

6 $$B\begin{bmatrix} c_1 \\ c_2 \\ \vdots \\ c_n \end{bmatrix}$$

where c_1, c_2, . . ., c_n are the coordinates of $\bar{u}$ and the columns of B are the coordinates of $T\bar{u}_1$, $T\bar{u}_2$, . . ., $T\bar{u}_n$ (all with respect to $\bar{u}_1$,$\bar{u}_2$, . . .,$\bar{u}_n$).

The matrix B is called the **matrix** of T **relative to** $\bar{u}_1$, $\bar{u}_2$, . . ., $\bar{u}_n$.

EXAMPLE 1 Suppose, relative to the standard basis, T is the linear transformation of R^2 into R^2 given by

7 $$T\bar{u} = A\bar{u}, \qquad A = \begin{bmatrix} 3 & -1 \\ 2 & 0 \end{bmatrix}$$

The vectors

$$\bar{u}_1 = \begin{bmatrix} 1 \\ 1 \end{bmatrix} \qquad \bar{u}_2 = \begin{bmatrix} 1 \\ -1 \end{bmatrix}$$

are independent, and hence are a basis for R^2. Let us find the matrix of T relative to the basis $\bar{u}_1$, $\bar{u}_2$. Call this matrix B. The first column of B consists of the coordinates of $T\bar{u}_1$ relative to $\bar{u}_1$, $\bar{u}_2$.

We have

$$T\bar{u}_1 = \begin{bmatrix} 3 & -1 \\ 2 & 0 \end{bmatrix} \begin{bmatrix} 1 \\ 1 \end{bmatrix} = \begin{bmatrix} 2 \\ 2 \end{bmatrix}$$

so that $T\bar{u}_1 = 2\bar{u}_1 = 2\bar{u}_1 + 0 \cdot \bar{u}_2$. Thus the first column of B is $\begin{bmatrix} 2 \\ 0 \end{bmatrix}$. Similarly

$$T\bar{u}_2 = A\bar{u}_2 = \begin{bmatrix} 4 \\ 2 \end{bmatrix} = 3 \cdot \begin{bmatrix} 1 \\ 1 \end{bmatrix} + 1 \cdot \begin{bmatrix} 1 \\ -1 \end{bmatrix} = 3 \cdot \bar{u}_1 + 1 \cdot \bar{u}_2$$

so the second column of B is $\begin{bmatrix} 3 \\ 1 \end{bmatrix}$. Hence

$$B = \begin{bmatrix} 2 & 3 \\ 0 & 1 \end{bmatrix}$$

If $\bar{u}$ is given in terms of $\bar{u}_1$ and $\bar{u}_2$, say

$$\bar{u} = -3\bar{u}_1 + \tfrac{7}{2}\bar{u}_2$$

then

$$B \begin{bmatrix} -3 \\ \tfrac{7}{2} \end{bmatrix} = \begin{bmatrix} \tfrac{9}{2} \\ \tfrac{7}{2} \end{bmatrix}$$

so

$$T\bar{u} = \tfrac{9}{2}\bar{u}_1 + \tfrac{7}{2}\bar{u}_2$$

EXAMPLE 2 It is sometimes possible to choose a basis so as to simplify the description of a transformation which was defined geometrically. For example, suppose P is the projection along $\bar{v} = \frac{1}{3}\,(1,2,2)$. Any vector parallel to $\bar{v}$

is left fixed by P while any vector orthogonal to $\bar{v}$ is sent into $\bar{0}$ by P. Let us translate this into a matrix statement. Put $\bar{w}_1 = \bar{v}$ and choose two independent vectors $\bar{w}_2$ and $\bar{w}_3$ which are orthogonal to $\bar{v}$. One way to do this is to choose $\bar{w}_2$ and $\bar{w}_3$ so that $\bar{w}_1$, $\bar{w}_2$, and $\bar{w}_3$ are orthogonal, which can be done as follows: Choose any $\bar{w}_2$ satisfying

$$\bar{w}_1 \cdot \bar{w}_2 = 0 \qquad \bar{w}_2 \neq \bar{0}$$

then choose $\bar{w}_3$ to satisfy

$$\bar{w}_1 \cdot \bar{w}_3 = 0 \qquad \bar{w}_2 \cdot \bar{w}_3 = 0 \qquad \bar{w}_3 \neq \bar{0}$$

For example,

$$\bar{w}_2 = (0,1,-1) \qquad \bar{w}_3 = (-4,1,1)$$

is one such choice. The definition of P gives

$$\begin{aligned} P\bar{w}_1 &= P\bar{v} = (\bar{v} \cdot \bar{v})\bar{v} = \bar{v} \\ P\bar{w}_2 &= (\bar{w}_2 \cdot \bar{v})\bar{v} = \bar{0} \\ P\bar{w}_3 &= (\bar{w}_3 \cdot \bar{v})\bar{v} = \bar{0} \end{aligned}$$

so the matrix of P relative to the bases $\bar{w}_1$, $\bar{w}_2$, $\bar{w}_3$ is

$$\begin{bmatrix} 1 & 0 & 0 \\ 0 & 0 & 0 \\ 0 & 0 & 0 \end{bmatrix}$$

This is a much simpler form than the matrix of P relative to the standard basis (see page 144).

DISCUSSION

If T is a given linear transformation, then there should be some relationship between its matrix relative to one basis and its matrix relative to another. Let us describe this in the case when one of these is the standard basis for R^2. Suppose A is the matrix of T relative to the standard basis and B is the matrix of T relative to another basis $\bar{u}_1$, $\bar{u}_2$, where

$$A = \begin{bmatrix} A_{11} & A_{12} \\ A_{21} & A_{22} \end{bmatrix} \qquad B = \begin{bmatrix} B_{11} & B_{12} \\ B_{21} & B_{22} \end{bmatrix}$$

Let P be the matrix whose columns are $\bar{u}_1$, $\bar{u}_2$. It will now be shown that B and A are related by:

8 $$B = P^{-1}AP$$

The proof of this goes as follows.

We know that $P\begin{bmatrix}1\\0\end{bmatrix}$ is the first column of P, so $P\begin{bmatrix}1\\0\end{bmatrix} = \bar{u}_1$. Thus

$$P^{-1}AP\begin{bmatrix}1\\0\end{bmatrix} = P^{-1}A\bar{u}_1$$

We also know that (see formulas **3** and **4**) $A\bar{u}_1 = T\bar{u}_1$ and $T\bar{u}_1 = B_{11}\bar{u}_1 + B_{21}\bar{u}_2$, so

$$\begin{aligned}P^{-1}AP\begin{bmatrix}1\\0\end{bmatrix} &= P^{-1}T\bar{u}_1\\ &= P^{-1}(B_{11}\bar{u}_1 + B_{21}\bar{u}_2)\\ &= B_{11}P^{-1}\bar{u}_1 + B_{21}P^{-1}\bar{u}_2\end{aligned}$$

Since

$$P\begin{bmatrix}1\\0\end{bmatrix} = \bar{u}_1 \qquad \text{and} \qquad P\begin{bmatrix}0\\1\end{bmatrix} = \bar{u}_2$$

we can multiply by P^{-1} to obtain

$$P^{-1}\bar{u}_1 = \begin{bmatrix}1\\0\end{bmatrix} \qquad \text{and} \qquad P^{-1}\bar{u}_2 = \begin{bmatrix}0\\1\end{bmatrix}$$

Substitution of these gives

$$\begin{aligned}P^{-1}AP\begin{bmatrix}1\\0\end{bmatrix} &= B_{11}P^{-1}\bar{u}_1 + B_{21}P^{-1}\bar{u}_2\\ &= B_{11}\begin{bmatrix}1\\0\end{bmatrix} + B_{21}\begin{bmatrix}0\\1\end{bmatrix}\\ &= \begin{bmatrix}B_{11}\\B_{21}\end{bmatrix}\end{aligned}$$

This shows that the first column of $P^{-1}AP$ has the same entries as the first column of B. A similar calculation of

$$P^{-1}AP\begin{bmatrix}0\\1\end{bmatrix}$$

establishes that the second columns also have the same entries and completes the proof of formula **8**.

We state a general version of this result, the proof of which is omitted:

9 Suppose $\bar{u}_1, \bar{u}_2, \ldots, \bar{u}_n$ is a basis for V and that $\bar{v}_1, \bar{v}_2, \ldots, \bar{v}_n$ is also a basis for V. Suppose T is a linear transformation from V into V whose matrices with respect to these bases are, respectively, A and B. Let P be the matrix whose columns consist of the coordinates of each of the vectors $\bar{v}_1, \bar{v}_2, \ldots, \bar{v}_n$ with respect to $\bar{u}_1, \bar{u}_2, \ldots, \bar{u}_n$. Let Q be the matrix whose columns are the coordinates of each of the vectors $\bar{u}_1, \bar{u}_2, \ldots, \bar{u}_n$ with respect to $\bar{v}_1, \bar{v}_2, \ldots, \bar{v}_n$. Then

$$QP = PQ = I$$

so that

$$Q = P^{-1} \qquad \text{and} \qquad P = Q^{-1}$$

Furthermore,

$$B = P^{-1}AP \qquad \text{and} \qquad A = Q^{-1}BQ$$

EXAMPLE 3 Let us verify formula **8** for the operator T of Example **1**. We have

$$A = \begin{bmatrix} 3 & -1 \\ 2 & 0 \end{bmatrix} \qquad B = \begin{bmatrix} 2 & 3 \\ 0 & 1 \end{bmatrix} \qquad \bar{u}_1 = \begin{bmatrix} 1 \\ 1 \end{bmatrix} \qquad \bar{u}_2 = \begin{bmatrix} 1 \\ -1 \end{bmatrix}$$

Now put

$$P = \begin{bmatrix} 1 & 1 \\ 1 & -1 \end{bmatrix} \qquad \text{so that} \qquad P^{-1} = \begin{bmatrix} \frac{1}{2} & \frac{1}{2} \\ \frac{1}{2} & -\frac{1}{2} \end{bmatrix}$$

Therefore,

$$\begin{aligned} P^{-1}AP &= P^{-1}\begin{bmatrix} 3 & -1 \\ 2 & 0 \end{bmatrix}\begin{bmatrix} 1 & 1 \\ 1 & -1 \end{bmatrix} \\ &= P^{-1}\begin{bmatrix} 2 & 4 \\ 2 & 2 \end{bmatrix} \\ &= \begin{bmatrix} \frac{1}{2} & \frac{1}{2} \\ \frac{1}{2} & -\frac{1}{2} \end{bmatrix}\begin{bmatrix} 2 & 4 \\ 2 & 2 \end{bmatrix} \\ &= \begin{bmatrix} 2 & 3 \\ 0 & 1 \end{bmatrix} \\ &= B \end{aligned}$$

EXAMPLE 4 The above concepts extend with little change to other vector spaces, such as the space V of all polynomial functions of degree not exceeding 2, discussed in Example 5, page 83.

The polynomials $\bar{u}_1 = 1$, $\bar{u}_2 = x$, and $\bar{u}_3 = x^2$ are a basis for V. We shall find the matrix of D, the differentiation operator, relative to this basis. Since $D1 = 0$, $Dx = 1$, and $Dx^2 = 2x$, we have

$$D\bar{u}_1 = 0 \cdot \bar{u}_1 + 0 \cdot \bar{u}_2 + 0 \cdot \bar{u}_3$$
$$D\bar{u}_2 = 1 \cdot \bar{u}_1 + 0 \cdot \bar{u}_2 + 0 \cdot \bar{u}_3$$
$$D\bar{u}_3 = 0 \cdot \bar{u}_1 + 2 \cdot \bar{u}_2 + 0 \cdot \bar{u}_3$$

The matrix D_0 of D relative to $\bar{u}_1$, $\bar{u}_2$, and $\bar{u}_3$ has as its columns the coordinates of $D\bar{u}_1$, $D\bar{u}_2$, and $D\bar{u}_3$ with respect to $\bar{u}_1$, $\bar{u}_2$, and $\bar{u}_3$. Thus

$$D_0 = \begin{bmatrix} 0 & 1 & 0 \\ 0 & 0 & 2 \\ 0 & 0 & 0 \end{bmatrix}$$

Note that

$$D_0^2 = \begin{bmatrix} 0 & 0 & 2 \\ 0 & 0 & 0 \\ 0 & 0 & 0 \end{bmatrix} \quad \text{and} \quad D_0^3 = \begin{bmatrix} 0 & 0 & 0 \\ 0 & 0 & 0 \\ 0 & 0 & 0 \end{bmatrix}$$

which are, respectively, the matrices of D^2 and D^3 relative to $\bar{u}_1$, $\bar{u}_2$, and $\bar{u}^3$. We can verify this easily, for

$$D^2\bar{u}_1 = 0 = 0 \cdot \bar{u}_1 + 0 \cdot \bar{u}_2 + 0 \cdot \bar{u}_3$$
$$D^2\bar{u}_2 = 0 = 0 \cdot \bar{u}_1 + 0 \cdot \bar{u}_2 + 0 \cdot \bar{u}_3$$
$$D^2\bar{u}_3 = 2 = 2 \cdot \bar{u}_1 + 0 \cdot \bar{u}_2 + 0 \cdot \bar{u}_3$$

and

$$D^3\bar{u}_1 = D^3\bar{u}_2 = D^3\bar{u}_3 = 0$$

Note that D is *not* invertible, for the columns of D_0 are dependent.

Let us examine the matrix of D relative to another basis. Recall that

$$f_1 = \frac{(x-1)(x-2)}{2} \qquad f_2 = \frac{x(x-2)}{-1} \qquad f_3 = \frac{x(x-1)}{2}$$

are also a basis for V and that for any f in V (see the Lagrange interpolation formula, page 91)

10 $$f = f(0)f_1 + f(1)f_2 + f(2)f_3$$

To find the matrix of D relative to f_1, f_2, and f_3 note that

$$f_1 = x^2/2 - 3x/2 + 1 \qquad f_2 = -x^2 + 2x \qquad f_3 = x^2/2 - x/2$$

so that

11 $$Df_1 = x - \tfrac{3}{2} \qquad Df_2 = -2x + 2 \qquad Df_3 = x - \tfrac{1}{2}$$

To find the matrix of D relative to f_1, f_2, and f_3 we need to find the coordinates of Df_1, Df_2, and Df_3 relative to f_1, f_2, and f_3. Formula **10** tells us that the coordinates of any f, relative to f_1, f_2, and f_3, are $f(0)$, $f(1)$, and $f(2)$. Therefore, setting $x = 0$, 1, and 2 in each expression in **11** we find that

$$(-\tfrac{3}{2},-\tfrac{1}{2},\tfrac{1}{2}) \qquad (2,0,-2) \qquad (-\tfrac{1}{2},\tfrac{1}{2},\tfrac{3}{2})$$

are the respective coordinates of Df_1, Df_2, and Df_3 relative to f_1, f_2, and f_3. The matrix D_1 of D relative to this basis has these as its columns:

$$D_1 = \begin{bmatrix} -\frac{3}{2} & 2 & -\frac{1}{2} \\ -\frac{1}{2} & 0 & \frac{1}{2} \\ \frac{1}{2} & -2 & \frac{3}{2} \end{bmatrix}$$

The result **9** tells us that

$$D_1 = P^{-1}D_0P \qquad \text{and} \qquad D_0 = Q^{-1}D_1Q$$

where the columns of P are, respectively, the coordinates of f_1, f_2, f_3 with respect to $\bar{u}_1, \bar{u}_2, \bar{u}_3$ and the columns of Q are, respectively, the coordinates of $\bar{u}_1, \bar{u}_2, \bar{u}_3$ with respect to f_1, f_2, f_3.

We have

$$\begin{aligned} f_1 &= \frac{x^2}{2} - \frac{3x}{2} + 1 = 1 \cdot \bar{u}_1 - \tfrac{3}{2}\bar{u}_2 + \tfrac{1}{2}\bar{u}_3 \\ f_2 &= -x^2 + 2x \qquad = 0 \cdot \bar{u}_1 + 2 \cdot \bar{u}_2 - \bar{u}_3 \\ f_3 &= \frac{x^2}{2} - \frac{x}{2} \qquad = 0 \cdot \bar{u}_1 - \tfrac{1}{2}\bar{u}_2 + \tfrac{1}{2}\bar{u}_3 \end{aligned}$$

and (using formula **10** for each of 1, x, and x^2)

$$\begin{aligned} \bar{u}_1 &= 1 \cdot f_1 + 1 \cdot f_2 + 1 \cdot f_3 \\ \bar{u}_2 &= 0 \cdot f_1 + 1 \cdot f_2 + 2 \cdot f_3 \\ \bar{u}_3 &= 0 \cdot f_1 + 1 \cdot f_2 + 4 \cdot f_3 \end{aligned}$$

Thus

$$P = \begin{bmatrix} 1 & 0 & 0 \\ -\frac{3}{2} & 2 & -\frac{1}{2} \\ \frac{1}{2} & -1 & \frac{1}{2} \end{bmatrix} \qquad Q = \begin{bmatrix} 1 & 0 & 0 \\ 1 & 1 & 1 \\ 1 & 2 & 4 \end{bmatrix}$$

The student can then verify that $QP = PQ = I$, $D_1 = P^{-1}D_0P$, and $D_0 = Q^{-1}D_1Q$.

REMARK The results of this section can be extended to cover the case of linear transformations defined for vectors in one vector space V with values in a second vector space W. In this case a matrix description can be given relative to selection of a basis for V and a basis for W. Because the more interesting and useful theory occurs when V and W are the same, we have confined our discussion to this case.

EXERCISES

1 For each of the following matrices A suppose T is defined by $T\bar{u} = A\bar{u}$ for $\bar{u}$ in R^2. Proceed as in Example **1** to find the matrix B of T relative to $\bar{u}_1 = (3,1)$ and $\bar{u}_2 = (5,2)$.

a $A = \begin{bmatrix} 28 & -75 \\ 10 & -27 \end{bmatrix}$ **b** $A = \begin{bmatrix} -2 & 9 \\ -1 & 4 \end{bmatrix}$

2 Suppose T is defined for $\bar{u}$ in R^3 by $T\bar{u} = A\bar{u}$, where

$$A = \begin{bmatrix} 8 & 9 & 9 \\ 3 & 2 & 3 \\ -9 & -9 & -10 \end{bmatrix}$$

a Find the matrix B of T relative to $(-1,1,0)$, $(-1,0,1)$, and $(-3,-1,3)$.
b What is the matrix of T^6 relative to this basis?

3 **a** For A and B as in Exercise **1a** find an invertible matrix P such that $B = P^{-1}AP$. (See Example **3**.)
b For A and B as in Exercise **2** find an invertible matrix P such that $B = P^{-1}AP$.

4 For each of the operators T of Exercise **1** find the matrix of each of the following, relative to (3,1) and (5,2).

a T^3 **b** $T + I$
c $T^6 - T^4 + T^2$ **d** T^{-1}

5 What is the relation between the matrix of T relative to $\bar{u}_1$, $\bar{u}_2$ and the matrix of T relative to $\bar{u}_2$, $\bar{u}_1$?

6 What is the matrix of the identity operator I relative to the basis $\bar{u}_1, \bar{u}_2, \ldots, \bar{u}_n$ for V? What is the matrix of the zero operator relative to this basis? The matrix of aI?

7 Suppose V is the vector space of polynomials of degree not exceeding 3. Find the matrix of the differentiation operator D with respect to the basis $1, x, x^2$, and x^3.

8 For V and D as in Exercise **7** find the matrix of D^2, D^3, and D^4 with respect to $1, x, x^2$, and x^3.

9 For V and D as in Exercise **7**, find the matrix of D relative to the basis

$$\bar{v}_1 = \frac{(x-1)(x-2)(x-3)}{-6}, \qquad \bar{v}_2 = \frac{x(x-2)(x-3)}{2}$$

$$\bar{v}_3 = \frac{x(x-1)(x-3)}{-2}, \qquad \bar{v}_4 = \frac{x(x-1)(x-2)}{6}$$

(See Exercise **10** on page 92.)

10 Suppose V is as in Example **4** and that T is the operator defined by

$$Tf = \frac{1}{x}\int_0^x f(t)\,dt \qquad \text{for } f \text{ in } V$$

a Find $T(x^2 + 1)$.
b Find $T(1 - x)$.
c Find the matrix A of T relative to 1, x, and x^2.
d Find the inverse of the matrix of part **c** and use it to describe $T^{-1}1$, $T^{-1}x$, and $T^{-1}x^2$.

11 For V and T as in Exercise **10** find the matrix B of T relative to the basis f_1, f_2, f_3 of Example **4**.

12 Suppose A, B, and P each have two rows and two columns and that $B = P^{-1}AP$. Show that there is a linear transformation T from R^2 into R^2 such that A and B are each the matrix of T with respect to a basis. (Hint: Let $T\bar{u} = A\bar{u}$ for $\bar{u}$ in R^2 and show that B is the matrix of T with respect to the columns of P.)

13 For the matrices A and B of Exercises **10** and **11**, proceed as in Example **4** to find P and Q such that $QP = PQ = I$, $B = P^{-1}AP$, and $A = Q^{-1}BQ$.

□ 14 The results of this section hold for complex vector spaces. Suppose T is defined for $\bar{u}$ in C^2 by

$$T\bar{u} = A\bar{u} \qquad \text{where } A = \begin{bmatrix} 7+4i & -12+24i \\ -1+2i & -7-3i \end{bmatrix}$$

a Find the matrix B of T relative to the basis $(4i, -1)$ and $(-3, -i)$.
b Find an invertible matrix P such that $B = P^{-1}AP$.

SECTION 2 Calculations with Similar Matrices

We say that the matrix B is **similar** to the matrix A if there is an invertible matrix P such that

1 $$B = P^{-1}AP$$

Similarity is related to coordinate change as follows: Suppose T is the linear transformation defined by $T\bar{u} = A\bar{u}$ for $\bar{u}$ in R^n, so that A is the matrix of T with respect to the standard basis. If B is the matrix of T with respect to another basis, then formula **8**, page 241, tells us that B is similar to A. It is also true that if B is similar to A, then B is the matrix of T with respect to some other basis for R^n. (See Exercise **12**, page 247.)

Our purpose in this section is to gain facility in calculating with formula **1**. We first note that

2 If B is similar to A, then A is similar to B.

If P is invertible and $B = P^{-1}AP$, we can solve for A by suitably multiplying by P and P^{-1} as follows:

$$\begin{aligned} PBP^{-1} &= P(P^{-1}AP)P^{-1} \\ &= (PP^{-1})A(PP^{-1}) \\ &= A \end{aligned}$$

If we put $Q = P^{-1}$, so that $Q^{-1} = P$, we can then write

3 $$A = Q^{-1}BQ$$

and thus establish statement **2**.

The above procedure indicates a basic trick for calculating with formula **1**, namely, the cancellation effect of multiplying P by P^{-1} or P^{-1} by P. As another example, suppose $B = P^{-1}AP$. Then

$$\begin{aligned} B^2 &= (P^{-1}AP)(P^{-1}AP) \\ &= (P^{-1}A)(PP^{-1})(AP) \\ &= (P^{-1}A)I(AP) \\ &= P^{-1}AAP \\ &= P^{-1}A^2P \end{aligned}$$

Notice that the six factors of $(P^{-1}AP)(P^{-1}AP)$ collapse to the four factors of $P^{-1}A^2P$ because of the cancellation of $P^{-1}P$. This cancellation appears again in the calculation of other powers of B. For example,

$$\begin{aligned} B^3 &= (P^{-1}AP)(P^{-1}AP)(P^{-1}AP) \\ &= (P^{-1}A)(PP^{-1})A(PP^{-1})(AP) \\ &= P^{-1}A^3P \end{aligned}$$

In general, for any positive integer k:

4 $$B^k = P^{-1}A^kP \qquad \text{if } B = P^{-1}AP$$

From this it follows that

5 If B is similar to A, then B^k is similar to A^k.

EXAMPLE 1 Let us verify these formulas for

$$A = \begin{bmatrix} 1 & 1 \\ 0 & 1 \end{bmatrix} \quad \text{and} \quad P = \begin{bmatrix} 1 & 1 \\ 1 & -1 \end{bmatrix}$$

Then (the student should verify these calculations)

$$P^{-1} = \begin{bmatrix} \frac{1}{2} & \frac{1}{2} \\ \frac{1}{2} & -\frac{1}{2} \end{bmatrix} \quad \text{and} \quad P^{-1}AP = \begin{bmatrix} \frac{3}{2} & -\frac{1}{2} \\ \frac{1}{2} & \frac{1}{2} \end{bmatrix}$$

Therefore if

$$B = \begin{bmatrix} \frac{3}{2} & -\frac{1}{2} \\ \frac{1}{2} & \frac{1}{2} \end{bmatrix}$$

then B is similar to A.

We shall verify directly that A is also similar to B. Put

$$Q = \begin{bmatrix} \frac{1}{2} & \frac{1}{2} \\ \frac{1}{2} & -\frac{1}{2} \end{bmatrix} \quad \text{so that} \quad Q^{-1} = \begin{bmatrix} 1 & 1 \\ 1 & -1 \end{bmatrix}$$

It follows that

$$\begin{aligned} Q^{-1}BQ &= Q^{-1}\begin{bmatrix} \frac{3}{2} & -\frac{1}{2} \\ \frac{1}{2} & \frac{1}{2} \end{bmatrix}\begin{bmatrix} \frac{1}{2} & \frac{1}{2} \\ \frac{1}{2} & -\frac{1}{2} \end{bmatrix} \\ &= Q^{-1}\begin{bmatrix} \frac{1}{2} & 1 \\ \frac{1}{2} & 0 \end{bmatrix} \\ &= \begin{bmatrix} 1 & 1 \\ 1 & -1 \end{bmatrix}\begin{bmatrix} \frac{1}{2} & 1 \\ \frac{1}{2} & 0 \end{bmatrix} \\ &= \begin{bmatrix} 1 & 1 \\ 0 & 1 \end{bmatrix} \\ &= A \end{aligned}$$

To verify formula **4** for $k = 2$, note that

$$B^2 = \begin{bmatrix} 2 & -1 \\ 1 & 0 \end{bmatrix} \qquad \text{and} \qquad A^2 = \begin{bmatrix} 1 & 2 \\ 0 & 1 \end{bmatrix}$$

Then,

$$\begin{aligned} P^{-1}A^2P &= P^{-1}\begin{bmatrix} 1 & 2 \\ 0 & 1 \end{bmatrix} \begin{bmatrix} 1 & 1 \\ 1 & -1 \end{bmatrix} \\ &= P^{-1}\begin{bmatrix} 3 & -1 \\ 1 & -1 \end{bmatrix} \\ &= \begin{bmatrix} \frac{1}{2} & \frac{1}{2} \\ \frac{1}{2} & -\frac{1}{2} \end{bmatrix} \begin{bmatrix} 3 & -1 \\ 1 & -1 \end{bmatrix} \\ &= \begin{bmatrix} 2 & -1 \\ 1 & 0 \end{bmatrix} \end{aligned}$$

The powers of A are easy to find. Direct calculation gives

$$A^3 = \begin{bmatrix} 1 & 3 \\ 0 & 1 \end{bmatrix}, \qquad A^4 = \begin{bmatrix} 1 & 4 \\ 0 & 1 \end{bmatrix}$$

In general, we obtain the formula

$$A^k = \begin{bmatrix} 1 & k \\ 0 & 1 \end{bmatrix}$$

We can now use formula **4** to find B^k, for

$$\begin{aligned} B^k = P^{-1}A^kP &= P^{-1}\begin{bmatrix} 1 & k \\ 0 & 1 \end{bmatrix} \begin{bmatrix} 1 & 1 \\ 1 & -1 \end{bmatrix} \\ &= P^{-1}\begin{bmatrix} 1+k & 1-k \\ 1 & -1 \end{bmatrix} \\ &= \begin{bmatrix} \frac{1}{2} & \frac{1}{2} \\ \frac{1}{2} & -\frac{1}{2} \end{bmatrix} \begin{bmatrix} 1+k & 1-k \\ 1 & -1 \end{bmatrix} \\ &= \begin{bmatrix} 1 + k/2 & -k/2 \\ k/2 & 1 - k/2 \end{bmatrix} \end{aligned}$$

EXAMPLE 2 If B is similar to A, and A is a diagonal matrix, the use of formula 4 often simplifies calculation of large powers of B. For example, suppose

$$A = \begin{bmatrix} 2 & 0 \\ 0 & 1 \end{bmatrix} \quad \text{and} \quad P = \begin{bmatrix} 2 & -1 \\ -1 & 1 \end{bmatrix}$$

Then

$$P^{-1} = \begin{bmatrix} 1 & 1 \\ 1 & 2 \end{bmatrix} \quad \text{and} \quad P^{-1}AP = \begin{bmatrix} 3 & -1 \\ 2 & 0 \end{bmatrix}$$

Therefore, if $B = P^{-1}AP$, we know from formula 4 that $B^k = P^{-1}A^kP$. For example,

$$A^6 = \begin{bmatrix} 64 & 0 \\ 0 & 1 \end{bmatrix}$$

so that

$$\begin{aligned} B^6 &= P^{-1}A^6P \\ &= \begin{bmatrix} 1 & 1 \\ 1 & 2 \end{bmatrix} \begin{bmatrix} 64 & 0 \\ 0 & 1 \end{bmatrix} \begin{bmatrix} 2 & -1 \\ -1 & 1 \end{bmatrix} \\ &= \begin{bmatrix} 127 & -63 \\ 126 & -62 \end{bmatrix} \end{aligned}$$

This method of calculating B^6 is certainly easier than direct calculation.

DISCUSSION Some further consequences of the similarity relationship follow. Suppose B is similar to A and A is invertible. We then have $B = P^{-1}AP$, and it follows that

$$\begin{aligned} (P^{-1}A^{-1}P)B &= (P^{-1}A^{-1}P)(P^{-1}AP) \\ &= (P^{-1}A^{-1})(PP^{-1})(AP) \\ &= (P^{-1}A^{-1})(AP) \\ &= P^{-1}(A^{-1}A)P \\ &= P^{-1}P = I \end{aligned}$$

Thus $P^{-1}A^{-1}P$ *must* be the inverse of B, and this fact shows that

6 $$B^{-1} = P^{-1}A^{-1}P$$

Recall (see Example 3, page 171) that if $q(\lambda)$ is the polynomial

$$q(\lambda) = a_k\lambda^k + a_{k-1}\lambda^{k-1} + \cdots + a_1\lambda + a_0$$

then

$$q(A) = a_kA^k + a_{k-1}A^{k-1} + \cdots + a_1A + a_0I$$

Suppose B is similar to A, that is, there is a matrix P such that $B = P^{-1}AP$. Then

$$q(B) = a_kB^k + a_{k-1}B^{k-1} + \cdots + a_1B + a_0I$$

Now substitute, using formula **4** and the fact that $I = P^{-1}IP$ to write this as

$$q(B) = a_kP^{-1}A^kP + a_{k-1}P^{-1}A^{k-1}P + \cdots + a_1P^{-1}AP + a_0P^{-1}IP$$

It is now possible to factor out P^{-1} and P to obtain

$$q(B) = P^{-1}(a_kA^k + a_{k-1}A^{k-1} + \cdots + a_1A + a_0I)P$$

In other words,

7 $$q(B) = P^{-1}q(A)P \qquad \text{if } B = P^{-1}AP$$

so that, in particular,

8 If B is similar to A, then $q(B)$ is similar to $q(A)$ for any polynomial $q(\lambda)$.

EXAMPLE 3 Let us verify formula **6** for the matrices of Example 1. We have

$$A = \begin{bmatrix} 1 & 1 \\ 0 & 1 \end{bmatrix} \qquad P = \begin{bmatrix} 1 & 1 \\ 1 & -1 \end{bmatrix} \qquad P^{-1} = \begin{bmatrix} \frac{1}{2} & \frac{1}{2} \\ \frac{1}{2} & -\frac{1}{2} \end{bmatrix}$$

and

$$B = P^{-1}AP = \begin{bmatrix} \frac{3}{2} & -\frac{1}{2} \\ \frac{1}{2} & \frac{1}{2} \end{bmatrix}$$

Simple calculation shows that

$$A^{-1} = \begin{bmatrix} 1 & -1 \\ 0 & 1 \end{bmatrix} \qquad \text{and} \qquad B^{-1} = \begin{bmatrix} \frac{1}{2} & \frac{1}{2} \\ -\frac{1}{2} & \frac{3}{2} \end{bmatrix}$$

We have

$$\begin{aligned} P^{-1}A^{-1}P &= P^{-1}\begin{bmatrix} 1 & -1 \\ 0 & 1 \end{bmatrix}\begin{bmatrix} 1 & 1 \\ 1 & -1 \end{bmatrix} \\ &= P^{-1}\begin{bmatrix} 0 & 2 \\ 1 & -1 \end{bmatrix} \\ &= \begin{bmatrix} \frac{1}{2} & \frac{1}{2} \\ \frac{1}{2} & -\frac{1}{2} \end{bmatrix}\begin{bmatrix} 0 & 2 \\ 1 & -1 \end{bmatrix} \\ &= \begin{bmatrix} \frac{1}{2} & \frac{1}{2} \\ -\frac{1}{2} & \frac{3}{2} \end{bmatrix} \\ &= B^{-1} \end{aligned}$$

EXAMPLE 4 If A is diagonal and q is a polynomial, $q(A)$ is simple to calculate. For example, if $A = \begin{bmatrix} 2 & 0 \\ 0 & 1 \end{bmatrix}$ and $q(\lambda) = \lambda^5 - 3\lambda^2 - 2$, then

$$\begin{aligned} q(A) &= A^5 - 3A^2 - 2I \\ &= \begin{bmatrix} 32 & 0 \\ 0 & 1 \end{bmatrix} - 3\begin{bmatrix} 4 & 0 \\ 0 & 1 \end{bmatrix} - 2\begin{bmatrix} 1 & 0 \\ 0 & 1 \end{bmatrix} \\ &= \begin{bmatrix} 18 & 0 \\ 0 & -4 \end{bmatrix} \end{aligned}$$

Note that $18 = q(2)$ and $-4 = q(1)$; therefore

$$q(A) = \begin{bmatrix} q(2) & 0 \\ 0 & q(1) \end{bmatrix}$$

If B is similar to A, we can then use formula 7 to find $q(B)$. For example, suppose

$$P = \begin{bmatrix} 2 & -1 \\ -1 & 1 \end{bmatrix} \quad \text{so that} \quad P^{-1} = \begin{bmatrix} 1 & 1 \\ 1 & 2 \end{bmatrix}$$

Denoting $P^{-1}AP$ by B, we have (see Example 2)

$$B = \begin{bmatrix} 3 & -1 \\ 2 & 0 \end{bmatrix}$$

Thus, formula 7 gives

$$\begin{aligned} q(B) &= P^{-1}q(A)P \\ &= \begin{bmatrix} 1 & -1 \\ 1 & 2 \end{bmatrix} \begin{bmatrix} 18 & 0 \\ 0 & -4 \end{bmatrix} \begin{bmatrix} 2 & -1 \\ -1 & 1 \end{bmatrix} \\ &= \begin{bmatrix} 32 & -14 \\ 44 & -26 \end{bmatrix} \end{aligned}$$

EXAMPLE 5 *Finding Roots of a Matrix.*

We can sometimes use the similarity relationship to find roots. For example, suppose we want to find $B^{1/2}$, that is, we are looking for a matrix C such that $C^2 = B$. Suppose

$$B = \begin{bmatrix} 3 & -1 \\ 2 & 0 \end{bmatrix}$$

We can, of course, put

$$C = \begin{bmatrix} a & b \\ c & d \end{bmatrix}$$

and determine a, b, c, and d from the equation $C^2 = B$. The resulting equations will be nonlinear and rather difficult to solve. In certain cases, a simple trick will find such a C. Recall that, from Example 2, $B = P^{-1}AP$, where

$$A = \begin{bmatrix} 2 & 0 \\ 0 & 1 \end{bmatrix} \quad \text{and} \quad P = \begin{bmatrix} 2 & -1 \\ -1 & 1 \end{bmatrix}$$

We can then obtain solutions to $C^2 = B$ by obtaining solutions to $D^2 = A$ and using the similarity relationship. For example, if

$$D = \begin{bmatrix} \sqrt{2} & 0 \\ 0 & 1 \end{bmatrix}$$

then certainly $D^2 = A$. Thus, if we put $C = P^{-1}DP$ the cancellation effect gives

$$\begin{aligned} C^2 &= (P^{-1}DP)(P^{-1}DP) \\ &= (P^{-1}D)(P^{-1}P)(DP) \\ &= (P^{-1}D)(DP) \\ &= P^{-1}D^2P = P^{-1}AP \\ &= B \end{aligned}$$

The matrix C is given by

$$C = P^{-1}DP = \begin{bmatrix} 2\sqrt{2} - 1 & -\sqrt{2} + 1 \\ 2\sqrt{2} - 2 & -\sqrt{2} + 2 \end{bmatrix}$$

Therefore we have found a matrix C such that $C^2 = B$.

There are, of course, at least three other such matrices, obtained by using this process with each of the following in place of D.

$$\begin{bmatrix} -\sqrt{2} & 0 \\ 0 & 1 \end{bmatrix} \quad \begin{bmatrix} \sqrt{2} & 0 \\ 0 & -1 \end{bmatrix} \quad \begin{bmatrix} -\sqrt{2} & 0 \\ 0 & -1 \end{bmatrix}$$

EXAMPLE 6 *The Determinant of a Linear Transformation.*

We can use the properties of the determinant to show that similar matrices have the same determinant, for if $B = P^{-1}AP$, then $\det B = \det(P^{-1}AP)$. Now, the facts that the determinant of a product is the product of the determinants, and that scalars commute give us

$$\begin{aligned} \det B &= \det P^{-1} \det A \det P \\ &= \det A \det P^{-1} \det P \end{aligned}$$

Since $\det P^{-1} \det P = \det P^{-1}P = \det I = 1$ we must have $\det B = \det A$.

Suppose T is a linear transformation from V into V and that A is the matrix of T with respect to some basis for V. We then define $\det T$ by

$$\det T = \det A$$

The matrix A depends upon the choice of basis for V, but the number $\det T$ does not depend upon this basis choice, for if B is the matrix of T with respect to some other basis, we know that B is similar to A. Thus the above shows that $\det B = \det A$.

EXERCISES

1 Suppose $B = P^{-1}AP$. Why is it that P^{-1} and P do not necessarily cancel each other in this formula?

2 Suppose

$$A = \begin{bmatrix} 1 & 1 & 0 \\ 0 & 1 & 1 \\ 0 & 0 & 1 \end{bmatrix} \quad \text{and} \quad P = \begin{bmatrix} 1 & 1 & 1 \\ 2 & 3 & 3 \\ 1 & 0 & 1 \end{bmatrix}$$

a Find P^{-1}.
b Find $B = P^{-1}AP$.
c Verify that $B^2 = P^{-1}A^2P$.
d For $Q = P^{-1}$ verify that $A = Q\ ^{1}BQ$.
e Verify that $B^{-1} = P^{-1}A^{-1}P$.
f For $q(\lambda) = \lambda^3 - 3\lambda + 4$ verify that $q(B) = P^{-1}q(A)P$.

3 Suppose

$$B = \begin{bmatrix} 8 & 9 & 9 \\ 3 & 2 & 3 \\ -9 & -9 & -10 \end{bmatrix} \quad \text{and} \quad Q = \begin{bmatrix} -1 & -1 & -3 \\ 1 & 0 & -1 \\ 0 & 1 & 3 \end{bmatrix}$$

a Find Q^{-1}.
b Find $A = Q^{-1}BQ$.
c Put $P = Q^{-1}$ and verify that $B = P^{-1}AP$.
d Verify that $B^2 = P^{-1}A^2P$.
e Find B^6 by using formula **4**.
f For $q(\lambda) = 3\lambda^4 - \lambda^2 + 2\lambda$ find $q(A)$ and then use formula **7** to find $q(B)$.

4 Suppose

$$B = \begin{bmatrix} 19 & -12 \\ 24 & -15 \end{bmatrix} \quad \text{and} \quad Q = \begin{bmatrix} 3 & 2 \\ 4 & 3 \end{bmatrix}$$

a Show that $Q^{-1}BQ = \begin{bmatrix} 3 & 0 \\ 0 & 1 \end{bmatrix}$.

b Use the technique of Example **5** to find a matrix C such that $C^2 = B$.

c Find a matrix C_1 such that $C_1^3 = B$.

5 Suppose

$$A = \begin{bmatrix} a & 0 & 0 \\ 0 & b & 0 \\ 0 & 0 & c \end{bmatrix}$$

and that B is similar to A. What is det B?

6 Suppose A is similar to I. What is A?

7 Suppose $q(\lambda)$ is a polynomial such that $q(A) = 0$. If B is similar to A, what is $q(B)$?

8 If

$$A = \begin{bmatrix} 0 & 1 & 1 \\ 0 & 0 & 1 \\ 0 & 0 & 0 \end{bmatrix}$$

and B is similar to A:

a What is B^3?

b Show that $(I + B)^{-1} = I - B + B^2$.

9 Suppose

$$A = \begin{bmatrix} a & 0 & 0 \\ 0 & b & 0 \\ 0 & 0 & c \end{bmatrix}$$

a If $q(\lambda) = a_3\lambda^3 + a_2\lambda^2 + a_1\lambda + a_0$, show that

$$q(A) = \begin{bmatrix} q(a) & 0 & 0 \\ 0 & q(b) & 0 \\ 0 & 0 & q(c) \end{bmatrix}$$

b If $q(\lambda) = (\lambda - a)(\lambda - b)(\lambda - c)$ use the information of part **a** to find $q(A)$.

c If $q(\lambda) = (\lambda - a)(\lambda - b)(\lambda - c)$ and B is similar to A, what is $q(B)$?

10 Suppose D is the differentiation operator on the vector space of polynomials of degree not exceeding two. By finding the matrix of D with respect to a basis, show that det $D = 0$.

11 a Show that A is similar to A.

b Show that if A is similar to B and B is similar to C, then A is similar to C.

12 Suppose $B = P^{-1}AP$ and $C = P^{-1}DP$.

a Show that BC is similar to AD.

b Show that if $BC = CB$, then $AD = DA$.

13 Suppose A is similar to B. Show that the transpose matrix A^t is similar to B^t.

14 Suppose A is invertible. Show that AB is similar to BA. Give an example which shows that AB might not be similar to BA if A is not invertible.

□ 15 The *trace* of A is the sum of the diagonal entries of A. Establish each of the following for matrices with two rows and two columns.

a Trace AB = trace BA.

b Similar matrices have the same trace.

c There do not exist matrices A and B such that $AB - BA = I$.

d For a linear transformation T, the trace of T is the trace of the matrix of T relative to a basis. Does this depend upon the basis?

□ 16 Suppose

$$A = \begin{bmatrix} 1+i & 0 \\ 0 & -1 \end{bmatrix} \quad \text{and} \quad P = \begin{bmatrix} -i & 3 \\ 1 & 4i \end{bmatrix}$$

a Find P^{-1} and $B = P^{-1}AP$.

b Use the similarity relationship to find B^8.

SECTION 3 Characteristic Vectors

If V is a finite-dimensional vector space and T is a linear transformation from V into V, then, as was shown in Section 1, T can be represented, in terms of coordinates with respect to a given basis, as multiplication by a matrix A. The choice of a different basis results in a different matrix B, which is similar to A. In the next two sections, we shall discuss the problem of finding, if possible, a basis such that the matrix B is diagonal. In this section a concept is presented that will be useful in those later discussions.

We say that a vector $\bar{v}$ is a **characteristic vector** for T and that λ is a **characteristic value** of T if

$$T\bar{v} = \lambda\bar{v} \quad \text{and} \quad \bar{v} \neq \bar{0}$$

In other words, a nonzero vector $\bar{v}$ is a characteristic vector if $T\bar{v}$ is a *multiple* of $\bar{v}$.

This concept gives a useful formulation of the problem of finding a diagonal matrix representation for T, for

1 Suppose T is a linear transformation from V into V and that the matrix B of T with respect to the basis $\bar{u}_1, \bar{u}_2, \ldots, \bar{u}_n$ is diagonal. Then each $\bar{u}_i$ is a characteristic vector for T, and the diagonal entries of B are the corresponding characteristic values.

To simplify the proof of this result assume that $n = 2$. The columns of the matrix B are the coordinates of $T\bar{u}_1$ and $T\bar{u}_2$ with respect to the basis $\bar{u}_1$, $\bar{u}_2$. (See formula **6**, page 239.) Therefore

$$B = \begin{bmatrix} B_{11} & B_{12} \\ B_{21} & B_{22} \end{bmatrix} \quad \text{where} \quad \begin{aligned} T\bar{u}_1 &= B_{11}\bar{u}_1 + B_{21}\bar{u}_2 \\ T\bar{u}_2 &= B_{12}\bar{u}_1 + B_{22}\bar{u}_2 \end{aligned}$$

Thus, if B is diagonal, then $B_{12} = B_{21} = 0$, and therefore

2 $$T\bar{u}_1 = B_{11}\bar{u}_1 \quad \text{and} \quad T\bar{u}_2 = B_{22}\bar{u}_2$$

The vectors $\bar{u}_1$ and $\bar{u}_2$ are independent and hence nonzero. We therefore conclude that $\bar{u}_1$ and $\bar{u}_2$ are characteristic vectors, with corresponding characteristic values B_{11} and B_{22}, the diagonal entries of B.

The converse of statement **1** is also true, because

3 If V has a basis $\bar{u}_1, \bar{u}_2, \ldots, \bar{u}_n$ consisting of characteristic vectors for T, the matrix of T with respect to this basis is diagonal.

To simplify the proof of this, assume that $n = 2$. Suppose $\bar{u}_1$ and $\bar{u}_2$ are a basis for V consisting of characteristic vectors for T. Therefore $T\bar{u}_1$ is a multiple of $\bar{u}_1$ and $T\bar{u}_2$ is a multiple of $\bar{u}_2$, so we can write

$$T\bar{u}_1 = a\bar{u}_1 \quad \text{and} \quad T\bar{u}_2 = b\bar{u}_2$$

Rewrite this in the form

$$T\bar{u}_1 = a\bar{u}_1 + 0\bar{u}_2 \quad \text{and} \quad T\bar{u}_2 = 0\bar{u}_1 + b\bar{u}_2$$

Thus the matrix of T with respect to $\bar{u}_1$ and $\bar{u}_2$ is the diagonal matrix

$$\begin{bmatrix} a & 0 \\ 0 & b \end{bmatrix}$$

Some examples will now be given to illustrate these concepts, after which a method for finding characteristic vectors and characteristic values will be discussed.

EXAMPLE 1 Geometric methods can be used to find characteristic vectors and characteristic values for rotations, reflections, and projections. These make use of the fact that a nonzero vector $\bar{v}$ is a characteristic vector for T if and only if $T\bar{v}$ is a multiple of $\bar{v}$.

Suppose T is a counterclockwise rotation in R^2 through the angle θ and that θ is *not* a multiple of π. If $\bar{v} \neq \bar{0}$, $T\bar{v}$ cannot be a multiple of $\bar{v}$, as shown in Figure 1.

Figure 1

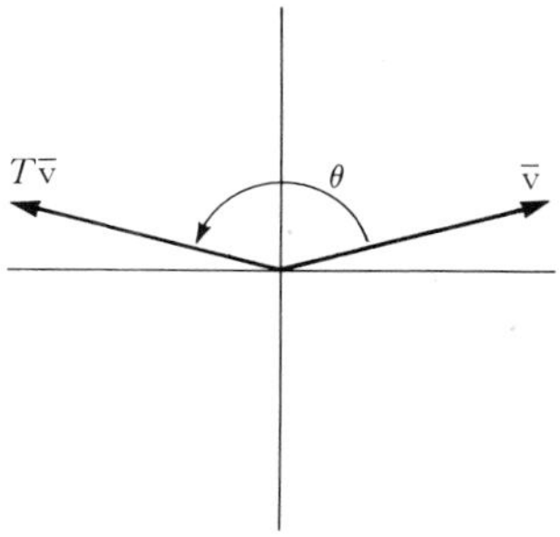

Figure 2

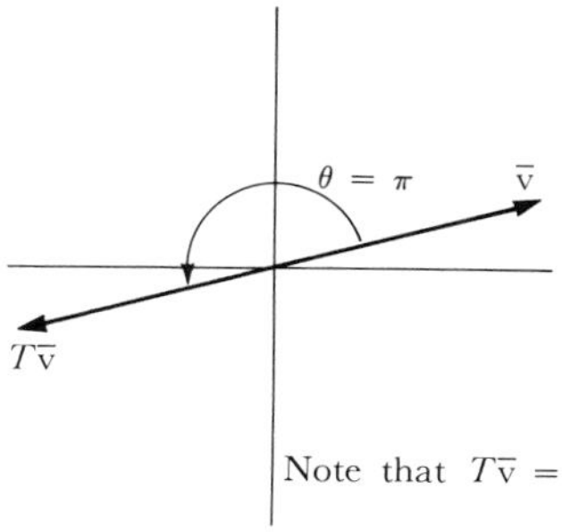

Note that $T\bar{v} = -\bar{v}$.

This fact establishes that

> A rotation through θ has no characteristic vectors, if θ is not a multiple of π.

Suppose $\theta = \pi$. Then, as Figure 2 indicates, for *every* nonzero vector $\bar{v}$ we have $T\bar{v} = -\bar{v}$. Since this is also true if θ is any odd multiple of π, we know that

> If θ is an odd multiple of π, then every nonzero vector is a characteristic vector belonging to the characteristic value $\lambda = -1$.

If θ is an even multiple of π, then T is just the identity operator. Thus for every nonzero vector $\bar{v}$ we have $T\bar{v} = \bar{v}$. In other words,

> If θ is an even multiple of π then every nonzero vector is a characteristic vector belonging to the characteristic value $\lambda = 1$.

EXAMPLE 2 Suppose T is the projection in R^2 onto the nonzero vector $\bar{w}$. A vector parallel to $\bar{w}$ is left fixed by T; that is, if $\bar{v}$ is a multiple of $\bar{w}$, $T\bar{v} = \bar{v}$. This shows that every nonzero multiple of $\bar{w}$ is a characteristic vector belonging to the characteristic value $\lambda = 1$.

If $\bar{v}$ is orthogonal to $\bar{w}$, then $T\bar{v} = \bar{0}$. Since $\bar{0} = 0\bar{v}$, this statement shows that every nonzero vector $\bar{v}$ which is orthogonal to $\bar{w}$ is a characteristic vector belonging to the characteristic value $\lambda = 0$.

If $\bar{v}$ is neither parallel nor orthogonal to $\bar{w}$, then, as Figure 3 indicates, $T\bar{v}$ is *not* parallel to $\bar{v}$, and such a vector $\bar{v}$ cannot be a characteristic vector.

In summary, we have shown that

> The projection T onto $\bar{w}$ has the characteristic values $\lambda = 0$ and $\lambda = 1$. The nonzero multiples of $\bar{w}$ are the characteristic vectors belonging to $\lambda = 1$, and the nonzero vectors orthogonal to $\bar{w}$ are the characteristic vectors belonging to $\lambda = 0$.

Figure 3

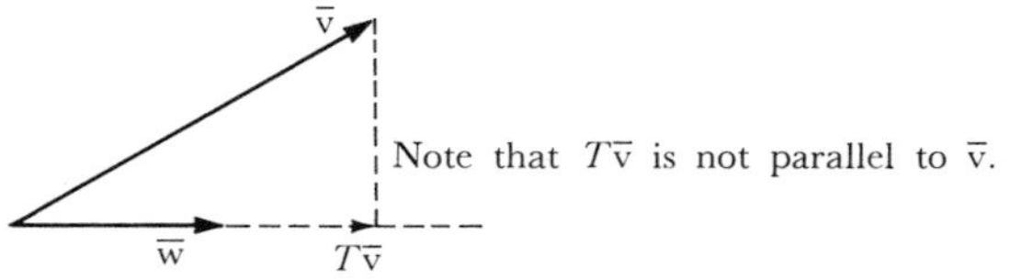
v̄
Note that $T\bar{v}$ is not parallel to $\bar{v}$.
w̄
$T\bar{v}$

DISCUSSION

We extend our terminology to matrices by saying that a vector $\bar{v}$ (written as a column matrix) is a **characteristic vector** for A belonging to the **characteristic value** λ if

$$A\bar{v} = \lambda\bar{v} \qquad \text{and} \qquad \bar{v} \neq \bar{0}$$

Since $I\bar{v} = \bar{v}$, we can rewrite the equation $A\bar{v} = \lambda\bar{v}$ as

$$(\lambda I - A)\bar{v} = \bar{0}$$

Therefore, if $\bar{v}$ is a characteristic vector for A belonging to λ, it follows that $\bar{v}$ is a nonzero solution to $(\lambda I - A)\bar{u} = \bar{0}$; that is, $\bar{v}$ is a nonzero vector in the null space of $\lambda I - A$. Such a $\bar{v}$ can exist only if $\lambda I - A$ is not invertible (Theorem **11**, page 188). We therefore have established that

4 If $\bar{v}$ is a characteristic vector for A belonging to λ, $\lambda I - A$ is not invertible.

If $\lambda I - A$ is not invertible and $\bar{v}$ is a nonzero vector such that $(\lambda I - A)\bar{v} = \bar{0}$, we can then rewrite this equation to conclude that $A\bar{v} = \lambda\bar{v}$ and $\bar{v} \neq \bar{0}$. This shows that

5 If $\lambda I - A$ is not invertible, any nonzero vector $\bar{v}$ in the null space of $\lambda I - A$ is a characteristic vector belonging to the characteristic value λ.

We know from Theorem **14** that $\lambda I - A$ is not invertible if and only if $\det(\lambda I - A) = 0$. The function

$$f(\lambda) = \det(\lambda I - A)$$

is called the **characteristic polynomial** of A. In summary:

THEOREM 15

The characteristic values of A are the roots of the characteristic polynomial $f(\lambda) = \det(\lambda I - A)$. If λ is a root of this polynomial, then any nonzero vector in the null space of $\lambda I - A$ is a characteristic vector belonging to λ.

We have previously noted that f is indeed a polynomial whose degree is the size of A. (See Example **4**, page 222.)

EXAMPLE 3 Theorem **15** provides a method for finding the characteristic values and corresponding characteristic vectors for a matrix. For example, suppose

$$A = \begin{bmatrix} 3 & -1 \\ 2 & 0 \end{bmatrix}$$

Form the matrix

$$\lambda I - A = \lambda \begin{bmatrix} 1 & 0 \\ 0 & 1 \end{bmatrix} - \begin{bmatrix} 3 & -1 \\ 2 & 0 \end{bmatrix} = \begin{bmatrix} \lambda - 3 & 1 \\ -2 & \lambda \end{bmatrix}$$

and take its determinant to obtain the characteristic polynomial of A:

$$\begin{aligned} f(\lambda) &= \det(\lambda I - A) \\ &= (\lambda - 3)\lambda + 2 \\ &= \lambda^2 - 3\lambda + 2 \\ &= (\lambda - 2)(\lambda - 1) \end{aligned}$$

Since the roots of $f(\lambda)$ are $\lambda = 2$ and $\lambda = 1$, Theorem **15** tells us that these are the characteristic values of A. We then find the corresponding characteristic vectors by finding the nonzero vectors in the null spaces of $2I - A$ and $1I - A$. We have

$$2I - A = 2\begin{bmatrix} 1 & 0 \\ 0 & 1 \end{bmatrix} - \begin{bmatrix} 3 & -1 \\ 2 & 0 \end{bmatrix} = \begin{bmatrix} -1 & 1 \\ -2 & 2 \end{bmatrix}$$

$$1I - A = 1\begin{bmatrix} 1 & 0 \\ 0 & 1 \end{bmatrix} - \begin{bmatrix} 3 & -1 \\ 2 & 0 \end{bmatrix} = \begin{bmatrix} -2 & 1 \\ -2 & 1 \end{bmatrix}$$

which reduce, respectively, to

$$\begin{bmatrix} 1 & -1 \\ 0 & 0 \end{bmatrix} \quad \text{and} \quad \begin{bmatrix} 1 & -\frac{1}{2} \\ 0 & 0 \end{bmatrix}$$

We see that

$$\bar{v}_1 = \begin{bmatrix} 1 \\ 1 \end{bmatrix} \quad \text{and} \quad \bar{v}_2 = \begin{bmatrix} 1 \\ 2 \end{bmatrix}$$

are respective bases for the null spaces of $2I - A$ and $1I - A$. Therefore,

> The characteristic values of A are $\lambda = 2$ and $\lambda = 1$. The nonzero multiples of $\bar{v}_1$ are the characteristic vectors belonging to $\lambda = 2$, while the nonzero multiples of $\bar{v}_2$ are the characteristic vectors belonging to $\lambda = 1$.

This information will be used in the next section to show that A is similar to

$$\begin{bmatrix} 2 & 0 \\ 0 & 1 \end{bmatrix}$$

EXAMPLE 4 In using Theorem **15** it is necessary to calculate $\det(\lambda I - A)$. This can be done using the definition of determinant. (See formula **2**, page 217.) The formulas of Exercise **10**, page 230, are, however, worth remembering:

6 If $$A = \begin{bmatrix} a & b \\ c & d \end{bmatrix}$$

then $\det(\lambda I - A) = \lambda^2 - (a + d)\lambda + ad - bc$

If $$A = \begin{bmatrix} A_{11} & A_{12} & A_{13} \\ A_{21} & A_{22} & A_{23} \\ A_{31} & A_{32} & A_{33} \end{bmatrix} \text{ then}$$

7 $$\det(\lambda I - A) = \lambda^3 - (A_{11} + A_{22} + A_{33})\lambda^2 + \left\{\det\begin{bmatrix} A_{11} & A_{12} \\ A_{21} & A_{22} \end{bmatrix} + \det\begin{bmatrix} A_{11} & A_{13} \\ A_{31} & A_{33} \end{bmatrix} + \det\begin{bmatrix} A_{22} & A_{23} \\ A_{32} & A_{33} \end{bmatrix}\right\}\lambda - \det A$$

For example, if

$$A = \begin{bmatrix} 8 & 9 & 9 \\ 3 & 2 & 3 \\ -9 & -9 & -10 \end{bmatrix}$$

formula **7** gives

$$\det(\lambda I - A) = \lambda^3 - (8 + 2 - 10)\lambda^2 + \left\{\det\begin{bmatrix} 8 & 9 \\ 3 & 2 \end{bmatrix} + \det\begin{bmatrix} 8 & 9 \\ -9 & -10 \end{bmatrix} + \det\begin{bmatrix} 2 & 3 \\ -9 & -10 \end{bmatrix}\right\}\lambda - \det A$$

Calculation of these determinants gives

$$\det(\lambda I - A) = \lambda^3 - 3\lambda - 2 = (\lambda + 1)^2(\lambda - 2)$$

Theorem **15** then tells us that the characteristic values of A are $\lambda = -1$ and $\lambda = 2$. Corresponding characteristic vectors are found by finding the nonzero vectors in the null spaces of $-1I - A$ and $2I - A$.

EXAMPLE 5 The characteristic values of an upper (or lower) triangular matrix are easy to find. For example, if

$$A = \begin{bmatrix} 2 & 0 & 1 & 2 \\ 0 & 2 & -1 & 3 \\ 0 & 0 & -3 & 1 \\ 0 & 0 & 0 & 4 \end{bmatrix}$$

then

$$f(\lambda) = \det(\lambda I - A) = \det \begin{bmatrix} \lambda - 2 & 0 & -1 & -2 \\ 0 & \lambda - 2 & 1 & -3 \\ 0 & 0 & \lambda + 3 & -1 \\ 0 & 0 & 0 & \lambda - 4 \end{bmatrix}$$

The determinant of an upper triangular matrix is the product of the diagonal entries. (See property **4c**, page 218.) Therefore

$$f(\lambda) = (\lambda - 2)(\lambda - 2)(\lambda + 3)(\lambda - 4)$$

so that the characteristic values are 2, -3, and 4.

In general,

8 The characteristic values of an upper (or lower) triangular matrix are the diagonal entries of the matrix.

EXERCISES

1 For each of the following operators T describe the characteristic values and vectors. Figures may be helpful.
   - **a** T is reflection in R^2 through the line through $\overline{w}$.
   - **b** T is projection in R^2 *orthogonal* to $\overline{w}$.
   - **c** T is reflection in R^3 through the line through $\overline{w}$.
   - **d** T is projection in R^3 onto $\overline{w}$.
   - **e** T is projection in R^3 orthogonal to $\overline{w}$.
   - **f** T is counterclockwise rotation in R^2 through $\pi/4$ followed by reflection in the x-axis.

2 Suppose $Df = f'$. Show that every real number is a characteristic value of D. [Hint: Calculate $D(e^{ax})$.]

3 Show that $\begin{bmatrix} 1 \\ 1 \end{bmatrix}$ and $\begin{bmatrix} 1 \\ -1 \end{bmatrix}$ are characteristic vectors for $A = \begin{bmatrix} 2 & 1 \\ 1 & 2 \end{bmatrix}$

What are the corresponding characteristic values?

4 Use Theorem **15** and formulas **6**, **7**, or **8** to find the characteristic polynomial and the characteristic values for each of the following matrices.

a $\begin{bmatrix} \frac{1}{2} & \frac{1}{2} \\ \frac{1}{2} & \frac{1}{2} \end{bmatrix}$ **b** $\begin{bmatrix} 3 & 1 \\ 1 & 3 \end{bmatrix}$ **c** $\begin{bmatrix} 3 & 1 & 0 \\ 1 & 3 & 0 \\ 0 & 0 & 2 \end{bmatrix}$

d $\begin{bmatrix} 5 & 1 & 1 \\ -3 & 1 & -3 \\ -2 & -2 & 2 \end{bmatrix}$ **e** $\begin{bmatrix} 1 & 2 & 1 \\ 0 & 2 & 1 \\ 0 & 0 & 3 \end{bmatrix}$ **f** $\begin{bmatrix} 0 & 1 & 2 & 1 \\ 0 & 0 & 1 & 3 \\ 0 & 0 & 0 & 1 \\ 0 & 0 & 0 & 0 \end{bmatrix}$

5 For each of the matrices of Exercise **4** find at least one characteristic vector for each characteristic value.

6 Show that $\begin{bmatrix} 1 & -2 \\ 1 & -1 \end{bmatrix}$ has no real characteristic values. (See also Exercise **11**.)

7 **a** Show that if $A\bar{v} = \lambda\bar{v}$, then $A^2\bar{v} = \lambda^2\bar{v}$.

b Suppose $\lambda_1, \lambda_2, \ldots, \lambda_k$ are the characteristic values of A. What are the characteristic values of A^2?

c Suppose

$$A = \begin{bmatrix} 1 & 2 & 1 \\ 0 & -1 & 0 \\ 0 & 0 & 3 \end{bmatrix}$$

What are the characteristic values of A^2? of A^3?

8 **a** Show that if $A\bar{v} = \lambda\bar{v}$, then

$$(A^3 - 3A^2 + A - 2I)\bar{v} = (\lambda^3 - 3\lambda^2 + \lambda - 2)\bar{v}$$

b Suppose $q(\lambda)$ is a polynomial and $\lambda_1, \lambda_2, \ldots, \lambda_k$ are the characteristic values of A. What are the characteristic values of $q(A)$? (Hint: See part **a**.)

9 **a** Show that similar matrices have the same characteristic polynomial. [Hint: $\lambda I - P^{-1}AP = P^{-1}(\lambda I - A)P$.]

b How would you define the characteristic polynomial of a linear transformation T from V into V? Does your definition depend upon the choice of basis for V?

10 Suppose A is similar to

$$\begin{bmatrix} 2 & 0 & 0 \\ 0 & -1 & 0 \\ 0 & 0 & 3 \end{bmatrix}$$

What are the characteristic values of A? (Hint: See Exercise **9a**.)

□ **11** Show that $\begin{bmatrix} 2 \\ 1-i \end{bmatrix}$ and $\begin{bmatrix} 2 \\ 1+i \end{bmatrix}$ are complex characteristic vectors for $\begin{bmatrix} 1 & -2 \\ 1 & -1 \end{bmatrix}$. What are the corresponding characteristic values? Compare this with Exercise **6**.

□ **12** Find the characteristic polynomial, the complex characteristic values, and at least one corresponding characteristic vector for each of the following.

a $\begin{bmatrix} 2-i & 2 \\ 1+3i & -1+3i \end{bmatrix}$ **b** $\begin{bmatrix} 0 & -1 \\ 1 & 0 \end{bmatrix}$

□ **13** An important theorem, known as the Cayley-Hamilton Theorem, asserts that if $f(\lambda)$ is the characteristic polynomial of A, $f(A)$ is the zero matrix. Verify that this is so for

$$A = \begin{bmatrix} 2 & -1 \\ 1 & 3 \end{bmatrix}$$

SECTION 4 Matrices Similar to Diagonal Matrices

We now turn to the problem of finding, for a given matrix A, an invertible matrix P such that $B = P^{-1}AP$ is a diagonal matrix. If T is a linear transformation from V into V such that A is the matrix of T with respect to some basis, the problem is simply to find another basis for which the matrix of T is diagonal. (See statement **9**, page 243.) We have seen that this is just the same as the problem of finding a basis for V consisting of characteristic vectors for T. (See statements **1** and **3** of the previous section.) For matrices this translates to:

1 A matrix A with n rows and n columns is similar to a diagonal matrix if and only if R^n has a basis consisting of characteristic vectors for A.

If R^n has such a basis, and we let P be the matrix with these basis vectors as columns, P is invertible and $P^{-1}AP$ is a diagonal matrix whose diagonal entries are the corresponding characteristic values of A.

Not every matrix is similar to a diagonal matrix (see Example **2**, below), and hence not every linear transformation has a diagonal representation. Knowledge of the roots of the characteristic polynomial is not enough to decide whether a given matrix is similar to a diagonal matrix. If these roots are real and distinct, however, the matrix can be diagonalized, as the following theorem implies. The proof can be found in the Appendix.

THEOREM 16 Suppose A has n rows and n columns and its characteristic polynomial has n *distinct*, real roots $a_1, a_2, \ldots, a_n$. If $\bar{v}_1, \bar{v}_2, \ldots, \bar{v}_n$ are corresponding characteristic vectors, these vectors are independent and hence are a basis for R^n.

If A satisfies the conditions of Theorem **16**, we can let P be the matrix with $\bar{v}_1, \bar{v}_2, \ldots, \bar{v}_n$ as columns and then $P^{-1}AP$ will be the diagonal matrix with diagonal entries $a_1, a_2, \ldots, a_n$.

We now give some examples which indicate a systematic procedure for diagonalizing a matrix or a transformation.

EXAMPLE 1 Let us show how to find an invertible matrix P such that $B = P^{-1}AP$ is diagonal, where

$$A = \begin{bmatrix} 22 & 20 \\ -25 & -23 \end{bmatrix}$$

First find the characteristic polynomial (using formula **6**, page 262):

$$\begin{aligned} f(\lambda) &= \det(\lambda I - A) \\ &= \lambda^2 - (22 - 23)\lambda + \det A \\ &= \lambda^2 + \lambda - 6 \\ &= (\lambda - 2)(\lambda + 3) \end{aligned}$$

This gives the characteristic values $\lambda = 2$ and $\lambda = -3$. Now find one characteristic vector for each characteristic value by finding a nonzero vector in the null spaces of $2I - A$ and $-3I - A$. We have

$$2I - A = \begin{bmatrix} -20 & -20 \\ 25 & 25 \end{bmatrix} \quad \text{and} \quad -3I - A = \begin{bmatrix} -25 & -20 \\ 25 & 20 \end{bmatrix}$$

which reduce, respectively, to

$$\begin{bmatrix} 1 & 1 \\ 0 & 0 \end{bmatrix} \quad \text{and} \quad \begin{bmatrix} 1 & \frac{4}{5} \\ 0 & 0 \end{bmatrix}$$

Thus

$$\bar{v}_1 = \begin{bmatrix} 1 \\ -1 \end{bmatrix} \quad \text{and} \quad \bar{v}_2 = \begin{bmatrix} -4 \\ 5 \end{bmatrix}$$

are, respectively, nonzero vectors in the null space of $2I - A$ and $-3I - A$, and consequently,

2 $$A\bar{v}_1 = 2\bar{v}_1 \quad \text{and} \quad A\bar{v}_2 = -3\bar{v}_2$$

The vectors $\bar{v}_1$ and $\bar{v}_2$ are easily seen to be independent (or we can apply Theorem **16** to reach the same conclusion) and hence are a basis for R^2 consisting of characteristic vectors for A. If we let P be the matrix whose columns are $\bar{v}_1$ and $\bar{v}_2$, then

$$P = \begin{bmatrix} 1 & -4 \\ -1 & 5 \end{bmatrix} \quad \text{and a calculation gives} \quad P^{-1} = \begin{bmatrix} 5 & 4 \\ 1 & 1 \end{bmatrix}$$

Since the columns of P are $\bar{v}_1$ and $\bar{v}_2$, relations **2** tell us that the columns of AP are $2\bar{v}_1$ and $-3\bar{v}_2$. We know that

$$P^{-1}\bar{v}_1 = \begin{bmatrix} 1 \\ 0 \end{bmatrix} \quad \text{and} \quad P^{-1}\bar{v}_2 = \begin{bmatrix} 0 \\ 1 \end{bmatrix}$$

Therefore,

$$P^{-1}(2\bar{v}_1) = \begin{bmatrix} 2 \\ 0 \end{bmatrix} \quad \text{and} \quad P^{-1}(-3\bar{v}_2) = \begin{bmatrix} 0 \\ -3 \end{bmatrix}$$

In other words,

$$P^{-1}AP = \begin{bmatrix} 2 & 0 \\ 0 & -3 \end{bmatrix}$$

Note that the diagonal entries of $P^{-1}AP$ are the characteristic values corresponding to $\bar{v}_1$ and $\bar{v}_2$.

Interchanging the columns of P will interchange the diagonal entries. For example, if we put

$$P = \begin{bmatrix} -4 & 1 \\ 5 & -1 \end{bmatrix}$$

we have (the student should verify this)

$$P^{-1}AP = \begin{bmatrix} -3 & 0 \\ 0 & 2 \end{bmatrix}$$

EXAMPLE 2 It may not be possible to find enough characteristic vectors for A to form a basis. This may occur for two reasons: the characteristic polynomial may have complex roots, or there may be an insufficient number of independent characteristic vectors (which *may* occur when the characteristic polynomial has repeated roots). For example, if

$$A = \begin{bmatrix} 1 & -2 \\ 1 & -1 \end{bmatrix}$$

the characteristic polynomial is $f(\lambda) = \det(\lambda I - A) = \lambda^2 - (1 - 1)\lambda + \det A = \lambda^2 + 1$, which has no real roots. Hence there are no characteristic vectors in R^2. Thus, from property **1**, there does not exist a *real* invertible matrix P such that $P^{-1}AP$ is diagonal. One can, however, find a complex invertible matrix P such that $P^{-1}AP$ is diagonal by proceeding, as in Example **1**, using complex matrices. (See Exercise **11**.) Suppose

$$A_1 = \begin{bmatrix} 1 & 1 \\ 0 & 1 \end{bmatrix}$$

Since its characteristic polynomial is $f(\lambda) = \det(\lambda I - A_1) = (\lambda - 1)^2$, $\lambda = 1$ is the only characteristic value for A_1. We have

$$1I - A_1 = \begin{bmatrix} 0 & -1 \\ 0 & 0 \end{bmatrix} \quad \text{which reduces to} \quad \begin{bmatrix} 0 & 1 \\ 0 & 0 \end{bmatrix}$$

Thus, the nonzero vectors in the null space of $1I - A_1$ are the nonzero multiples of

$$\bar{v} = \begin{bmatrix} 1 \\ 0 \end{bmatrix}$$

We have shown that

> The only characteristic vectors for A_1 are the nonzero multiples of $\bar{v} = \begin{bmatrix} 1 \\ 0 \end{bmatrix}$.

Since $\begin{bmatrix} 1 \\ 0 \end{bmatrix}$ is not a basis for R^2, statement **1** tells us that

$$A_1 = \begin{bmatrix} 1 & 1 \\ 0 & 1 \end{bmatrix}$$

is not similar to a diagonal matrix.

EXAMPLE 3 Let us look at a three-dimensional example. Suppose

$$A = \begin{bmatrix} 1 & 0 & 1 \\ 0 & 2 & 3 \\ 0 & 0 & -1 \end{bmatrix}$$

The characteristic values of A are the diagonal entries $\lambda = 1, 2, -1$, since A is upper triangular. (See statement **8**, page 263.) From Theorem **16** we deduce that A is similar to

$$B = \begin{bmatrix} 1 & 0 & 0 \\ 0 & 2 & 0 \\ 0 & 0 & -1 \end{bmatrix}$$

To find an invertible matrix P such that $B = P^{-1}AP$ requires finding one nonzero vector in the null space of each of $I - A$, $2I - A$, and $-I - A$. We have

$$I - A = \begin{bmatrix} 0 & 0 & -1 \\ 0 & -1 & -3 \\ 0 & 0 & 2 \end{bmatrix} \qquad 2I - A = \begin{bmatrix} 1 & 0 & -1 \\ 0 & 0 & -3 \\ 0 & 0 & 3 \end{bmatrix}$$

$$-I - A = \begin{bmatrix} -2 & 0 & -1 \\ 0 & -3 & -3 \\ 0 & 0 & 0 \end{bmatrix}$$

which reduce, respectively, to

$$\begin{bmatrix} 0 & 1 & 0 \\ 0 & 0 & 1 \\ 0 & 0 & 0 \end{bmatrix} \quad \begin{bmatrix} 1 & 0 & 0 \\ 0 & 0 & 1 \\ 0 & 0 & 0 \end{bmatrix} \quad \begin{bmatrix} 1 & 0 & \frac{1}{2} \\ 0 & 1 & 1 \\ 0 & 0 & 0 \end{bmatrix}$$

These give the respective nonzero solutions

$$\begin{bmatrix} 1 \\ 0 \\ 0 \end{bmatrix} \quad \begin{bmatrix} 0 \\ 1 \\ 0 \end{bmatrix} \quad \begin{bmatrix} -1 \\ -2 \\ 2 \end{bmatrix}$$

Therefore, if we put

$$P = \begin{bmatrix} 1 & 0 & -1 \\ 0 & 1 & -2 \\ 0 & 0 & 2 \end{bmatrix} \qquad \text{then} \qquad P^{-1} = \begin{bmatrix} 1 & 0 & \frac{1}{2} \\ 0 & 1 & 1 \\ 0 & 0 & \frac{1}{2} \end{bmatrix}$$

and $B = P^{-1}AP$. (The student should, of course, verify that these calculations are correct.)

EXAMPLE 4 Theorem **16** applies only if the roots of the characteristic polynomial are real and distinct. If some of the roots are complex, as in Example **2**, the matrix is not similar to a real diagonal matrix. If some roots are repeated, the matrix may or may not be similar to a diagonal matrix. For example, the matrices

$$\begin{bmatrix} 1 & 1 \\ 0 & 1 \end{bmatrix} \qquad \text{and} \qquad \begin{bmatrix} 1 & 0 \\ 0 & 1 \end{bmatrix}$$

have the same characteristic polynomial, $(\lambda - 1)^2$. The latter matrix is diagonal, while the former is not similar to a diagonal matrix. (See Example **2**.)

In general a matrix is similar to a diagonal matrix if for each root of the characteristic polynomial of multiplicity k we can find k independent characteristic vectors. (The proof of this fact is omitted.) For example, consider the matrix

$$A = \begin{bmatrix} 8 & 9 & 9 \\ 3 & 2 & 3 \\ -9 & -9 & -10 \end{bmatrix}$$

We saw in Example **4**, page 262, that its characteristic polynomial is $(\lambda + 1)^2(\lambda - 2)$: thus the characteristic roots are $\lambda = -1$ and $\lambda = 2$. Note that

$$-I - A = \begin{bmatrix} -9 & -9 & -9 \\ -3 & -3 & -3 \\ 9 & 9 & 9 \end{bmatrix} \qquad \text{which reduces to} \qquad \begin{bmatrix} 1 & 1 & 1 \\ 0 & 0 & 0 \\ 0 & 0 & 0 \end{bmatrix}$$

Thus, the vectors

$$\bar{v}_1 = \begin{bmatrix} -1 \\ 1 \\ 0 \end{bmatrix} \qquad \text{and} \qquad \bar{v}_2 = \begin{bmatrix} -1 \\ 0 \\ 1 \end{bmatrix}$$

are a basis for the null space of $-I - A$ and are therefore independent characteristic vectors belonging to $\lambda = -1$.

By reducing the matrix $2I - A$ we find the characteristic vector (belonging to $\lambda = 2$)

$$\bar{v}_3 = \begin{bmatrix} -3 \\ -1 \\ 3 \end{bmatrix}$$

The student can show that $\bar{v}_1$, $\bar{v}_2$, and $\bar{v}_3$ are independent. Therefore, if we put

$$P = \begin{bmatrix} -1 & -1 & -3 \\ 1 & 0 & -1 \\ 0 & 1 & 3 \end{bmatrix} \quad \text{then} \quad P^{-1}AP = \begin{bmatrix} -1 & 0 & 0 \\ 0 & -1 & 0 \\ 0 & 0 & 2 \end{bmatrix}$$

EXAMPLE 5 To find a diagonal matrix representation for a linear transformation T we can proceed directly by some means to find a basis consisting of characteristic vectors; or we can first select a basis, then find the matrix A of T with respect to this basis, and, finally, apply the above methods, as shown in Example **6**, below.

For example, if T is a counterclockwise rotation in R^2 through the angle θ, then, as we saw in Example **1** of the previous section, T has *no* characteristic values if θ is not a multiple of π. Therefore, unless θ is a multiple of π, T has *no* diagonal matrix representation. If θ is an even multiple of π, $T = I$, and its matrix, with respect to *any* basis, is the identity matrix (which is certainly diagonal). If θ is an odd multiple of π, then $T = -I$, and consequently the matrix of T with respect to any basis is $-I$.

Suppose T is projection in R^2 onto the nonzero vector $\bar{w}$. Let $\bar{w}_1$ be a nonzero vector orthogonal to $\bar{w}$. We then know that $T\bar{w} = \bar{w}$, $T\bar{w}_1 = \bar{0}$, and that $\bar{w}$ and $\bar{w}_1$ are independent. Thus $\bar{w}$ and $\bar{w}_1$ form a basis for R^2. The matrix of T with respect to $\bar{w}$ and $\bar{w}_1$ is

$$\begin{bmatrix} 1 & 0 \\ 0 & 0 \end{bmatrix}$$

EXAMPLE 6 Suppose V is the vector space of polynomials of degree not exceeding two and D is the differentiation operator on V. The matrix of D with respect to the basis 1, x, x^2 is (see Example **4**, page 244)

$$D_0 = \begin{bmatrix} 0 & 1 & 0 \\ 0 & 0 & 2 \\ 0 & 0 & 0 \end{bmatrix}$$

This matrix has only one characteristic root, namely $\lambda = 0$ (see property **8**, page 263), and $0I - D_0$ reduces to

$$\begin{bmatrix} 0 & 1 & 0 \\ 0 & 0 & 1 \\ 0 & 0 & 0 \end{bmatrix}$$

so that every characteristic vector of D_0 is a nonzero multiple of

$$\begin{bmatrix} 1 \\ 0 \\ 0 \end{bmatrix}$$

Since this is not a basis for R^3, we conclude that D_0 is *not* similar to a diagonal matrix. Therefore, the differentiation operator D on V has *no* diagonal matrix representation.

EXERCISES

1 For each of the following matrices A find an invertible matrix P such that $P^{-1}AP$ is a diagonal matrix B.

a $\begin{bmatrix} \frac{1}{2} & \frac{1}{2} \\ \frac{1}{2} & \frac{1}{2} \end{bmatrix}$ **b** $\begin{bmatrix} 3 & 1 \\ 1 & 3 \end{bmatrix}$ **c** $\begin{bmatrix} 3 & 1 & 0 \\ 1 & 3 & 0 \\ 0 & 0 & 2 \end{bmatrix}$

d $\begin{bmatrix} 1 & 2 & 1 \\ 0 & 2 & 1 \\ 0 & 0 & 3 \end{bmatrix}$ **e** $\begin{bmatrix} 5 & 1 & 1 \\ -3 & 1 & -3 \\ -2 & -2 & 2 \end{bmatrix}$ **f** $\begin{bmatrix} 3 & 11 & 3 \\ 0 & -4 & -3 \\ 2 & 6 & 1 \end{bmatrix}$

2 For the matrix A of Example **3**, Section 3, find an invertible matrix P such that $P^{-1}AP$ is diagonal.

3 Show that $\begin{bmatrix} 0 & 1 & 1 \\ 0 & 0 & 2 \\ 0 & 0 & 1 \end{bmatrix}$ *is not* similar to a diagonal matrix.

4 Show that

$$A = \begin{bmatrix} a & d & e \\ 0 & b & f \\ 0 & 0 & c \end{bmatrix}$$

is similar to a diagonal matrix if $(a - b)(b - c)(a - c) \neq 0$.

5 Find A^{10} if:

a $A = \begin{bmatrix} 2 & 1 \\ 1 & 2 \end{bmatrix}$ **b** $A = \begin{bmatrix} 2 & -1 & 1 \\ 3 & -2 & 3 \\ 3 & -1 & 0 \end{bmatrix}$

(Hint: First find a diagonal matrix similar to A.)

6 Find a matrix C such that $C^2 = A$, where

$$A = \begin{bmatrix} 2 & 4 & 2 \\ -7 & -12 & -5 \\ 25 & 40 & 15 \end{bmatrix}$$

7 For each of the operators T of Exercise **1**, page 263, find a diagonal matrix which is the matrix of T with respect to some basis.

8 Suppose V is the set of all functions of the form $ae^x + be^{-x}$. Find the matrix of the differentiation operator D with respect to the basis e^x, e^{-x}. Why does this not contradict the results of Example **6**?

9 Suppose V is the vector space of Example **6** and T is the operation that replaces x by $2x - 1$; that is, $T(a + bx + cx^2) = a + b(2x - 1) + c(2x - 1)^2$.

a Show that T is linear.

b Find the matrix of T with respect to $1, x, x^2$.

c Can you deduce from this matrix that there is a basis for which the matrix of T is diagonal?

d Find three independent polynomials in V which are characteristic vectors for T.

10 Show that if B is diagonal and $B^2 = 0$ then $B = 0$. Deduce from this that if A is similar to a diagonal matrix and $A^2 = 0$ then $A = 0$.

□ **11** Find a *complex* invertible matrix P such that $P^{-1}AP$ is diagonal where

$$A = \begin{bmatrix} 1 & -2 \\ 1 & -1 \end{bmatrix}$$

□ **12** Show that if $A^2 = A$, then A is similar to a diagonal matrix with 1's and 0's on the diagonal. [Hint: Show that if $\bar{u}_1, \bar{u}_2, \ldots, \bar{u}_m$ are a basis for the null space of A, and $\bar{v}_1, \bar{v}_2, \ldots, \bar{v}_k$ are a basis for the null space of $I - A$, these vectors form a basis consisting of characteristic vectors. The relation $I = A + (I - A)$ is useful.]

SECTION 5 Symmetric Matrices

Theorem **16** gives a condition which guarantees that a matrix is similar to a diagonal matrix. This theorem asserts that a matrix is similar to a diagonal matrix if its characteristic polynomial has distinct real roots. If the roots are real and some are repeated, then, as shown in Example **4** of the previous section, further effort is required to see whether there are enough independent characteristic vectors to give a basis. This procedure is of limited use in higher dimensions, because determination of the roots of polynomials of large degree is, in general, quite difficult. Fortunately, there are theorems which, from simple conditions on the entries of the matrix, assert that the matrix is similar to a diagonal matrix. For example, if a matrix is upper (or lower) triangular, with distinct diagonal entries, it is similar to a diagonal matrix. (See Exercise **4** of the previous section.)

A much deeper theorem asserts that if a matrix is symmetric, then it is similar to a diagonal matrix. The known proofs of this theorem all require a more sophisticated analysis than we can give at this point, for it must be shown that the characteristic polynomial of a symmetric matrix has *only* real roots and that for each root of multiplicity k, one can find k independent characteristic vectors. The common occurrence of symmetric matrices in applications makes this result particularly important.

Recall that the transpose A^t of A is the matrix obtained from A by interchanging the rows and columns of A. (See page 198.) We say that A is **symmetric** if $A^t = A$. For example,

$$A = \begin{bmatrix} 2 & 1 \\ 1 & 3 \end{bmatrix} \qquad \text{is symmetric since} \qquad A^t = \begin{bmatrix} 2 & 1 \\ 1 & 3 \end{bmatrix} = A$$

while

$$A = \begin{bmatrix} 2 & 1 \\ 2 & 3 \end{bmatrix} \qquad \text{is } \textit{not} \text{ symmetric since} \qquad A^t = \begin{bmatrix} 2 & 2 \\ 1 & 3 \end{bmatrix} \neq A$$

For symmetric matrices we have the following theorem:

THEOREM 17 If A is symmetric, there is an invertible matrix P such that $B = P^{-1}AP$ is a diagonal matrix.

This theorem is usually called the *spectral theorem.* (A generalization of this result has applications to calculating atomic spectra.) A two-dimensional proof is given in Example **3** below. The reader is referred to the Bibliography for general proofs.

The spectral theorem tell us, in particular, that if a matrix is symmetric, its characteristic polynomial must have only real roots. Many methods for approximating these roots have been developed, modern high-speed computers being particularly useful in this task. (See the Bibliography for references to some of these methods.)

If A is symmetric we can actually find an orthogonal matrix P such that $P^{-1}AP$ is diagonal. (Recall that an orthogonal matrix is a matrix whose columns are orthonormal, page 199.) This result is a consequence of some elementary tricks with the dot product.

Suppose A has n rows and n columns. For each $\bar{u}$ and $\bar{v}$ in R^n we then have

1 $$(A\bar{u} \cdot \bar{v}) = \bar{u} \cdot (A^t\bar{v})$$

For example, suppose

$$A = \begin{bmatrix} a & b \\ c & d \end{bmatrix} \qquad \bar{u} = \begin{bmatrix} u_1 \\ u_2 \end{bmatrix} \qquad \bar{v} = \begin{bmatrix} v_1 \\ v_2 \end{bmatrix}$$

Then

$$A\bar{u} = \begin{bmatrix} a & b \\ c & d \end{bmatrix} \begin{bmatrix} u_1 \\ u_2 \end{bmatrix} = \begin{bmatrix} au_1 + bu_2 \\ cu_1 + du_2 \end{bmatrix}$$

and therefore

2 $$(A\bar{u}) \cdot \bar{v} = (au_1 + bu_2)v_1 + (cu_1 + du_2)v_2$$

Also, it follows from

$$A^t\bar{v} = \begin{bmatrix} a & c \\ b & d \end{bmatrix} \begin{bmatrix} v_1 \\ v_2 \end{bmatrix} = \begin{bmatrix} av_1 + cv_2 \\ bv_1 + dv_2 \end{bmatrix}$$

that

$$\bar{u} \cdot (A^t\bar{v}) = u_1(av_1 + cv_2) + u_2(bv_1 + dv_2)$$

This is just a rewritten version of the right side of equation **2**; thus

$$(A\bar{u}) \cdot \bar{v} = \bar{u} \cdot (A^t\bar{v})$$

Let us use this to show that the following statement is true.

3 If A is symmetric and a_1 and a_2 are distinct characteristic values of A, with corresponding characteristic vectors $\bar{u}_1$ and $\bar{u}_2$, then $\bar{u}_1$ and $\bar{u}_2$ must be orthogonal.

By assumption, $a_1 \neq a_2$ and

4 $$A\bar{u}_1 = a_1\bar{u}_1 \qquad A\bar{u}_2 = a_2\bar{u}_2$$

Furthermore,

$$a_1(\bar{u}_1 \cdot \bar{u}_2) - a_2(\bar{u}_1 \cdot \bar{u}_2) = (a_1\bar{u}_1 \cdot \bar{u}_2) - (\bar{u}_1 \cdot a_2\bar{u}_2)$$

Substitute relations **4** into the right-hand side of this and obtain

$$a_1(\bar{u}_1 \cdot \bar{u}_2) - a_2(\bar{u}_1 \cdot \bar{u}_2) = (A\bar{u}_1 \cdot \bar{u}_2) - (\bar{u}_1 \cdot A\bar{u}_2)$$

Now use equation **1** and the assumption that $A = A^t$ to conclude that the right-hand side must be zero. Thus

$$(a_1 - a_2)(\bar{u}_1 \cdot \bar{u}_2) = a_1(\bar{u}_1 \cdot \bar{u}_2) - a_2(\bar{u}_1 \cdot \bar{u}_2) = 0$$

Therefore, since $a_1 \neq a_2$, we must have $\bar{u}_1 \cdot \bar{u}_2 = 0$.

In other words, we have shown that $\bar{u}_1$ and $\bar{u}_2$ must be orthogonal, and the proof of statement **3** is completed.

For a given symmetric matrix A, one can modify slightly the procedure of Section 4 to find an orthogonal matrix P such that $P^{-1}AP$ is diagonal. After finding the characteristic polynomial of A and its roots, we then find, for each characteristic value λ, an *orthonormal* basis for the null space of $\lambda I - A$. Theorem **17** guarantees that the collection of such characteristic vectors is then a basis for R^n, while statement **3** guarantees that these basis vectors are orthonormal. Thus the matrix P whose columns are these basis vectors is an orthogonal matrix such that $P^{-1}AP$ is diagonal. The fact that P is orthogonal makes the calculation of P^{-1} quite simple, for we know that

5 If P is orthogonal, then $P^{-1} = P^t$

EXAMPLE 1 An example will illustrate the calculation of an orthogonal matrix P such that $P^{-1}AP$ is diagonal. Suppose

$$A = \begin{bmatrix} 3 & 1 \\ 1 & 3 \end{bmatrix}$$

Then A is symmetric. The characteristic polynomial of A is

$$\begin{aligned} \det(\lambda I - A) &= \det\left\{\begin{bmatrix} \lambda & 0 \\ 0 & \lambda \end{bmatrix} - \begin{bmatrix} 3 & 1 \\ 1 & 3 \end{bmatrix}\right\} \\ &= \det\begin{bmatrix} \lambda - 3 & -1 \\ -1 & \lambda - 3 \end{bmatrix} \\ &= (\lambda - 3)^2 - 1 \\ &= \lambda^2 - 6\lambda + 8 \\ &= (\lambda - 4)(\lambda - 2) \end{aligned}$$

so that the characteristic values are $\lambda = 4$, $\lambda = 2$. We now find orthonormal bases for the null spaces of $4I - A$ and $2I - A$. First,

$$4I - A = \begin{bmatrix} 4 & 0 \\ 0 & 4 \end{bmatrix} - \begin{bmatrix} 3 & 1 \\ 1 & 3 \end{bmatrix} = \begin{bmatrix} 1 & -1 \\ -1 & 1 \end{bmatrix}$$

which reduces to

$$\begin{bmatrix} 1 & -1 \\ 0 & 0 \end{bmatrix}$$

so that the vectors in the null space of $4I - A$ are those of the form $\begin{bmatrix} y \\ y \end{bmatrix}$. Clearly

$$\begin{bmatrix} 1 \\ 1 \end{bmatrix}$$

is a basis for the space. Therefore, if we normalize, then

$$\bar{u}_1 = \begin{bmatrix} 1/\sqrt{2} \\ 1/\sqrt{2} \end{bmatrix}$$

is an orthonormal basis for the null space of $4I - A$. For $\lambda = 2$, proceed in the same manner:

$$2I - A = \begin{bmatrix} 2 & 0 \\ 0 & 2 \end{bmatrix} - \begin{bmatrix} 3 & 1 \\ 1 & 3 \end{bmatrix} = \begin{bmatrix} -1 & -1 \\ -1 & -1 \end{bmatrix}$$

which reduces to $\begin{bmatrix} 1 & 1 \\ 0 & 0 \end{bmatrix}$ so that

$$\bar{u}_2 = \begin{bmatrix} -1/\sqrt{2} \\ 1/\sqrt{2} \end{bmatrix}$$

is an orthonormal basis for the null space of $2I - A$.

From our construction we know that $A\bar{u}_1 = 4\bar{u}_1$ and $A\bar{u}_2 = 2\bar{u}_2$ (a fact also easily checked by calculation). We know from statement 3 that $\bar{u}_1 \cdot \bar{u}_2 = 0$ (which is also easily checked by direct calculation), so that $\bar{u}_1$ and $\bar{u}_2$ are an orthonormal basis for R^2. Hence

$$P = \begin{bmatrix} 1/\sqrt{2} & -1/\sqrt{2} \\ 1/\sqrt{2} & 1/\sqrt{2} \end{bmatrix}$$

is necessarily an orthogonal matrix (since its columns were constructed to be orthonormal). It follows that

$$P^{-1} = P^t = \begin{bmatrix} 1/\sqrt{2} & 1/\sqrt{2} \\ -1/\sqrt{2} & 1/\sqrt{2} \end{bmatrix}$$

Then $B = P^{-1}AP$ is a diagonal matrix. Direct calculation of this product gives

$$B = \begin{bmatrix} 4 & 0 \\ 0 & 2 \end{bmatrix}$$

EXAMPLE 2 Let us look at a three-dimensional example. Suppose A is the symmetric matrix

$$A = \begin{bmatrix} 1 & 1 & 0 \\ 1 & 1 & 0 \\ 0 & 0 & 2 \end{bmatrix}$$

Note that

6 $$\lambda I - A = \begin{bmatrix} \lambda & 0 & 0 \\ 0 & \lambda & 0 \\ 0 & 0 & \lambda \end{bmatrix} - \begin{bmatrix} 1 & 1 & 0 \\ 1 & 1 & 0 \\ 0 & 0 & 2 \end{bmatrix} = \begin{bmatrix} \lambda - 1 & -1 & 0 \\ -1 & \lambda - 1 & 0 \\ 0 & 0 & \lambda - 2 \end{bmatrix}$$

so that

$$\begin{aligned} \det(\lambda I - A) &= (\lambda - 1)\det\begin{bmatrix} \lambda - 1 & 0 \\ 0 & \lambda - 2 \end{bmatrix} - (-1)\det\begin{bmatrix} -1 & 0 \\ 0 & \lambda - 2 \end{bmatrix} \\ &\quad + 0\det\begin{bmatrix} -1 & \lambda - 1 \\ 0 & 0 \end{bmatrix} \\ &= (\lambda - 1)^2(\lambda - 2) - (\lambda - 2) \\ &= ((\lambda - 1)^2 - 1)(\lambda - 2) \\ &= (\lambda^2 - 2\lambda + 1 - 1)(\lambda - 2) \\ &= (\lambda^2 - 2\lambda)(\lambda - 2) \\ &= \lambda(\lambda - 2)^2 \end{aligned}$$

Thus the characteristic values of A are $\lambda = 0$ and $\lambda = 2$. Using equation **6** we see that $0I - A$ is

$$\begin{bmatrix} -1 & -1 & 0 \\ -1 & -1 & 0 \\ 0 & 0 & -2 \end{bmatrix} \quad \text{which reduces to} \quad \begin{bmatrix} 1 & 1 & 0 \\ 0 & 0 & 1 \\ 0 & 0 & 0 \end{bmatrix}$$

Therefore, the null space of A consists of vectors of the form

$$\begin{bmatrix} -y \\ y \\ 0 \end{bmatrix} \qquad \text{so that} \qquad \bar{u}_1 = \begin{bmatrix} 1/\sqrt{2} \\ -1/\sqrt{2} \\ 0 \end{bmatrix}$$

is an orthonormal basis for the null space of $0I - A$.

Using equation **6** again, $2I - A$ is

$$\begin{bmatrix} 1 & -1 & 0 \\ -1 & 1 & 0 \\ 0 & 0 & 0 \end{bmatrix} \qquad \text{which reduces to} \qquad \begin{bmatrix} 1 & -1 & 0 \\ 0 & 0 & 0 \\ 0 & 0 & 0 \end{bmatrix}$$

so that the null space of $2I - A$ consists of vectors of the form

$$\begin{bmatrix} y \\ y \\ z \end{bmatrix} = y \begin{bmatrix} 1 \\ 1 \\ 0 \end{bmatrix} + z \begin{bmatrix} 0 \\ 0 \\ 1 \end{bmatrix}$$

Thus

$$\begin{bmatrix} 1 \\ 1 \\ 0 \end{bmatrix} \qquad \text{and} \qquad \begin{bmatrix} 0 \\ 0 \\ 1 \end{bmatrix}$$

are an orthogonal basis for this null space, and normalizing shows that

$$\bar{u}_2 = \begin{bmatrix} 1/\sqrt{2} \\ 1/\sqrt{2} \\ 0 \end{bmatrix} \qquad \text{and} \qquad \bar{u}_3 = \begin{bmatrix} 0 \\ 0 \\ 1 \end{bmatrix}$$

are an orthonormal basis for the null space of $2I - A$.

Our construction gives

$$A\bar{u}_1 = 0\bar{u}_1 \qquad A\bar{u}_2 = 2\bar{u}_2 \qquad A\bar{u}_3 = 2\bar{u}_3$$
$$|\bar{u}_1| = |\bar{u}_2| = |\bar{u}_3| = 1 \qquad \text{and} \qquad \bar{u}_2 \cdot \bar{u}_3 = 0$$

while property **3** gives

$$\bar{u}_1 \cdot \bar{u}_2 = 0 \qquad \text{and} \qquad \bar{u}_1 \cdot \bar{u}_3 = 0$$

(These can also be checked by direct calculation.) Therefore $\bar{u}_1$, $\bar{u}_2$, and $\bar{u}_3$ are an orthonormal basis consisting of characteristic vectors for A. Put

$$P = \begin{bmatrix} 1/\sqrt{2} & 1/\sqrt{2} & 0 \\ -1/\sqrt{2} & 1/\sqrt{2} & 0 \\ 0 & 0 & 1 \end{bmatrix}$$

Then, since this is an orthogonal matrix,

$$P^{-1} = P^t = \begin{bmatrix} 1/\sqrt{2} & -1/\sqrt{2} & 0 \\ 1/\sqrt{2} & 1/\sqrt{2} & 0 \\ 0 & 0 & 1 \end{bmatrix}$$

We know that $B = P^{-1}AP$ is diagonal, and its diagonal entries are the characteristic values of A corresponding to the columns of P. Thus

$$B = \begin{bmatrix} 0 & 0 & 0 \\ 0 & 2 & 0 \\ 0 & 0 & 2 \end{bmatrix}$$

which can be checked by calculating $P^{-1}AP$ directly.

EXAMPLE 3 We outline here a proof of the spectral theorem for symmetric matrices with two rows and two columns. This proof uses the quadratic formula and can, with difficulty, be generalized to larger matrices.

Suppose

$$A = \begin{bmatrix} a & b \\ b & c \end{bmatrix}$$

If $b = 0$, A is already diagonal. Thus we can assume $b \neq 0$. The characteristic polynomial is

$$\lambda^2 - (a + c)\lambda + ac - b^2$$

From the quadratic formula this polynomial will have distinct real roots if

7 $$(a + c)^2 - 4(ac - b^2) > 0$$

Let us show that if $b \neq 0$, inequality 7 will always be true, no matter what values a, b, and c have. Multiply out and regroup to obtain

$$\begin{aligned} (a + c)^2 - 4(ac - b^2) &= a^2 + 2ac + c^2 - 4ac + 4b^2 \\ &= a^2 - 2ac + c^2 + 4b^2 \\ &= (a - c)^2 + 4b^2 \end{aligned}$$

Thus $(a + c)^2 - 4(ac - b^2)$ can be written as the sum of two squares $(a - c)^2 + 4b^2$, one of which is not zero (since $b \neq 0$ by assumption). We conclude that inequality **7** must always be true if $b \neq 0$. Therefore the characteristic polynomial has distinct real roots, and we conclude from Theorem **16** that indeed A is similar to a diagonal matrix.

EXERCISES

1 For each of the following matrices A find an orthogonal matrix P such that $P^{-1}AP$ is a diagonal matrix B.

a $\begin{bmatrix} 2 & 1 \\ 1 & 2 \end{bmatrix}$ **b** $\begin{bmatrix} 2 & -1 \\ -1 & 2 \end{bmatrix}$ **c** $\begin{bmatrix} 1 & 1 & 1 \\ 1 & 1 & 1 \\ 1 & 1 & 1 \end{bmatrix}$

d $\begin{bmatrix} -8 & 5 & 4 \\ 5 & 3 & 1 \\ 4 & 1 & 0 \end{bmatrix}$ **e** $\begin{bmatrix} 4 & -2 & 2 \\ -2 & 1 & -1 \\ 2 & -1 & 1 \end{bmatrix}$

2 Find a matrix C such that $C^2 = \begin{bmatrix} 1 & 1 & 1 \\ 1 & 1 & 1 \\ 1 & 1 & 1 \end{bmatrix}$.

3 Suppose $A = \begin{bmatrix} 3 & 1 & 2 \\ 0 & 1 & 1 \\ 1 & 0 & 1 \end{bmatrix}$. Find A^t and verify that formula **1** is correct for A.

4 Suppose $\bar{v} = \frac{1}{\sqrt{2}}(1, -1)$ and P is projection onto $\bar{v}$. Show that the matrix of P relative to the standard basis is symmetric and find an orthonormal basis for R^2 consisting of characteristic vectors for P. (Hint: If $\bar{w} \cdot \bar{v} = 0$, then $P\bar{w} = 0$.)

5 Suppose $A = A^t$ and

$$A\begin{bmatrix} 1 \\ 1 \\ 1 \end{bmatrix} = \begin{bmatrix} 2 \\ 2 \\ 2 \end{bmatrix} \quad A\begin{bmatrix} 1 \\ -2 \\ 1 \end{bmatrix} = \begin{bmatrix} 0 \\ 0 \\ 0 \end{bmatrix} \quad A\begin{bmatrix} 1 \\ 0 \\ -1 \end{bmatrix} = \begin{bmatrix} -3 \\ 0 \\ 3 \end{bmatrix}$$

Find A.

(Hint: The vectors $\begin{bmatrix} 1 \\ 1 \\ 1 \end{bmatrix}$, $\begin{bmatrix} 1 \\ -2 \\ 1 \end{bmatrix}$, $\begin{bmatrix} 1 \\ 0 \\ -1 \end{bmatrix}$ are orthogonal. Find an orthogonal matrix P such that $\begin{bmatrix} 2 & 0 & 0 \\ 0 & 0 & 0 \\ 0 & 0 & -3 \end{bmatrix} = P^{-1}AP$. Then calculate A.)

6 Show that A^tA, AA^t, and $A + A^t$ are symmetric by finding the transpose of each.

7 Find A^{10} if $A = \begin{bmatrix} 2 & -1 \\ -1 & 2 \end{bmatrix}$.

8 Show that if A is symmetric and p is a real polynomial, $p(A)$ is symmetric.

9 Show that if $A = A^t$ and $A^2 = 0$ then $A = 0$. (Hint: See Exercise **10**, page 273. A simple alternative proof can be given by using equation **1** to calculate $A^2\bar{u} \cdot \bar{u}$.)

10 Show that if $A^2 = I$ and $A = A^t$, then A is similar to a matrix with diagonal entries $+1$ or -1.

□ **11** There are complex number versions of Theorem **17**. For $z = a + ib$, a and b real, we define the *conjugate* $z^* = a - ib$. For complex A we define the *conjugate* A^* as the matrix obtained by conjugating the entries of A. The *transpose conjugate* (or *adjoint*) is $(A^*)^t$. A matrix A is *unitary* if $(A^*)^t = A^{-1}$; *self-adjoint* if $A = (A^*)^t$; and *normal* if $(A^*)^tA = A(A^*)^t$.

a For $A = \begin{bmatrix} 2 - i & i \\ 1 + i & 3i \end{bmatrix}$ find A^* and $(A^*)^t$.

b For $A = \begin{bmatrix} i/\sqrt{2} & 1/\sqrt{2} \\ -1/\sqrt{2} & -i/\sqrt{2} \end{bmatrix}$ verify that A is unitary.

c For $A = \begin{bmatrix} 2 & 1 - i \\ 1 + i & 0 \end{bmatrix}$ verify that A is self-adjoint.

d For $A = \begin{bmatrix} 0 & 2 + i \\ -2 + i & 0 \end{bmatrix}$ verify that A is normal.

A complex number version of Theorem **17** can be stated as follows.

e If A is self-adjoint (or normal), there is a unitary matrix P such that $P^{-1}AP$ is diagonal. If A is self-adjoint, this diagonal matrix has only real entries.

The conditions that P be unitary are the conditions that the columns of P be orthonormal with respect to the dot product of Exercise **17**, page 116.

f For $A = \begin{bmatrix} 1 & i \\ -i & 1 \end{bmatrix}$, find a unitary matrix P such that $P^{-1}AP$ is diagonal.

g For $A = \begin{bmatrix} 0 & -1 \\ 1 & 0 \end{bmatrix}$, find a unitary matrix P such that $P^{-1}AP$ is diagonal.

SECTION 6 Reduction of Quadratics

The spectral theorem (Theorem **17**) has many applications in mathematics. One such application, that of reduction of quadratics to standard form, is discussed in this section.

Let us begin with the general quadratic equation, with no linear terms, in two unknowns:

1 $$ax^2 + bxy + cy^2 = d$$

Put

$$A = \begin{bmatrix} a & \frac{b}{2} \\ \frac{b}{2} & c \end{bmatrix} \quad \text{and} \quad \bar{u} = \begin{bmatrix} x \\ y \end{bmatrix}$$

Then A is symmetric and

$$A\bar{u} \cdot \bar{u} = \begin{bmatrix} ax + \frac{b}{2}y \\ \frac{b}{2}x + cy \end{bmatrix} \cdot \begin{bmatrix} x \\ y \end{bmatrix} = ax^2 + bxy + cy^2$$

so we can rewrite equation **1** as

2 $$A\bar{u} \cdot \bar{u} = d$$

The matrix A is called the **matrix of the form** $ax^2 + bxy + cy^2$. Note that its diagonal entries are the coefficients of x^2 and y^2, while its non-diagonal entries are each half the coefficient of the term xy.

Now apply the spectral theorem to obtain an orthogonal matrix P and a diagonal matrix B such that $B = P^{-1}AP$. Solve this for A to obtain $A = PBP^{-1}$. Substituting this in equation **2** gives

3 $$PBP^{-1}\bar{u} \cdot \bar{u} = d$$

We have seen how to shift a matrix across the dot product. For any matrix A and any $\bar{u}$ and $\bar{v}$ we have $A\bar{u} \cdot \bar{v} = \bar{u} \cdot A^t\bar{v}$. (See formula **1**, page 275.) Use this to shift P across the dot product

$$PBP^{-1}\bar{u} \cdot \bar{u} = BP^{-1}\bar{u} \cdot P^t\bar{u}$$

Since P is orthogonal we know that $P^t = P^{-1}$. Hence equation **3** becomes

$$BP^{-1}\bar{u} \cdot P^{-1}\bar{u} = d$$

Put

$$B = \begin{bmatrix} b_1 & 0 \\ 0 & b_2 \end{bmatrix} \quad \text{and} \quad P^{-1}\bar{u} = \begin{bmatrix} x_1 \\ y_1 \end{bmatrix}$$

Therefore, equation **3** can now be written as

$$\begin{bmatrix} b_1 & 0 \\ 0 & b_2 \end{bmatrix} \begin{bmatrix} x_1 \\ y_1 \end{bmatrix} \cdot \begin{bmatrix} x_1 \\ y_1 \end{bmatrix} = d$$

which, after carrying out these products, gives

4 $$b_1 x_1^2 + b_2 y_1^2 = d$$

Thus, relative to the x_1, y_1 coordinate system we can rewrite equation **1** in the form **4**, in which no "xy" term appears. We can easily graph equation **4** in the x_1, y_1 system and thereby obtain the graph of equation **1**.

EXAMPLE 1 Let us use the above method to describe the graph of

5 $$3x^2 + 2xy + 3y^2 = 1$$

Put

$$A = \begin{bmatrix} 3 & 1 \\ 1 & 3 \end{bmatrix} \quad \text{and} \quad \bar{u} = \begin{bmatrix} x \\ y \end{bmatrix}$$

and apply the methods of the previous section to find P and B.

The characteristic polynomial of A is

$$\begin{aligned} f(\lambda) &= \det(\lambda I - A) \\ &= \det \begin{bmatrix} \lambda - 3 & -1 \\ -1 & \lambda - 3 \end{bmatrix} \\ &= (\lambda - 3)^2 - 1 \\ &= \lambda^2 - 6\lambda + 8 \\ &= (\lambda - 2)(\lambda - 4) \end{aligned}$$

Therefore, the characteristic roots are $\lambda = 2$, $\lambda = 4$.

The matrices

$$2I - A = \begin{bmatrix} -1 & -1 \\ -1 & -1 \end{bmatrix} \quad \text{and} \quad 4I - A = \begin{bmatrix} 1 & -1 \\ -1 & 1 \end{bmatrix}$$

reduce respectively to

$$\begin{bmatrix} 1 & 1 \\ 0 & 0 \end{bmatrix} \quad \text{and} \quad \begin{bmatrix} 1 & -1 \\ 0 & 0 \end{bmatrix}$$

which give the respective characteristic vectors (normalized to have length 1)

$$\bar{u}_1 = \begin{bmatrix} 1/\sqrt{2} \\ -1/\sqrt{2} \end{bmatrix} \quad \text{and} \quad \bar{u}_2 = \begin{bmatrix} 1/\sqrt{2} \\ 1/\sqrt{2} \end{bmatrix}$$

With

$$P = \begin{bmatrix} 1/\sqrt{2} & 1/\sqrt{2} \\ -1/\sqrt{2} & 1/\sqrt{2} \end{bmatrix}$$

we have

$$P^{-1} = P^t = \begin{bmatrix} 1/\sqrt{2} & -1/\sqrt{2} \\ 1/\sqrt{2} & 1/\sqrt{2} \end{bmatrix}$$

(since P is orthogonal); thus

$$B = P^{-1}AP = \begin{bmatrix} 2 & 0 \\ 0 & 4 \end{bmatrix}$$

Therefore, we can rewrite equation **5** as

6 $$2x_1^2 + 4y_1^2 = 1 \quad \text{where} \quad \begin{bmatrix} x_1 \\ y_1 \end{bmatrix} = P^{-1}\begin{bmatrix} x \\ y \end{bmatrix}$$

To graph this equation in the (x_1,y_1) coordinate system, recall that x_1 and y_1 are the coordinates of (x,y) relative to the basis $\bar{u}_1$ and $\bar{u}_2$. (See Section 4, Chapter 4.) Since the graph of equation **6** is an ellipse, we obtain the graph shown in Figure 4.

Of course, the matrices P and B are not unique. For example, if we put

$$P = \begin{bmatrix} 1/\sqrt{2} & 1/\sqrt{2} \\ 1/\sqrt{2} & -1/\sqrt{2} \end{bmatrix} \quad \text{so that} \quad P^{-1} = \begin{bmatrix} 1/\sqrt{2} & 1/\sqrt{2} \\ 1/\sqrt{2} & -1/\sqrt{2} \end{bmatrix}$$

we obtain

$$B = P^{-1}AP = \begin{bmatrix} 4 & 0 \\ 0 & 2 \end{bmatrix}$$

Figure 4 *The graph of* $3x^2 + 2xy + 3y^2 = 1$.

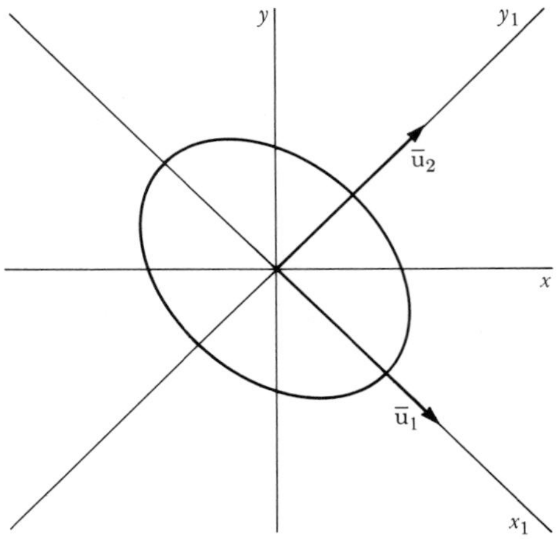

and thus equation **5** can be written as

$$7 \quad 4x_2^2 + 2y_2^2 = 1 \qquad \text{where} \qquad \begin{bmatrix} x_2 \\ y_2 \end{bmatrix} = \begin{bmatrix} 1/\sqrt{2} & 1/\sqrt{2} \\ 1/\sqrt{2} & -1/\sqrt{2} \end{bmatrix} \begin{bmatrix} x \\ y \end{bmatrix}$$

The pair (x_2,y_2) is the set of coordinates of (x,y) relative to the basis $\bar{u}_2$, $\bar{u}_1$. While the graph of equation **7** in the x_2, y_2 system is not the same as the graph of equation **6** in the x_1, y_1 system, these two equations give the same graph in the x, y system (as we should expect). For, merely by labeling the $\bar{u}_2$ axis as the x_2 axis and the $\bar{u}_1$ axis as the y_2 axis, we obtain the same figure shown in Figure 4.

EXAMPLE 2 The graph of the equation in Example **1** is an ellipse. Let us show that the graph of

$$-3x^2 + 8xy + 3y^2 = 10$$

is an hyperbola. Put

$$A = \begin{bmatrix} -3 & 4 \\ 4 & 3 \end{bmatrix}$$

The characteristic polynomial of A is

$$f(\lambda) = \lambda^2 - 25 = (\lambda - 5)(\lambda + 5)$$

and characteristic vectors corresponding to $\lambda = 5$, $\lambda = -5$ are, respectively,

$$\bar{u}_1 = \begin{bmatrix} \frac{1}{\sqrt{5}} \\ \frac{2}{\sqrt{5}} \end{bmatrix} \qquad \bar{u}_2 = \begin{bmatrix} \frac{-2}{\sqrt{5}} \\ \frac{1}{\sqrt{5}} \end{bmatrix}$$

With P having these as columns, we have

$$P^{-1}AP = \begin{bmatrix} 5 & 0 \\ 0 & -5 \end{bmatrix}$$

In coordinates (x_1,y_1) with respect to $\bar{u}_1$, $\bar{u}_2$ our equation is

$$5x_1^2 - 5y_1^2 = 10$$

The graph of this is the hyperbola shown in Figure 5.

Figure 5 *The graph of* $-3x^2 + 8xy + 3y^2 = 10$.

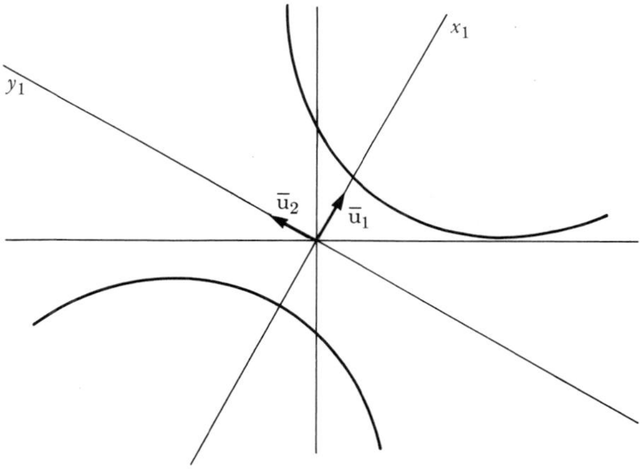

EXAMPLE 3 It is possible to determine the nature of the graph of

8 $$ax^2 + bxy + cy^2 = d$$

from the determinant of the matrix of the form:

$$A = \begin{bmatrix} a & \frac{b}{2} \\ \frac{b}{2} & c \end{bmatrix}$$

To see that this is so, let b_1, b_2 be the characteristic values of A, corresponding to the orthonormal set $\bar{u}_1$, $\bar{u}_2$ of characteristic vectors, so that, in terms of coordinates (x_1,y_1) with respect to $\bar{u}_1$, $\bar{u}_2$, the equation **8** is

9 $$b_1x_1^2 + b_2y_1^2 = d$$

If $d > 0$, the graph of this is

A ellipse if $b_1 > 0$ and $b_2 > 0$
A hyperbola if b_1 and b_2 have opposite signs
Empty if $b_1 < 0$ and $b_2 < 0$

In other words, with $B = \begin{bmatrix} b_1 & 0 \\ 0 & b_2 \end{bmatrix}$ we have $\det B = b_1b_2$

so the equation **9** represents

An ellipse or the empty set if $\det B > 0$
A hyperbola if $\det B < 0$

Since $B = P^{-1}AP$ and

$$\begin{aligned} \det B &= \det (P^{-1}AP) = \det P^{-1} \det A \det P \\ &= \det A \det P^{-1} \det P \\ &= \det A \det P^{-1}P \\ &= \det A \end{aligned}$$

our results can be summarized as follows:

10 The graph of the equation $ax^2 + bxy + cy^2 = d$, $d > 0$, is an ellipse or empty if $\det A > 0$, and is a hyperbola if $\det A < 0$, where A is the matrix of the form $ax^2 + bxy + cy^2$.

For example $\det \begin{bmatrix} -2 & 5 \\ 5 & 1 \end{bmatrix} = -27 < 0$ so the graph of

$$-2x^2 + 10xy + y^2 = 3$$

must be a hyperbola.

The case when $d < 0$ can be treated by first multiplying through equation **8** by -1, while the case $d = 0$ is left to Exercise **3**.

EXAMPLE 4 The above methods extend to quadratic expressions in more than two variables. Consider

11 $$ax^2 + by^2 + cz^2 + dxy + exz + fyz = g$$

The matrix of the form $ax^2 + by^2 + cz^2 + dxy + exz + fyz$ is the symmetric matrix:

$$A = \begin{bmatrix} a & \frac{d}{2} & \frac{e}{2} \\ \frac{d}{2} & b & \frac{f}{2} \\ \frac{e}{2} & \frac{f}{2} & c \end{bmatrix}$$

The expression **11** is then

$$(A\bar{u}) \cdot \bar{u} = g \qquad \text{where} \qquad \bar{u} = \begin{bmatrix} x \\ y \\ z \end{bmatrix}$$

Now one can find an orthogonal matrix P such that $P^{-1}AP$ is diagonal. The columns of P; $\bar{u}_1$, $\bar{u}_2$, $\bar{u}_3$, will be an orthonormal basis and if (x_1, y_1, z_1) are the coordinates of $\bar{u}$ with respect to this basis, the equation **11** takes the form

$$b_1x_1^2 + b_2x_2^2 + b_3x_3^2 = g$$

where b_1, b_2, b_3 are the diagonal entries of $P^{-1}AP$.

EXERCISES

1 Use the method of Example **1** to eliminate the xy term and use this information to draw a graph of each of the following.
  a $2x^2 + 2xy + 2y^2 = 1$ **b** $-3x^2 + 8xy + 3y^2 = \sqrt{5}$
  c $xy = 1$ **d** $3x^2 + 6xy + 11y^2 = 12$

2 Which of the following describe ellipses? Hyperbolas?
  a $x^2 + 3xy + y^2 = 1$ **b** $2x^2 + 4xy = -3$
  c $6x^2 + xy + y^2 = 3$ **d** $2x^2 + 3xy + 4y^2 = 10$

3 **a** Give an example of a quadratic equation whose matrix has a positive determinant yet whose graph is empty.
  b What is the graph of $ax^2 + bxy + cy^2 = 0$? Be sure to discuss all the possible cases.

c What is the graph of $ax^2 + bxy + cy^2 = 1$ in the case where the determinant of the matrix of the form is 0? Discuss all possibilities.

4 Use the method of Example **1** to eliminate the cross product terms in each of the following by first suitably choosing a symmetric matrix A so that the left-hand side is of the form $A\bar{u} \cdot \bar{u}$.

a $2x^2 + 2xy + 2y^2 - 3z^2 = 1$

b $-5x^2 - 5y^2 - 2z^2 + 14xy + 8xz + 8xy = 6$

(Hint: -12 is a root of $\lambda^3 + 12\lambda^2 - 36\lambda - 432 = 0$.)

5 Suppose the graph of $ax^2 + bxy + cy^2 = d$ is an hyperbola for some $d > 0$. Show, using result **10**, that for any other nonzero value of d the graph is still a hyperbola.

6 Show, using the result **10**, that the graph of $ax^2 + bxy + cy^2 = 1$ is an ellipse if $a > 0$ and $b^2 - 4ac < 0$ and a hyperbola if $b^2 - 4ac > 0$.

7 As Figure 4 shows, the two points on the graph of

$$3x^2 + 2xy + 3y^2 = 1$$

which are furthest from the origin are the places where the graph crosses the $\bar{u}_1$-axis, $\bar{u}_1$ being a characteristic vector belonging to the smaller characteristic value $b_1 = 2$.

a What points on the graph of

$$2x^2 + 2xy + 2y^2 = 1$$

are furthest from the origin? Closest to the origin?

b What points on the graph of

$$-3x^2 + 8xy + 3y^2 = 10$$

are closest to the origin? Show that there are points on its graph arbitrarily far from the origin.

8 a What are the characteristic values of $A = \begin{bmatrix} a & b/2 \\ b/2 & c \end{bmatrix}$?

b Which characteristic value is largest?

c What is the shortest distance from the origin to the graph of

$$ax^2 + bxy + cy^2 = 1.$$

9 Find the maximum value of $ax^2 + bxy + cy^2$ subject to the condition $x^2 + y^2 = 1$. {Hint: Find the matrix A of the form, its characteristic values b_1, b_2 and corresponding orthonormal characteristic vectors $\bar{u}_1$, $\bar{u}_2$, then show that

$$[A(a_1\bar{u}_1 + a_2\bar{u}_2)] \cdot (a_1\bar{u}_1 + a_2\bar{u}_2) = a_1^2 b_1 + a_2^2 b_2.\}$$

10 Show that if $A^t = A$ and $(A\bar{u} \cdot \bar{u}) = ax^2 + bxy + cy^2$, where $\bar{u} = \begin{bmatrix} x \\ y \end{bmatrix}$, then $A = \begin{bmatrix} a & b/2 \\ b/2 & c \end{bmatrix}$. {Hint: Put $A = \begin{bmatrix} A_{11} & A_{12} \\ A_{21} & A_{22} \end{bmatrix}$ and solve the identity $(A\bar{u}) \cdot \bar{u} = ax^2 + bxy + cy^2$ for the entries of A.}

CHAPTER 6

LINEAR DIFFERENTIAL OPERATORS

The operation of differentiation is a linear operation, that is, the derivative of the sum is the sum of the derivatives, and the derivative of a scalar multiple is that multiple of the derivative. This chapter will show how the methods of linear algebra can be combined with elementary calculus to solve differential equations of the form $p(D)y = f$, where f is a given function, D is the differentiation operator, and p is a polynomial.

Section 1 discusses the first-order homogeneous equation $(D - aT)y = 0$. Higher-order homogeneous equations are discussed in Sections 2, 3, and 4, and Section 5 discusses the nonhomogeneous case. Some elementary applications are presented in Section 6.

SECTION 1 First-Order Linear Operators

In this section we describe the solutions to equations of the form

$$a_1 y' + a_2 y = f$$

We first discuss the case where $f = 0$, $a_1 = 1$, and a_2 is constant. Our equation can then be rewritten as $y' - ay = 0$ by replacing a_2 by $-a$. In operator form this equation is

$$(D - aI)y = 0$$

so the problem of solving the equation $y' - ay = 0$ is the same as the problem of describing the null space of $D - aI$.

Recall from calculus that $De^x = e^x$; therefore, the chain rule gives $De^{ax} = ae^{ax}$ for any real number a. Thus, if L is the operator $L = D - aI$, it follows that $Le^{ax} = 0$.

In other words,

1 e^{ax} is in the null space of $D - aI$.

We shall show, in fact, that e^{ax} is a *basis* for the null space of $D - aI$ by using further results from calculus. In other words, it will be seen that

2 If y is in the null space of $D - aI$, then y is a constant multiple of e^{ax}.

To do this we use a trick: Suppose $(D - aI)y = 0$. Put $z = e^{-ax}y$. The product rule then gives

$$\begin{aligned} Dz &= -ae^{-ax}y + e^{-ax}y' \\ &= e^{-ax}(y' - ay) \\ &= e^{-ax}(D - aI)y \\ &= 0 \end{aligned}$$

Thus, since the derivative of z is zero, we know that (from a theorem of calculus) z must be constant; that is, there is a constant c such that $e^{-ax}y = c$. Solving for y gives $y = ce^{ax}$. This proves statement **2**.

We usually summarize statements **1** and **2** by saying that $y = ce^{ax}$ is the **general solution** to $y' - ay = 0$. This means that no matter what value c has, $y = ce^{ax}$ is a solution to this equation, and, conversely, any solution to this equation must be of the form ce^{ax} for some choice of c.

EXAMPLE 1 Consider the equation

3 $$y' = 2y$$

It can be written as

$$Ly = 0 \qquad \text{where} \qquad L = D - 2I$$

so that its general solution is $y = ce^{2x}$.

Suppose we are looking for the solution to equation **3** that satisfies the additional condition

4 $$y(3) = 4$$

We know that if there is a function that satisfies both equation **3** and equation **4**, it must be of the form

5 $$y = ce^{2x}$$

Therefore, we merely need to choose c so that property **4** holds. Putting $x = 3$ in equation **5** and using equation **4**, we have $4 = ce^6$, from which we obtain $c = 4e^{-6}$.

Our conclusion is that $y = 4e^{2x-6}$ is the *only* function that satisfies equation **3** and condition **4**.

EXAMPLE 2 Consider the equation

6 $$y' + 3y = 0$$

We can write this as $Ly = 0$, where $L = D + 3I$, and consequently have the general solution

7 $$y = ce^{-3x}$$

The solution to equation **6** that satisfies

8 $$y(-1) = -2$$

can be found, as in Example **1**, by setting $x = -1$ in the general solution **7** and calculating c, using condition **8**. We have $-2 = ce^3$, and, consequently, $c = -2e^{-3}$. Hence $y = -2e^{-3x-3}$ is the solution to equation **6** that satisfies condition **8**.

Problems of this type, for which one is to find a function which satisfies a differential equation and for which the function (and, possibly, one or more of its derivatives) has a specified value at a point, are known as **initial value problems**. This terminology arises from common physical situations in which the functions are functions of t (denoting time) and the conditions are specified for $t = 0$. (See Section 6, below.)

EXAMPLE 3 ☐ *Inhomogeneous Equations.*
A useful trick will enable us to solve the inhomogeneous problem

9 $$y' - ay = f$$

where f is assumed to be continuous. The trick is to multiply both sides of this equation by a function so chosen that the left-hand side is the derivative of a known function. Such a function is e^{-ax}. For if we multiply both sides of equation **9** by e^{-ax} we obtain

$$y'e^{-ax} - aye^{-ax} = e^{-ax}f$$

The left-hand side is just $D(ye^{-ax})$, as can be seen by calculating $D(ye^{-ax})$ using the product rule. Thus

$$D(ye^{-ax}) = e^{-ax}f$$

Let g be an integral of $e^{-ax}f$; that is, g is any function whose derivative is $e^{-ax}f$. The equation can then be rewritten as $D(ye^{-ax}) = Dg$, and therefore $ye^{-ax}g$ must be constant. Thus $ye^{-ax} = g + c$, which can be rewritten as

10 $$y = e^{ax}g + ce^{ax}$$

This is the general solution to equation **9**. For example, consider the equation

11 $$y' + 2y = x$$

Multiplying this equation by e^{2x} gives $y'e^{2x} + 2ye^{2x} = xe^{2x}$, which can be rewritten as

12 $$D(ye^{2x}) = xe^{2x}$$

To find an integral of xe^{2x} integrate by parts, with $u = x/2$ and $dv = 2e^{2x}\,dx$. Thus $du = \frac{1}{2}\,dx$ and $v = e^{2x}$, and we have

$$\int xe^{2x}\,dx = \frac{x}{2}e^{2x} - \tfrac{1}{2}\int e^{2x}\,dx$$
$$= \frac{x}{2}e^{2x} - \tfrac{1}{4}e^{2x}$$

Therefore, since we can rewrite equation **12** as

$$D(ye^{2x}) = D\left(\frac{xe^{2x}}{2} - \frac{e^{2x}}{4}\right)$$

there must be a constant c such that

$$ye^{2x} = \frac{xe^{2x}}{2} - \frac{e^{2x}}{4} + c$$

Multiplying by e^{-2x} then gives the general solution to equation **11**:

13 $$y = \frac{x}{2} - \frac{1}{4} + ce^{-2x}$$

If we are looking for the solution that satisfies the initial condition $y(0) = 2$, merely substitute $x = 0$ in the general solution **13** to obtain $2 = \frac{0}{2} - \frac{1}{4} + ce^{-0}$. Thus $c = \frac{9}{4}$, and we have

$$y = \frac{x}{2} - \frac{1}{4} + \frac{9}{4}e^{-2x}$$

This is the solution to the initial value problem:

$$y' + 2y = x,\ y(0) = 2$$

EXAMPLE 4 □ *First-Order Continuous Coefficient Operators.*
The trick used in Example **3** can be extended to give the general solution to $y' + p(x)y = f(x)$, if p and f are continuous. The trick is to multiply both sides by

$$e^{q(x)} \qquad \text{where} \qquad q = \int p(x)\,dx$$

so that the equation can be written as

$$D(ye^{q(x)}) = e^{q(x)}f(x)$$

For example, consider

14 $$y' + xy = x$$

Here we have $p(x) = x$ so that $q(x) = \int x\,dx = x^2/2$. Thus multiplying by $e^{x^2/2}$ yields

$$y'e^{x^2/2} + xye^{x^2/2} = xe^{x^2/2}$$

which can be rewritten as

$$D(ye^{x^2/2}) = xe^{x^2/2}$$

Since the function $xe^{x^2/2}$ is the derivative of $e^{x^2/2}$, we can rewrite this as

$$D(ye^{x^2/2}) = D(e^{x^2/2})$$

Therefore, there is a constant c such that

$$ye^{x^2/2} = e^{x^2/2} + c$$

and it follows that the general solution to equation **14** is

$$y = 1 + ce^{-x^2/2}$$

EXERCISES

1 Find the general solution to each of the following problems.
 a $y' - y = 0$ **b** $y' + \sqrt{5}y = 0$
 c $3y' + 2y = 0$ **d** $2y' = y$
 e $y' + ey = 0$

2 For each of the following find the function y that satisfies both the differential equation and the initial condition.
 a $y' = 2y,\ y(0) = 1$ **b** $y' + 2y = 0,\ y(0) = 1$
 c $2y' = y,\ y(1) = 2$ **d** $4y' + 5y = 0,\ y(2) = 3$

3 How many solutions does $y' = ay$ have that pass through 0 at 0?

4 Suppose $a < 0$ and y is a solution to the problem $y' = ay,\ y(0) = b$. What happens to this solution as $x \to \infty$? as $x \to -\infty$? What happens to the solution as $x \to \infty$ if $a > 0$?

☐ 5 Use the method of Example **3** to find the general solution to
 a $y' = 2y - x$ **b** $y' + 3y = e^x$
 c $y' = \sin x - y$ **d** $y' - y = x^2 + x$

☐ 6 For each of the equations of Exercise **4** find the solution that satisfies $y(0) = 1$ and the solution that satisfies $y(1) = 2$.

☐ 7 Use the method of Example **4** to find the general solution to

a $y' + x^2y = 0$ **b** $y' + \frac{1}{x}y = x^2$

c $y' + y \log x = x^{-x}$ **d** $y' + y \sin x = 0$

☐ 8 For each equation in Exercise **7** find the solution that satisfies $y(1) = 1$.

☐ 9 The results **1** and **2** can be derived for complex numbers a by suitably defining e^{ax} and Df for complex functions f. If $a = c + id$, where c and d are real, and $i^2 = -1$, we define e^a by

$$e^a = e^c(\cos d + i \sin d)$$

If $f(x) = f_1(x) + if_2(x)$, where f_1 and f_2 are real, we define

$$Df = Df_1 + i\,Df_2$$

a Show that $De^{ax} = ae^{ax}$ for complex a.

b Show that $D(fg) = f\,Dg + g\,Df$ for complex functions f and g. (Hint: Write $f = f_1 + if_2$ and $g = g_1 + ig_2$, where f_1, f_2, g_1, and g_2 are real functions.)

c Show that if $f = f_1 + if_2$, where f_1 and f_2 are real and have continuous derivatives, and if $Df = 0$, there is a complex number α such that $f = \alpha$.

d Derive property **2** for complex numbers a. (Hint: Proceed as in the derivation of property **2**, using **b** and **c**.)

SECTION 2 Second-Order Linear Operators

The differential equation

1 $$y'' + by' + cy = 0$$

can be written as $Ly = 0$, where

2 $$L = D^2 + bD + cI$$

It will now be shown how to solve equation **1** by suitably factoring expression **2**.

Consider the polynomial

3 $$p(\lambda) = \lambda^2 + b\lambda + c$$

This polynomial is usually called the **characteristic polynomial** of L. Its roots are called the **characteristic roots** of L. Exercise **6**, below, shows how this terminology is related to that used for matrices in Chapter 5.

As we know from polynomial theory, polynomial **3** may have two distinct real roots, precisely one real root, or two complex roots. Each of these cases leads to a different form for the general solution to equation **1**.

The roots of p will be denoted by α and β. From the quadratic formula we can write

$$\alpha = \frac{-b + \sqrt{b^2 - 4c}}{2} \qquad \beta = \frac{-b - \sqrt{b^2 - 4c}}{2}$$

The condition that α and β are real and distinct is simply the condition that $b^2 - 4c > 0$, while the condition that $\alpha = \beta$ is the condition $b^2 - 4c = 0$, and the condition that α and β are complex is the condition $b^2 - 4c < 0$.

We can then factor p as $p(\lambda) = (\lambda - \alpha)(\lambda - \beta)$. This factorization can be used to factor expression **2**. We have

$$\begin{aligned}(D - \alpha I)(D - \beta I) &= D^2 - \alpha ID - \beta DI + \alpha\beta I^2 \\ &= D^2 - (\alpha + \beta)D + \alpha\beta I\end{aligned}$$

since $ID = DI = D$ and $I^2 = I$. We also have

$$\begin{aligned}(D - \beta I)(D - \alpha I) &= D^2 - \beta ID - \alpha DI + \beta\alpha I^2 \\ &= D^2 - (\alpha + \beta)D + \alpha\beta I\end{aligned}$$

In each case the right-hand side is $p(D)$, since $p(\lambda) = \lambda^2 - (\alpha + \beta)\lambda + \alpha\beta$. We therefore know that $D - \alpha I$ and $D - \beta I$ *commute* and that

4 $$L = (D - \alpha I)(D - \beta I) = (D - \beta I)(D - \alpha I)$$

Case 1: α and β real, $\alpha \neq \beta$.
Write equation **1** as $Ly = 0$, where $L = D^2 + bD + cI$. Then express L as a product of $D - \alpha I$ and $D - \beta I$, in either order, by using formula **4**.

$$L = (D - \alpha I)(D - \beta I) = (D - \beta I)(D - \alpha I)$$

The results of Section 1 tell us that

$$(D - \alpha I)e^{\alpha x} = 0 \qquad \text{and} \qquad (D - \beta I)e^{\beta x} = 0$$

Combining these with each expression of formula **4** we have

$$\begin{aligned}Le^{\alpha x} &= (D - \beta I)(D - \alpha I)e^{\alpha x} = (D - \beta I)0 = 0 \\ Le^{\beta x} &= (D - \alpha I)(D - \beta I)e^{\beta x} = (D - \alpha I)0 = 0\end{aligned}$$

That is,

5 $e^{\alpha x}$ and $e^{\beta x}$ are in the null space of L.

These functions are independent, for if $c_1e^{\alpha x} + c_2e^{\beta x} = 0$, we can differentiate to obtain $c_1\alpha e^{\alpha x} + c_2\beta e^{\beta x} = 0$. Setting $x = 0$ in each of these gives the two equations

$$c_1 + c_2 = 0$$
$$c_1\alpha + c_2\beta = 0$$

The determinant of the matrix of coefficients of this system is

$$\det\begin{bmatrix} 1 & 1 \\ \alpha & \beta \end{bmatrix} = \beta - \alpha$$

which, by assumption, is *not* zero, for $\beta \neq \alpha$. We conclude that we must have $c_1 = c_2 = 0$ and, therefore, that $e^{\alpha x}$ and $e^{\beta x}$ are indeed independent.

In Section 4 it will be shown that the null space of L has dimension *at most* 2. Since two independent vectors in this null space have been found, we conclude that

The functions $e^{\alpha x}$ and $e^{\beta x}$ are a basis for the null space of

6 $$L = (D - \alpha I)(D - \beta I)$$

if α and β are real and $\alpha \neq \beta$.

We often express this result by saying that

$$y = c_1e^{\alpha x} + c_2e^{\beta x}$$

is the **general solution** to $Ly = 0$. This means that, for any choice of c_1 and c_2, this function is a solution to $Ly = 0$, and that every solution to $Ly = 0$ is of this form, for some c_1 and c_2.

Case 2: α and β real, $\alpha = \beta$.
In this case, write (using formula **4**)

$$L = (D - \alpha I)^2$$

Since $(D - \alpha I)e^{\alpha x} = 0$, $e^{\alpha x}$ must be in the null space of L. A useful and elementary trick enables us to find a second, independent solution to $Ly = 0$. Since the product rule gives

7 $$D(xe^{\alpha x}) = e^{\alpha x} + \alpha xe^{\alpha x}$$

it follows that

$$(D - \alpha I)(xe^{\alpha x}) = e^{\alpha x}$$

Hence, applying $D - \alpha I$ again, we have

$$
\begin{aligned}
(D - \alpha I)^2(xe^{\alpha x}) &= (D - \alpha I)[(D - \alpha I)(xe^{\alpha x})] \\
&= (D - \alpha I)e^{\alpha x} \\
&= 0
\end{aligned}
$$

Therefore, $xe^{\alpha x}$ is *also* in the null space of L.

The functions $e^{\alpha x}$ and $xe^{\alpha x}$ are independent, for if

$$c_1 e^{\alpha x} + c_2 x e^{\alpha x} = 0$$

differentiating gives

$$c_1 \alpha e^{\alpha x} + c_2(e^{\alpha x} + \alpha x e^{\alpha x}) = 0$$

Setting $x = 0$ in each of these gives the two equations

$$
\begin{aligned}
c_1 &= 0 \\
c_1 \alpha + c_2 &= 0
\end{aligned}
$$

Thus, c_1 and c_2 must be zero.

We shall show in Section 4 that the dimension of the null space of L is at most 2; hence

8 The functions $e^{\alpha x}$ and $xe^{\alpha x}$ are a basis for the null space of $L = (D - \alpha I)^2$. Thus, the general solution to $Ly = 0$ is

$$c_1 e^{\alpha x} + c_2 x e^{\alpha x}$$

Case 3: α and β complex.

We could use the complex function discussion of Exercise **9** of the previous section to obtain complex solutions of the form given in statement **6**.

Instead, a generalization of result **7** will be used to describe the general solutions in terms of real functions. This will be discussed in the next section.

EXAMPLE 1 Consider the equation $y'' - y = 0$. The associated differential operator is $L = D^2 - I$, whose characteristic polynomial is $p(\lambda) = \lambda^2 - 1 = (\lambda - 1)(\lambda + 1)$. Since the roots are real and unequal, we apply statement **6** to obtain the general solution

$$y = c_1 e^{x} + c_2 e^{-x}$$

EXAMPLE 2 Consider the equation $y'' + 4y' - y = 0$, whose associated differential operator is $L = D^2 + 4D - I$. The characteristic polynomial is $p(\lambda) =$

$\lambda^2 + 4\lambda - 1$. The quadratic formula gives the roots

$$\frac{-4 \pm \sqrt{20}}{2} = -2 \pm \sqrt{5}$$

Since these roots are real and unequal, the general solution is

$$y = c_1 e^{(-2+\sqrt{5})x} + c_2 e^{(-2-\sqrt{5})x}$$

EXAMPLE 3 The associated differential operator for $y'' + 8y' + 16y = 0$ is $L = D^2 + 8D + 16I$. Therefore, the characteristic polynomial is $p(\lambda) = \lambda^2 + 8\lambda + 16 = (\lambda + 4)^2$. Since the roots are real and equal, we apply statement **8** to obtain the general solution

$$y = c_1 e^{-4x} + c_2 x e^{-4x}$$

EXAMPLE 4 Suppose we wish to solve the initial value problem

9
$$\begin{aligned} y'' - y' - 6y &= 0 \\ y(0) &= 1 \\ y'(0) &= -1 \end{aligned}$$

The associated differential operator is $L = D^2 - D - 6I$, whose characteristic polynomial is $p(\lambda) = \lambda^2 - \lambda - 6 = (\lambda - 3)(\lambda + 2)$. Since the roots are real and distinct, we apply result **6** to obtain the general solution

$$y = c_1 e^{3x} + c_2 e^{-2x} \qquad \text{to} \qquad y'' - y' - 6y = 0$$

By differentiating we have

$$y' = 3c_1 e^{3x} - 2c_2 e^{-2x}$$

Setting $x = 0$ in each of these and using the conditions that $y(0) = 1$ and $y'(0) = -1$ gives the two equations

$$\begin{aligned} c_1 + c_2 &= 1 \\ 3c_1 - 2c_2 &= -1 \end{aligned}$$

This system has the unique solution $c_1 = \frac{1}{5}$, $c_2 = \frac{4}{5}$. Thus the solution to the initial value problem **9** is

$$y = \tfrac{1}{5}e^{3x} + \tfrac{4}{5}e^{-2x}$$

EXAMPLE 5 Consider the initial value problem

10
$$\begin{aligned} y'' + 2y' + y &= 0 \\ y(1) &= 0 \\ y'(1) &= 2 \end{aligned}$$

The equation $y'' + 2y' + y = 0$ is the same as $Ly = 0$, where $L = D^2 + 2D + I$.

The characteristic polynomial of L is $p(\lambda) = \lambda^2 + 2\lambda + 1 = (\lambda + 1)^2$. Since the root $\lambda = -1$ is repeated, we obtain the general solution $y = c_1e^{-x} + c_2xe^{-x}$, to $y'' + 2y' + y = 0$. Differentiation gives

$$y' = -c_1e^{-x} + c_2(e^{-x} - xe^{-x})$$

Thus setting $x = 1$ gives the equations

$$0 = c_1e^{-1} + c_2e^{-1}$$
$$2 = -c_1e^{-1} + c_2(e^{-1} - e^{-1})$$

The unique solution to this system is $c_1 = -2e$, $c_2 = 2e$, and, therefore, the solution to initial value problem **10** is

$$y = -2e^{-x+1} + 2xe^{-x+1}$$

EXAMPLE 6 Our method can be extended to equations in which the leading coefficient is a nonzero number other than 1. For example, the system

11 $$10y'' + 13y' - 3y = 0$$

has the same solutions as the system

$$y'' + \tfrac{13}{10}y' - \tfrac{3}{10}y = 0$$

The characteristic polynomial of the latter system is

$$p(\lambda) = \lambda^2 + \tfrac{13}{10}\lambda - \tfrac{3}{10}$$

which has the same roots as

12 $$q(\lambda) = 10\lambda^2 + 13\lambda - 3$$

Thus, to solve equation **11**, form its characteristic polynomial **12**, find its roots, and write down the general solution as before.

The roots are

$$\frac{-13 \pm \sqrt{169 - 4(-3)(10)}}{20} = \frac{-13 \pm 17}{20}$$

which give $\frac{1}{5}$ and $-\frac{3}{2}$. Since these are real and distinct, the general solution to equation **11** is

$$y = c_1e^{(1/5)x} + c_2e^{(-3/2)x}$$

EXERCISES

1 By first giving the associated differential operator L and then factoring the characteristic polynomial, find the general solution to each of the following.

a $y'' - 5y' - 6y = 0$ **b** $y'' - 5y' + 6y = 0$
c $y'' + 4y' + 4y = 0$ **d** $y'' + 3y' + y = 0$
e $y'' = 0$ **f** $y'' - 2y = 0$
g $y'' - 6y' + 9y = 0$ **h** $y'' - 2y' - 2y = 0$

2 Solve each of the following initial value problems.

a $\begin{aligned} y'' - 4y' + 4y &= 0 \\ y(0) &= 0 \\ y'(0) &= -1 \end{aligned}$ **b** $\begin{aligned} y'' + 3y' + 2y &= 0 \\ y(0) &= 2 \\ y'(0) &= 5 \end{aligned}$

c $\begin{aligned} y'' - 4y' + 4y &= 0 \\ y(3) &= -1 \\ y'(3) &= 2 \end{aligned}$ **d** $\begin{aligned} y'' - 3y' + 2y &= 0 \\ y(10) &= 1 \\ y'(10) &= 0 \end{aligned}$

3 Suppose y is a solution to $Ly = 0$, where $L = (D - \alpha I)(D - \beta I)$. Use the form of the general solution to answer each of the following.

a If α and β are both negative what is $\lim_{x\to\infty} y$? If y is not the zero function does the limit $\lim_{x\to-\infty} y$ exist?

b If $\alpha > 0$ and $\beta < 0$, what must y be in order that $\lim_{x\to\infty} y$ exists? In order that $\lim_{x\to-\infty} y$ exists?

c If α and β are positive and y is not the zero function does $\lim_{x\to\infty} y$ exist? Does $\lim_{x\to-\infty} y$ exist?

4 Find the general solution to (see Example **6**)

a $3y'' - 17y' + 10y = 0$ **b** $4y'' + 4y' + y = 0$

5 Initial value problems will not, in general, have a unique solution if too many or not enough initial conditions are specified. For each of the following decide whether a solution exists and, if so, whether there is a unique solution.

a $y'' - 3y' + 2y = 0$
$y(0) = 1, y'(0) = 0, y''(0) = 0$

b $y'' = 0$
$y(0) = 1, y'(0) = 2, y''(0) = 0$

c $y'' - 2y' + y = 0$
$y(0) = 1$

□ 6 If $L = D^2 + bD + cI$, the differential equation $y'' + by' + cy = 0$ can be expressed as a system of differential equations, thereby associating a matrix with the operator L. Put $z = y'$, and $z' = y''$. Therefore, if the differential equation for y is used, two simultaneous equations for y, z, and their first derivatives are obtained.

a Let $\bar{u} = \begin{bmatrix} y' \\ z' \end{bmatrix}$ and $\bar{v} = \begin{bmatrix} y \\ z \end{bmatrix}$. Find a matrix A such that $\bar{u} = A\bar{v}$.

b Show that A has the characteristic polynomial $\lambda^2 + b\lambda + c$.

c If the characteristic polynomial of A has distinct real roots, we can find an invertible matrix P such that $B = P^{-1}AP$ is diagonal. Put $\bar{w} = P^{-1}\bar{v}$.

Show that the derivatives of the coordinates of $\bar{w}$ are the coordinates of $P^{-1}\bar{u}$ and that $A\bar{v} = \bar{u}$ can be written as $\dfrac{d\bar{w}}{dt} = B\bar{w}$.

d Assuming that the characteristic values of A are real and distinct, we see that the system $d\bar{w}/dt = B\bar{w}$ is easy to solve, for B is diagonal. Show that if $\bar{w}$ is a solution to this system $P\bar{w}$ is a solution to $A\bar{v} = \bar{u}$. Show that the first coordinate of $P\bar{w}$ is then a solution to $y'' + by' + cy = 0$. (This gives an alternative method for solving this equation.)

e Assuming that $\lambda^3 + a\lambda^2 + b\lambda + c = 0$ has three distinct real roots, extend the above methods to solve $y''' + ay'' + by' + cy = 0$. (Hint: First put $z = y'$ and $w = y''$.)

□ **7** Problems other than initial value problems are also common in applications. If the values of y, y', and so on, are specified at two points, the problem is called a *boundary value problem*. In this case we may have either no solution or more than one solution. Discuss the existence and uniqueness of solutions to

a $y'' = 0$, $y(0) = 0$, $y'(1) = 2$

b $y'' + 2y' + y = 0$, $y(0) = 0$, $y'(1) = 3$

c $y'' + 2y' + y = 0$, $y(0) = 0$, $y'(1) = 0$

□ **8** Case 3 can be solved by using the complex function discussion of Exercise **9**, page 296. For example, suppose $p(\lambda) = \lambda^2 + b\lambda + c$, with $b^2 - 4c < 0$. With $k = 2\sqrt{4c - b^2}$ and $a = -b/2$, the roots are $a \pm ik$.

a Show that $e^{(a+ik)x}$ and $e^{(a-ik)x}$ are solutions to $y'' + by' + cy = 0$.

b Show that $e^{(a+ik)x}$ and $e^{(a-ik)x}$ are independent.

c Show that $\sin kx = (e^{ikx} - e^{-ikx})/2i$ and $\cos kx = (e^{ikx} + e^{-ikx})/2$.

d Use **a** and **c** to show that $e^{ax} \sin kx$ and $e^{ax} \cos kx$ are solutions to $y'' + by' + cy = 0$.

e Show that $c_1 e^{(a+ik)x} + c_2 e^{(a-ik)x}$ can be written as

$$e^{ax}(d_1 \sin kx + d_2 \cos kx) + ie^{ax}(d_3 \sin kx + d_4 \cos kx)$$

where d_1, d_2, d_3, and d_4 are real.
(Hint: If $c_1 = u_1 + iu_2$, $c_2 = v_1 + iv_2$, where u_1, u_2, v_1, v_2 are real, apply **c** with $d_1 = v_2 - u_2$, $d_2 = u_1 + v_1$, $d_3 = u_1 - v_1$, and $d_4 = u_2 + v_2$.)

f Find the general real solution to $y'' + 2y' + 2y = 0$.
(Hint: Use the real part of the solution of part **e**.)

SECTION 3 The Translation Principle

To complete the discussion of second-order homogeneous equations it is necessary to find the general solution to

1 $$y'' + by' + cy = 0$$

for the case in which $b^2 - 4c < 0$. This is the case in which the roots α and β of $p(\lambda) = \lambda^2 + b\lambda + c$ are complex. We first consider the case in which $b = 0$ and $c = 1$, that is, the equation $y'' + y = 0$. Recall the following two formulas from calculus:

$$(\sin x)' = \cos x, \qquad (\cos x)' = -\sin x$$

Differentiate again to obtain

$$(\sin x)'' = -\sin x, \qquad (\cos x)'' = -\cos x$$

In other words,

$\sin x$ and $\cos x$ are solutions to $y'' + y = 0$.

In fact we can show that

2 The functions $\sin x$ and $\cos x$ are a basis for the null space of $L = D^2 + I$, and therefore, the general solution to $y'' + y = 0$ is

$$y = c_1 \sin x + c_2 \cos x$$

To prove this it is necessary to show that $\sin x$ and $\cos x$ are independent *and* that they *span* the null space of $D^2 + I$. Suppose

$$c_1 \sin x + c_2 \cos x = 0$$

Differentiating gives

$$c_1 \cos x - c_2 \sin x = 0$$

and putting $x = 0$ in each of these gives

$$c_1 = c_2 = 0$$

which shows that, indeed, $\sin x$ and $\cos x$ are independent. The proof that they span the null space of L is outlined in Exercise **7**, below.

The chain rule gives $(\sin kx)' = k \cos kx$ and $(\cos kx)' = -k \sin kx$, so that $\sin kx$ and $\cos kx$ are solutions to $y'' + k^2y = 0$.

Arguments similar to those above can be used to show that

3 The general solution to $y'' + k^2y = 0$ is $c_1 \sin kx + c_2 \cos kx$.

The analysis of

4 $$y'' + by' + cy = 0 \qquad \text{where} \qquad b^2 - 4c < 0$$

is completed by using a translation principle to convert it into an equation of the form $y'' + k^2y = 0$. First complete the square on the characteristic polynomial

$$\begin{aligned} p(\lambda) &= \lambda^2 + b\lambda + c \\ &= \lambda^2 + b\lambda + \frac{b^2}{4} + c - \frac{b^2}{4} \\ &= \left(\lambda + \frac{b}{2}\right)^2 + c - \frac{b^2}{4} \end{aligned}$$

We know that $c - b^2/4 > 0$, for we have assumed that $b^2 - 4c < 0$. Therefore, we can write

5 $$p(\lambda) = (\lambda - a)^2 + k^2 \qquad \text{where} \qquad a = -b/2,\ k = \sqrt{c - b^2/2}$$

We have, then, the fact that

6 If y is a solution to $y'' + k^2y = 0$, then $e^{ax}y$ is a solution to $y'' + by' + cy = 0$. Conversely, if y is a solution to $y'' + by' + cy = 0$, then $e^{-ax}y$ is a solution to $y'' + k^2y = 0$.

Before establishing this result we deduce some of its consequences. The functions $\sin kx$ and $\cos kx$ are a basis for the solutions to $y'' + k^2y = 0$. Result **6** tells us that

7 $$e^{ax} \sin kx \qquad \text{and} \qquad e^{ax} \cos kx$$

are solutions to equation **4**. If

$$c_1 e^{ax} \sin kx + c_2 e^{ax} \cos kx = 0$$

we can multiply by e^{-ax} to obtain

$$c_1 \sin kx + c_2 \cos kx = 0$$

Thus, the independence of $\sin kx$ and $\cos kx$ guarantees that $c_1 = c_2 = 0$. In other words, the solutions **7** are independent.

Furthermore, they are a basis for the solutions to $y'' + by' + cy = 0$, for if y is a solution to this equation, then property **6** tells us that $e^{-ax}y$ is a solution to $y'' + k^2y = 0$. Statement **3** tells us, then, that there must be constants c_1 and c_2 such that

$$e^{-ax}y = c_1 \sin kx + c_2 \cos kx$$

so we must have

$$y = c_1e^{ax} \sin kx + c_2e^{ax} \cos kx$$

In summary:

8 The general solution to $y'' + by' + cy = 0$, where $b^2 - 4c < 0$, is $c_1e^{ax} \sin kx + c_2e^{ax} \cos kx$, where $a = -b/2$ and $k = \sqrt{c - b^2/4}$.

EXAMPLE 1 The equation $y'' + 3y = 0$ is of the form $y'' + k^2y = 0$, where $k = \sqrt{3}$. Apply property **3** to obtain the general solution

$$y = c_1 \sin \sqrt{3}x + c_2 \cos \sqrt{3}x$$

EXAMPLE 2 Consider the equation $y'' + 2y' + 4y = 0$. Here $p(\lambda) = \lambda^2 + 2\lambda + 4$ and $2^2 - 4 \cdot 4 < 0$. Therefore, completing the square gives

$$\begin{aligned} p(\lambda) &= (\lambda^2 + 2\lambda + 1) + (4 - 1) \\ &= (\lambda + 1)^2 + 3 \end{aligned}$$

We conclude from result **8** that the general solution is

$$y = c_1e^{-x} \sin \sqrt{3}x + c_2e^{-x} \cos \sqrt{3}x$$

EXAMPLE 3 The characteristic polynomial for $y'' - 6y' + 10y = 0$ is $p(\lambda) = \lambda^2 - 6\lambda + 10$. Completing the square gives

$$\begin{aligned} p(\lambda) &= (\lambda^2 - 6\lambda + 9) + (10 - 9) \\ &= (\lambda - 3)^2 + 1 \end{aligned}$$

Therefore, statement **8** gives the general solution

$$y = c_1e^{3x} \sin x + c_2e^{3x} \cos x$$

EXAMPLE 4 Suppose we wish to solve the initial value problem

$$\begin{aligned} y'' + 4y &= 0 \\ y\left(\frac{\pi}{4}\right) &= 1 \\ y'\left(\frac{\pi}{4}\right) &= -1 \end{aligned}$$

The general solution to $y'' + 4y = 0$ is $y = c_1 \sin 2x + c_2 \cos 2x$, whose derivative is $y' = 2c_1 \cos 2x - 2c_2 \sin 2x$. Setting $x = \pi/4$ and using the given conditions gives

$$1 = c_1 \sin \frac{\pi}{2} + c_2 \cos \frac{\pi}{2}$$

$$-1 = 2c_1 \cos \frac{\pi}{2} - 2c_2 \sin \frac{\pi}{2}$$

Thus $c_1 = 1$ and $c_2 = \frac{1}{2}$. This gives the solution to our problem:

$$y = \sin 2x + \tfrac{1}{2} \cos 2x$$

EXAMPLE 5 Suppose we wish to solve the initial value problem

$$\begin{aligned} y'' - 3y' + 3y &= 0 \\ y(0) &= -1 \\ y'(0) &= 0 \end{aligned}$$

The characteristic polynomial of the differential equation is

$$\begin{aligned} p(\lambda) &= \lambda^2 - 3\lambda + 3 \\ &= (\lambda^2 - 3\lambda + \tfrac{9}{4}) + (3 - \tfrac{9}{4}) \\ &= (\lambda - \tfrac{3}{2})^2 + \tfrac{3}{4} \end{aligned}$$

so the general solution is (using statement **8**)

$$y = c_1 e^{3x/2} \sin \frac{\sqrt{3}}{2} x + c_2 e^{3x/2} \cos \frac{\sqrt{3}}{2} x$$

We have

$$\begin{aligned} y' = {} & c_1 \left(\tfrac{3}{2} e^{3x/2} \sin \frac{\sqrt{3}}{2} x + \frac{\sqrt{3}}{2} e^{3x/2} \cos \frac{\sqrt{3}}{2} x \right) \\ & + c_2 \left(\tfrac{3}{2} e^{3x/2} \cos \frac{\sqrt{3}}{2} x - \frac{\sqrt{3}}{2} e^{3x/2} \sin \frac{\sqrt{3}}{2} x \right) \end{aligned}$$

Setting $x = 0$ in each of these, and using the initial conditions, we have

$$\begin{aligned} -1 &= c_2 \\ 0 &= \frac{\sqrt{3}}{2} c_1 + \tfrac{3}{2} c_2 \end{aligned}$$

so that $c_1 = \sqrt{3}$ and $c_2 = -1$. Hence our solution is

$$y = \sqrt{3} e^{3x/2} \sin \frac{\sqrt{3}}{2} x - e^{3x/2} \cos \frac{\sqrt{3}}{2} x$$

DISCUSSION Let us now turn to the proof of statement **6**. The associated operators for $y'' + by' + cy = 0$ and $y'' + k^2y = 0$ are, respectively,

$$L = D^2 + bD + cI \qquad \text{and} \qquad L_1 = D^2 + k^2I$$

The respective characteristic polynomials are

$$p(\lambda) = \lambda^2 + b\lambda + c \qquad \text{and} \qquad q(\lambda) = \lambda^2 + k^2$$

Formula **5** tells us that $p(\lambda) = q(\lambda - a)$; that is, $p(\lambda)$ is obtained from $q(\lambda)$ by *translation* by $\lambda - a$ (in other words, the replacement of λ by $\lambda - a$). Below it will be shown that

9 $$L(e^{ax}y) = e^{ax}L_1y$$

Result **6** is a consequence of formula **9**, for if y is a solution to $y'' + k^2y = 0$, then $L_1y = 0$, and therefore

$$L(e^{ax}y) = e^{ax}L_1y = 0$$

Thus $e^{ax}y$ is a solution to $y'' + by' + cy = 0$.

Conversely, if y is a solution to $y'' + by' + cy = 0$, then $Ly = 0$. Since $y = e^{ax}(e^{-ax}y)$, formula **9** gives

$$Ly = L[e^{ax}(e^{-ax}y)] = e^{ax}L_1(e^{-ax}y)$$

Therefore, $e^{ax}L_1(e^{-ax}y)$ must be the zero function. Since we know that e^{ax} is never zero, we conclude that $L_1(e^{-ax}y) = 0$, or, in other words, that $e^{-ax}y$ is a solution to $y'' + k^2y = 0$. This completes the proof of the fact that statement **6** is a consequence of formula **9**.

To prove formula **9** first apply the product and chain rules to obtain

$$D(e^{ax}y) = ae^{ax}y + e^{ax}Dy$$

Observe that $ae^{ax}y = (aI)e^{ax}y$, so we can subtract from each side to obtain

10 $$(D - aI)e^{ax}y = e^{ax}Dy$$

Equation **10** gives us

11 $$(D - aI)^2e^{ax}y = e^{ax}D^2y$$

for we have

$$\begin{aligned}(D - aI)^2e^{ax}y &= (D - aI)[(D - aI)e^{ax}y]\\ &= (D - aI)(e^{ax}Dy)\end{aligned}$$

Now apply formula **10** to this, with $e^{ax}y$ replaced by $e^{ax}Dy$, to obtain

$$(D - aI)^2 e^{ax}y = e^{ax}D^2y$$

This is the desired formula **11**. Returning to the proof of formula **9**, we observe that since $L = p(D)$ and $p(\lambda) = (\lambda - a)^2 + k^2$ we have

$$L = (D - aI)^2 + k^2I$$

Thus

$$L(e^{ax}y) = (D - aI)^2(e^{ax}y) + k^2Ie^{ax}y$$

Applying formula **11** gives

$$L(e^{ax}y) = e^{ax}D^2y + k^2Ie^{ax}y$$

Since $Ie^{ax}y = e^{ax}y = e^{ax}Iy$, we have

$$\begin{aligned} L(e^{ax}y) &= e^{ax}(D^2y + k^2Iy) \\ &= e^{ax}(D^2 + k^2I)y \\ &= e^{ax}L_1y \end{aligned}$$

This is just formula **9**. These results can be generalized to obtain the following theorem, which is known as the *translation principle*. The proof is left to Exercise **8**.

THEOREM 18 If $L_1 = \alpha_nD^n + \alpha_{n-1}D^{n-1} + \cdots + \alpha_1D + \alpha_0I$ and $L = \alpha_n(D - aI)^n + \alpha_{n-1}(D - aI)^{n-1} + \cdots + \alpha_1(D - aI) + \alpha_0I$ so that L is obtained from L_1 by replacing D by $D - aI$, then $L(e^{ax}y) = e^{ax}L_1y$.

In particular, if $c_1\varphi_1 + c_2\varphi_2 + \cdots + c_n\varphi_n$ is the general solution to $L_1y = 0$, then

$$e^{ax}(c_1\varphi_1 + c_2\varphi_2 + \cdots + c_n\varphi_n)$$

is the general solution to $Ly = 0$.

This theorem enables us to describe the null space of L once we know the null space of L_1. We shall find it useful in the next section. Example **6** indicates one such use of the translation principle.

EXAMPLE 6 The functions $1, x, x^2, \ldots, x^{n-1}$ are independent and are certainly solutions to $D^ny = 0$. Repeated integration of this equation shows that any solution must be a polynomial of degree $\leq n - 1$. Therefore, the null space of $L_1 = D^n$ has the basis $1, x, x^2, \ldots, x^{n-1}$. The translation principle tells us that the null space of $L = (D - aI)^n$ must have the basis

$$e^{ax}, xe^{ax}, \ldots, x^{n-1}e^{ax}$$

This is the general form of result **8** of the previous section.

We see that the general solution to

$$y'''' + 4y''' + 6y'' + 4y' + y = 0$$

is

$$y = (c_1 + c_2x + c_3x^2 + c_4x^3)e^{-x}$$

since the associated operator is

$$(D + I)^4 = D^4 + 4D^3 + 6D^2 + 4D + I$$

EXERCISES

1 Find the general solution to each of the following.
a $y'' + 4y = 0$ **b** $y'' = -by$
c $3y'' + 7y = 0$ **d** $2y'' = -5y$

2 Using the method of Examples **2** and **3** find the general solution to each of the following.
a $y'' + 2y' + 2y = 0$ **b** $y'' + 4y' + 5y = 0$
c $4y'' - 4y' + 5y = 0$ **d** $y'' - 8y' + 18y = 0$

3 Find the general solution to each of the following.
a $y'' + 6y' + 9y = 0$ **b** $y'' + 6y' + 10y = 0$
c $y'' + 6y' + 8y = 0$ **d** $y'' + 6y' - 10y = 0$

4 Solve each of the following initial value problems.

a $y'' + 4y = 0$, $y(0) = 0$, $y'(0) = 1$

b $4y'' + 9y = 0$, $y(\pi) = 1$, $y'(\pi) = 0$

c $y'' + 2y' + 2y = 0$, $y(\pi/2) = 1$, $y'(\pi/2) = -1$

d $y'' - 4y' + 6y = 0$, $y(0) = 0$, $y'(0) = 1$

5 Find a basis for the null space of each of the following by using the methods of Example **6**.
a $(D - 3I)^4$ **b** $(D + 2I)^6$

6 Using Theorem **18** find:
a $(D - 3I)^4(e^{3x} \log x)$ **b** $(D + 2I)^3(x^5e^{-2x})$
c $(D + I)^5e^{2x}$ **d** $(D - \sqrt{2}I)^{12}(e^{\sqrt{2}x} \sin x)$

7 This exercise outlines a proof that $\sin x$ and $\cos x$ span the null space of $D^2 + I$. Suppose y is in the null space of $D^2 + I$. Put

$$\begin{aligned} A &= y(0) \\ B &= y'(0) \\ z(x) &= y(x) - A \sin x - B \cos x \\ E(x) &= [z(x)]^2 + [z'(x)]^2 \end{aligned}$$

a Show that $z(0) = z'(0) = 0$.
b Show that $(D^2 + I)z = 0$.
c Show that $DE = 2z'(D^2 + I)z$.
d Deduce that E is constant.
e Use parts **a** and **d** to show that E is the zero function.
f Deduce that z is identically zero.
g Does part **f** show that y is a linear combination of $\sin x$ and $\cos x$?

8 Prove Theorem **18**. [Hint: First show that $(D - aI)^k(e^{ax}y) = e^{ax}D^k y$ for $k = 1, 2, \ldots, n$.]

☐ **9** Solve each equation of Exercises **1** and **2** by using the method of Exercise **8**, page 303.

SECTION 4 Higher-Order Linear Operators

The nature of the general solution to

1 $$y^{(n)} + \beta_{n-1}y^{(n-1)} + \cdots + \beta_1 y' + \beta_0 y = 0$$

depends upon the factorization of the characteristic polynomial

$$p(\lambda) = \lambda^n + \beta_{n-1}\lambda^{n-1} + \cdots + \beta_1\lambda + \beta_0$$

Using the theory of such factorization (see the Bibliography for a reference to a discussion of this), we can write

2 $$p(\lambda) = p_1(\lambda)^{n_1}p_2(\lambda)^{n_2} \cdots p_k(\lambda)^{n_k}$$

where each polynomial p_i is of the form $(\lambda - \alpha_i)$ or $\lambda^2 + b_i\lambda + c_i$ with $b_i^2 - 4c_i < 0$, and $p_i \neq p_j$ if $i \neq j$. Since each side of equation **2** must have the same degree, we have

3 $$n = n_1 \text{ degree } p_1 + n_2 \text{ degree } p_2 + \cdots + n_k \text{ degree } p_k$$

Suppose

$$p_i(\lambda) = \lambda - \alpha_i \qquad \text{and} \qquad n_i = 1$$

This factor contributes to the general solution of equation **1** the term $e^{\alpha_i x}$. If $n_i = 2$, we have the two solutions $e^{\alpha_i x}$ and $xe^{\alpha_i x}$.

In general, as shown in Example **6** of the previous section, we have the n_i independent solutions

4 $$e^{\alpha_i x}, xe^{\alpha_i x}, \ldots, x^{n_i - 1}e^{\alpha_i x}$$

contributed by the factor $p_i(\lambda)^{n_i} = (\lambda - \alpha_i)^{n_i}$.

If

$$p_i(\lambda) = \lambda^2 + b_i\lambda + c_i \qquad \text{with} \qquad b_i^2 - 4c_i < 0$$

we can complete the square to write

$$p_i(\lambda) = (\lambda - a_i)^2 + k_i^2$$

If $n_i = 1$, this factor contributes the two independent solutions

$$e^{a_i x} \sin k_i x \qquad \text{and} \qquad e^{a_i x} \cos k_i x$$

If $n_i = 2$, this factor contributes the four independent solutions (see Exercise **6**, below)

$$e^{a_i x} \sin k_i x \qquad e^{a_i x} \cos k_i x \qquad x e^{a_i x} \sin k_i x \qquad x e^{a_i x} \cos k_i x$$

In general, we obtain the $2n_i$ independent solutions

5
$$\begin{gathered} e^{a_i x} \sin k_i x,\ x e^{a_i x} \sin k_i x,\ \ldots,\ x^{n_i - 1} e^{a_i x} \sin k_i x \\ e^{a_i x} \cos k_i x,\ x e^{a_i x} \cos k_i x,\ \ldots,\ x^{n_i - 1} e^{a_i x} \cos k_i x \end{gathered}$$

contributed by the factor $p_i(\lambda)^{n_i} = (\lambda^2 + b_i\lambda + c_i)^{n_i}$.

This process—constructing solutions for each factor $p_i(\lambda)^{n_i}$—gives us n solutions to equation **1** (from formula **3**) which can be shown to be independent. To see that they are a basis for the solutions to equation **1** we refer to the following result, the proof of which is given in the Appendix.

THEOREM 19 Suppose S and T are linear transformations from V into V, *each* with a finite-dimensional null space. Then the null space of ST is finite-dimensional, and the dimension of the null space of ST *cannot exceed the sum* of the dimensions of the null space of S and the null space of T.

The associated operator for equation **1** is

$$L = D^n + \beta_{n-1}D^{n-1} + \cdots + \beta_1 D + \beta_0 I$$

We can then write

$$L = p(D) = p_1(D)^{n_1} p_2(D)^{n_2} \ldots p_k(D)^{n_k}$$

so that L is a product of powers of operators of either first or second order.

We have shown that if $p_i(\lambda)$ has degree 1 the null space of $p_i(D)$ has dimension 1. (See property **2**, page 291.) Exercise **7**, page 310, tells us that the null space of $p_i(D)$ has dimension 2 if $p_i(\lambda)$ is of the form $\lambda^2 + k^2$, while property **8**, page 306, tells us that this null space has dimension 2 if $p_i(\lambda) = \lambda^2 + b_i\lambda + c_i$ with $b_i^2 - 4c_i < 0$. Thus L is the product of powers of operators whose null spaces have dimension 1 or 2. We can conclude from Theorem **19** and formula **3** that

> The dimension of the null space of L cannot exceed n, the order of L.

Since we have constructed n independent solutions to equation **1**, we conclude that this null space has dimension at least n. In summary:

THEOREM 20 The dimension of the null space of

$$L = D^n + \beta_{n-1}D^{n-1} + \cdots + \beta_1 D + \beta_0 I$$

is n, the order of L.

EXAMPLE 1 Suppose we are looking for a basis for the null space of

$$L = (D - I)(D + 2I)^3(D^2 + 9I)^2D^4$$

The factor $D - I$ contributes e^x; the factor $(D + 2I)^3$ contributes e^{-2x}, xe^{-2x}, and x^2e^{-2x}; the factor $(D^2 + 9I)^2$ gives us $\sin 3x$, $\cos 3x$, $x \sin 3x$, and $x \cos 3x$; and D^4 gives 1, x, x^2, and x^3. Thus the general solution to $Ly = 0$ is

$$c_1e^x + c_2e^{-2x} + c_3xe^{-2x} + c_4x^2e^{-2x} + c_5 \sin 3x + c_6 \cos 3x$$
$$+ c_7x \sin 3x + c_8x \cos 3x + c_9 + c_{10}x + c_{11}x^2 + c_{12}x^3$$

EXAMPLE 2 Consider the equation

$$y^{(5)} + y^{(4)} - y^{(3)} - y^{(2)} = 0$$

The associated operator is $L = D^5 + D^4 - D^3 - D^2$, whose characteristic polynomial is

$$\lambda^5 + \lambda^4 - \lambda^3 - \lambda^2 = \lambda^2(\lambda^3 + \lambda^2 - \lambda - 1)$$
$$= \lambda^2(\lambda + 1)^2(\lambda - 1)$$

The factor λ^2 gives the two solutions 1, x; the factor $(\lambda + 1)^2$ gives the two solutions e^{-x}, xe^{-x}; and the factor $\lambda - 1$ gives the solution e^x. Thus the general solution is

$$y = c_1 + c_2x + c_3e^{-x} + c_4xe^{-x} + c_5e^x$$

EXAMPLE 3 To find the general solution to $Ly = 0$, where

$$L = (D^2 + 2D + 2I)^3(D^2 - I)$$

calculate the solutions corresponding to each factor of L. Completing the square for $D^2 + 2D + 2I$ gives

$$D^2 + 2D + 2I = (D + I)^2 + I$$

Therefore $(D^2 + 2D + 2I)^3$ gives the six functions $e^{-x} \sin x$, $e^{-x} \cos x$, $xe^{-x} \sin x$, $xe^{-x} \cos x$, $x^2e^{-x} \sin x$, and $x^2e^{-x} \cos x$. The factor $D^2 - I = (D + I)(D - I)$ gives us e^x and e^{-x}. The general solution is thus

$$\begin{aligned} c_1e^x + e^{-x}(c_2 \sin x + c_3 \cos x + c_4x \sin x \\ + c_5x \cos x + c_6x^2 \sin x + c_7x^2 \cos x + c_8) \end{aligned}$$

EXAMPLE 4 Suppose we wish to solve the initial value problem

$$y^{(5)} + y^{(4)} - y^{(3)} - y^{(2)} = 0$$

$$y(0) = 1,\ y'(0) = -1,\ y''(0) = 0,\ y'''(0) = 1,\ y''''(0) = 0$$

From Example 2 we know that the general solution to the differential equation is

$$y = c_1 + c_2x + c_3e^{-x} + c_4xe^{-x} + c_5e^x$$

so that

$$\begin{aligned} y' &= c_2 - c_3e^{-x} + c_4(e^{-x} - xe^{-x}) + c_5e^x \\ y'' &= c_3e^{-x} + c_4(-2e^{-x} + xe^{-x}) + c_5e^x \\ y''' &= -c_3e^{-x} + c_4(3e^{-x} - xe^{-x}) + c_5e^x \\ y'''' &= c_3e^{-x} + c_4(-4e^{-x} + xe^{-x}) + c_5e^x \end{aligned}$$

Setting $x = 0$ in each of these and using the initial conditions gives the following five equations:

$$\begin{aligned} 1 &= c_1 + c_3 + c_5 \\ -1 &= c_2 - c_3 + c_4 + c_5 \\ 0 &= c_3 - 2c_4 + c_5 \\ 1 &= -c_3 + 3c_4 + c_5 \\ 0 &= c_3 - 4c_4 + c_5 \end{aligned}$$

Solving these equations gives the unique solution $c_1 = 1$, $c_2 = -2$, $c_3 = -\frac{1}{2}$, $c_4 = 0$, and $c_5 = \frac{1}{2}$. Therefore, the solution is

$$y = 1 - 2x - \tfrac{1}{2}e^{-x} + \tfrac{1}{2}e^x$$

EXERCISES

1 By factoring the characteristic polynomial find the general solution to each of the following.

a $y'''' + 4y'' + 4y = 0$ **b** $y''' - 3y'' + 3y' - y = 0$

c $y^{(6)} - 3y^{(5)} - 4y^{(4)} = 0$ **d** $(y'' + 2y' + 2y)'' = 0$

e $y'''' + y = 0$ **f** $y''' - y = 0$

2 Find a basis for the null space of each of the following.

a $(D + 3I)^2(D^2 - I)^2(D^2 + I)$ **b** $D^5(D^2 + 4D + 5I)^3$

c $(D^2 - 3D - 4I)^2(D^2 + D + I)^2$ **d** $(D^4 - 6D^3 + 13D^2)^2$

3 Solve each of the following initial value problems.

a $y'''' + 4y'' + 4y = 0$
$y(0) = 1, y'(0) = 2, y''(0) = y'''(0) = 0$

b $y'''' - y = 0$
$y(0) = 0, y'(0) = 1, y''(0) = -1, y'''(0) = 1$

c $y''' + 3y'' + 3y' + y = 0$
$y(1) = 1, y'(1) = -1, y''(1) = 0$

d $(D^2 + 2D + 2I)^2(D - I)y = 0$
$y(0) = y'(0) = y''(0) = y'''(0) = 0, y''''(0) = 1$

4 Suppose K and L are constant coefficient linear differential operators.

a Show that KL is a constant coefficient linear differential operator. [Hint: There are polynomials p and q such that $K = p(D)$, $L = q(D)$.]

b Show that $KL = LK$. (Hint: Use the hint of part **a** and the fact that $pq = qp$.)

c Show that the order of KL is the sum of the orders of K and L.

5 Suppose that $A = \begin{bmatrix} 1 & 0 \\ 0 & 0 \end{bmatrix}$ and $B = \begin{bmatrix} 2 & 0 \\ 0 & 0 \end{bmatrix}$.

a What is the dimension of the null space of A? of B? of AB?

b Part **a** shows that the dimension of the null space of a product can be smaller than the sum of the dimensions of the null space of each factor. Show that if K and L are constant coefficient linear differential operators, then the dimension of the null space of KL *equals* the sum of the dimensions of the null spaces of K and L. (Hint: Use Theorem **20** and Exercise **4c**.)

6 **a** Show that $\sin kx$, $\cos kx$, $x \sin kx$, and $x \cos kx$ are solutions to

$$(D^2 + k^2I)^2y = 0$$

b Show that $\sin kx$, $\cos kx$, $x \sin kx$, and $x \cos kx$ are independent.

c Use the translation principle to show that $e^{ax} \sin kx$, $e^{ax} \cos kx$, $xe^{ax} \sin kx$, and $xe^{ax} \cos kx$ are solutions to $[(D - aI)^2 + k^2I]^2y = 0$.

d Show that the functions of part **c** are independent.

7 Proceed as in Exercise **6** to show that $e^{ax} \sin kx$, $e^{ax} \cos kx$, $xe^{ax} \sin kx$, $xe^{ax} \cos kx$, $x^2e^{ax} \sin kx$, and $x^2e^{ax} \cos kx$ are independent solutions to

$$[(D - aI)^2 + k^2I]^3y = 0$$

SECTION 5 The Method of Undetermined Coefficients

Let us now turn to the problem of finding the general form of solutions to $Ly = f$, where L is a constant coefficient linear differential operator. We first prove that

1 If $Ly_1 = f$, then the set of solutions to $Ly = f$ is precisely the set of functions of the form $y_1 + y_2$, where y_2 is in the null space of L.

For if y_2 is in the null space of L, then

$$L(y_1 + y_2) = Ly_1 + Ly_2 = f + 0 = f$$

Conversely, if $Ly = f$, and we let $y_2 = y - y_1$, then

$$Ly_2 = L(y - y_1) = Ly - Ly_1 = f - f = 0$$

We have shown that if $Ly = f$, then y must be of the form $y = y_1 + y_2$, where y_2 is in the null space of L, and have therefore completed the proof of statement **1**.

Suppose $\varphi_1, \varphi_2, \ldots, \varphi_n$ are a basis for the null space of L and that $Ly_1 = f$. The above result tells us that the solutions to $Ly = f$ are precisely the functions of the form

$$y_1 + c_1\varphi_1 + c_2\varphi_2 + \cdots + c_n\varphi_n$$

This is called the **general solution** to $Ly = f$. Observe that it is obtained by finding one solution, namely y_1, to $Ly = f$ and adding it to the general solution to the homogeneous equation $Ly = 0$.

Since the methods for finding the general solution to the homogeneous problem have already been developed, the discussion of constant coefficient linear differential equations can be completed by showing how to find *one* solution to the nonhomogeneous equation. A complicated but general method for finding such a solution for an arbitrary continuous function f is outlined in the exercises. We present here a useful and simple procedure that yields a solution when f is of a special form. To be precise, we assume that

2 There is a constant coefficient linear differential operator K, such that $Kf = 0$.

From our discussions in the previous sections about the form of the null space of such a K it is not too difficult to see that this is exactly the condition that f be a linear combination of products and derivatives of x,

e^{ax}, and sin kx. Thus, while the method to be presented does not apply to all functions f, it will cover a class of functions which frequently arise in practice.

If f satisfies condition **2** and if $Ly = f$, we have

$$(KL)y = K(Ly) = Kf = 0$$

In other words,

3 If f satisfies condition **2**, the solutions to $Ly = f$ are *all* in the null space of KL.

The operator KL is itself a constant coefficient linear differential operator (see Exercise **4**, page 315), so we can apply the methods previously learned to find a basis $\varphi_1, \varphi_2, \ldots, \varphi_n$ for the null space of KL. Then condition **3** tells us that we need only choose $c_1, c_2, \ldots, c_n$ so that

$$L(c_1\varphi_1 + c_2\varphi_2 + \cdots + c_n\varphi_n) = f$$

In other words, we merely calculate

$$L(c_1\varphi_1 + c_2\varphi_2 + \cdots + c_n\varphi_n)$$

and determine $c_1, c_2, \ldots, c_n$ so that the result equals f. For this reason, the method is known as the *method of undetermined coefficients.*

We have used the assumption that the equation $Ly = f$ does have a solution in order to know that we can actually find the desired coefficients. In the Appendix we shall prove a general result about linear transformations which will give the result that

4 If f satisfies condition **2**, there is a function y in the null space of KL such that $Ly = f$.

We demonstrate the use of this method with some examples.

EXAMPLE 1 To find a particular solution to

5 $$y'' + 2y = x^2 + 3$$

set $Ly = y'' + 2y$ and note that $(x^2 + 3)''' = 0$. Therefore, $x^2 + 3$ satisfies condition **2**, with $Ky = y'''$. We have

$$L = D^2 + 2I \qquad K = D^3$$

Hence

$$KL = D^3(D^2 + 2I)$$

Thus the null space of KL has the basis $1, x, x^2, \sin\sqrt{2}x, \cos\sqrt{2}x$. Now determine c_1, c_2, c_3, c_4, c_5 so that if $y = c_1 + c_2x + c_3x^2 + c_4 \sin\sqrt{2}x + c_5 \cos\sqrt{2}x$, then $Ly = x^2 + 3$. We have

$$Ly = L(c_1 + c_2x + c_3x^2) + L(c_4 \sin\sqrt{2}x + c_5 \cos\sqrt{2}x)$$

The second term is zero for any c_4 and c_5, since $\sin\sqrt{2}x$ and $\cos\sqrt{2}x$ are in the null space of L. Hence we need only choose c_1, c_2, c_3 so that

$$L(c_1 + c_2x + c_3x^2) = x^2 + 3$$

Since

$$L(c_1 + c_2x + c_3x^2) = 2c_1 + 2c_2x + 2c_3x^2 + 2c_3$$

we have the following equations (by equating coefficients of like powers):

$$2c_1 + 2c_3 = 3$$
$$2c_2 = 0$$
$$2c_3 = 1$$

The unique solution to this system is $c_1 = 1$, $c_2 = 0$, $c_3 = \frac{1}{2}$, which gives the function $y_1 = 1 + \frac{1}{2}x^2$ as a solution to equation **5**.

Since the general solution to $Ly = 0$ is

$$c_1 \sin\sqrt{2}x + c_2 \cos\sqrt{2}x$$

we apply property **1** to obtain the general solution to equation **5**,

$$y = 1 + \tfrac{1}{2}x^2 + c_1 \sin\sqrt{2}x + c_2 \cos\sqrt{2}x$$

EXAMPLE 2 Consider the equation

6 $$y'' + y = e^x$$

In this case $L = D^2 + I$, and, since $(D - I)e^x = 0$, we can choose $K = D - I$.

We have

$$KL = (D - I)(D^2 + I)$$

so a basis for the null space of KL is

$$e^x, \sin x, \cos x$$

Thus we want to find c_1, c_2, and c_3 such that

$$y = c_1e^x + c_2 \sin x + c_3 \cos x$$

is a particular solution to equation **6**. Apply L to obtain

$$Ly = c_1Le^x + c_2L \sin x + c_3L \cos x$$

Since

$$L \sin x = L \cos x = 0$$

the choice of c_2 and c_3 is arbitrary. Therefore we need only determine c_1 so that $c_1 Le^x = e^x$.

We have

$$Le^x = (e^x)'' + e^x = 2e^x$$

Therefore $c_1 Le^x = e^x$ gives $2c_1 e^x = e^x$. We can choose $c_1 = \frac{1}{2}$ to obtain the solution $y_1 = \frac{1}{2}e^x$.

The general solution to $Ly = 0$ is (since $L = D^2 + I$)

$$c_1 \sin x + c_2 \cos x$$

so the general solution to equation **6** is the sum of this and a particular solution:

$$y = \tfrac{1}{2}e^x + c_1 \sin x + c_2 \cos x$$

EXAMPLE 3 For the equation

7 $$y'' - y = e^x$$

put $L = D^2 - I$. Then choose $K = D - I$, since $(D - I)e^x = 0$. We have $KL = (D - I)^2(D + I)$; therefore, e^x, xe^x, and e^{-x} are a basis for the null space of KL. Thus we need to determine c_1, c_2, and c_3 so that

$$L(c_1 e^x + c_2 xe^x + c_3 e^{-x}) = e^x$$

Since

$$\begin{aligned} L(c_1 e^x + c_2 xe^x + c_3 e^{-x}) &= c_1 Le^x + c_2 Lxe^x + c_3 Le^{-x} \\ &= 0 + c_2 Lxe^x + 0 \end{aligned}$$

we need only choose c_2 so that

8 $$c_2 L(xe^x) = e^x$$

Since

$$L(xe^x) = (xe^x)'' - xe^x = 2e^x + xe^x - xe^x = 2e^x$$

equation **8** gives $2c_2 e^x = e^x$. Hence, if we choose $c_2 = \frac{1}{2}$ we obtain the particular solution $y_1 = \frac{1}{2}xe^x$. The general solution to $Ly = 0$ is $c_1 e^x + c_2 e^{-x}$. Applying property **1** we thus have the general solution to equation **7**:

$$y = \tfrac{1}{2}xe^x + c_1 e^x + c_2 e^{-x}$$

EXAMPLE 4 If L is a linear differential operator and $Lz_1 = f_1$, $Lz_2 = f_2$, then for any scalars a_1, a_2 we have

9 $$L(a_1z_1 + a_2z_2) = a_1Lz_1 + a_2Lz_2 = a_1f_1 + a_2f_2$$

We can use this principle to solve $Ly = f$, when f is a linear combination of functions which satisfy condition **2**. This principle is often called the *superposition principle*. For example, consider

10 $$y'' - y = 4x - 3xe^x$$

We can find particular solutions to each of the equations

$$y'' - y = x \qquad \text{and} \qquad y'' - y = xe^x$$

and then use the superposition principle to solve equation **10**. For $y'' - y = x$ put $L = D^2 - I$ and $K = D^2$, so that

$$KL = D^2(D^2 - I)$$

Thus there is a solution of the form

$$y_1 = c_1 + c_2x + c_3e^x + c_4e^{-x}$$

Apply L as follows:

$$\begin{aligned} Ly_1 &= c_1L1 + c_2Lx + c_3Le^x + c_4Le^{-x} \\ &= c_1L1 + c_2Lx \\ &= c_1[(1)'' - 1] + c_2[(x)'' - x] \\ &= -c_1 - c_2x \end{aligned}$$

The condition $Ly_1 = x$ gives the equation $-c_1 - c_2x = x$. We conclude that $c_1 = 0$ and $c_2 = -1$, so that $y_1 = -x$ is a solution to $y'' - y = x$.

Now find a solution y_2 to the equation $y'' - y = xe^x$. We have $L = D^2 - I$ and $K = (D - I)^2$, for $(D - I)^2(xe^x) = 0$. Thus

$$KL = (D - I)^2(D^2 - I) = (D - I)^3(D + I)$$

Therefore, there is a solution of the form

$$y_2 = c_1e^x + c_2xe^x + c_3x^2e^x + c_4e^{-x}$$

Apply L and use the fact that $Le^x = Le^{-x} = 0$; hence one need only choose c_2 and c_3 so that

$$c_2L(xe^x) + c_3L(x^2e^x) = xe^x$$

Since we have

$$\begin{aligned} L(xe^x) &= (xe^x)'' - xe^x = 2e^x \\ L(x^2e^x) &= (x^2e^x)'' - x^2e^x = 2e^x + 4xe^x \end{aligned}$$

we must therefore choose c_2 and c_3 so that

$$2c_2e^x + 2c_3e^x + 4c_3xe^x = xe^x$$

holds for all x. As e^x and xe^x are independent, this means that $c_3 = \frac{1}{4}$ and $c_2 = -c_3 = -\frac{1}{4}$, and it follows that $-\frac{1}{4}xe^x + \frac{1}{4}x^2e^x$ is a solution to $y'' - y = xe^x$. Since $-x$ is a solution to $y'' - y = x$, the superposition principle **9** shows that

$$-4x - 3(-\tfrac{1}{4}xe^x + \tfrac{1}{4}x^2e^x) = -4x + \tfrac{3}{4}e^x(x - x^2)$$

is a solution to equation **10**.

EXAMPLE 5 Suppose we wish to solve the initial value problem

11
$$\begin{aligned} y'' + y &= e^x \\ y(0) &= 1 \\ y'(0) &= 0 \end{aligned}$$

It was shown in Example **2** that the general solution to $y'' + y = e^x$ is

$$y = \tfrac{1}{2}e^x + c_1 \sin x + c_2 \cos x$$

Differentiating this gives

$$y' = \tfrac{1}{2}e^x + c_1 \cos x - c_2 \sin x$$

Set $x = 0$ in each of these and use the initial conditions to obtain

$$\begin{aligned} 1 &= \tfrac{1}{2} + c_2 \\ 0 &= \tfrac{1}{2} + c_1 \end{aligned}$$

so that $c_1 = -\frac{1}{2}$, $c_2 = \frac{1}{2}$. Therefore the solution to our initial value problem **11** is

$$y = \tfrac{1}{2}e^x - \tfrac{1}{2}\sin x + \tfrac{1}{2}\cos x$$

EXERCISES

1 Find a particular solution to each of the following.

a $y'' - 4y = xe^{-3x}$ **b** $y'' - 4y = e^{2x}$
c $y'' - 6y' + 8y = x^2 + 1$ **d** $y'' + 2y' + y = 2$
e $y'' + 2y' + y = xe^{-x}$ **f** $y'' + 4y' + 5y = x + \cos x$
g $y'' + 9y = 9x^2 + x \sin 3x$ **h** $y'''' - y = e^x + e^{-x} + \sin x$
i $y''' - y'' - 6y' = 1 + x + e^{-x}$ **j** $y'' - 2y' + y = 1 + xe^{2x}$

2 Solve each of the following problems.

a $y'' + 4y = e^x$, $y(\pi) = 1$, $y'(\pi) = 0$
b $y'' - y = e^{-x}$, $y(0) = 1$, $y'(0) = 0$
c $y'' - 3y' + 2y = 4x^2$, $y(0) = y'(0) = 0$
d $y'' + 2y' + y = e^{-x}$, $y(0) = y'(0) = 0$

3 For each of the following show that condition **2** is true by finding such a K.

a $x^2 \cos 2x + e^x$ **b** $e^{2x} \sin 3x + x^2 e^x$

c $\sin 2x + 3 \cos^2 x$ **d** $x^3 + 3x^2 + 2$

□ 4 There is a method, known as the method of *variation of parameters*, which gives a solution to $Ly = f$, once the null space of L is known. We illustrate this for $L = D^2 + bD + cI$. Let φ_1 and φ_2 be a basis for the null space of L. Let

12 $y = v_1\varphi_1 + v_2\varphi_2$, where v_1 and v_2 are functions which satisfy the condition

13 $v_1'\varphi_1 + v_2'\varphi_2 = 0$

a Show that $Ly = v_1 L\varphi_1 + v_2 L\varphi_2 + v_1'\varphi_1' + v_2'\varphi_2'$ (Hint: Calculate Dy and D^2y using equations **12** and **13**.)

b Show that if $v_1'\varphi_1' + v_2'\varphi_2' = f$, equation **12** gives a solution to $Ly = f$. (Hint: Use part **a**.)

c Show that if v_1 and v_2 satisfy

$$v_1' = \frac{\det\begin{bmatrix} 0 & \varphi_2 \\ f & \varphi_2' \end{bmatrix}}{\det\begin{bmatrix} \varphi_1 & \varphi_2 \\ \varphi_1' & \varphi_2' \end{bmatrix}}, \qquad v_2' = \frac{\det\begin{bmatrix} \varphi_1 & 0 \\ \varphi_1' & f \end{bmatrix}}{\det\begin{bmatrix} \varphi_1 & \varphi_2 \\ \varphi_1' & \varphi_2' \end{bmatrix}}$$

then $L(v\varphi_1 + v_2\varphi_2) = f$. [Hint: Use part **b**, equation **13**, and Cramer's rule (given in Example **9**, page 227).]

d Show that if

$$y = \varphi_1 \int \frac{(-f\varphi_2)}{\varphi_1\varphi_2' - \varphi_1'\varphi_2} + \varphi_2 \int \frac{f\varphi_1}{\varphi_1\varphi_2' - \varphi_1'\varphi_2}$$

then $Ly = f$. (Hint: Integrate the equation of part **c**.)

□ 5 Use the method of Exercise **4** to find a particular solution to

a $y'' + y = x$ **b** $y'' - y = \log x$

c $y'' - 3y' + 2y = \sqrt{x}$ **d** $y'' - 2y' + y = \sqrt{1 - x^2}$

□ 6 Let $\varphi_1, \varphi_2, \varphi_3$ be a basis for the null space of a third-order operator L. Put

$$\Phi = \det\begin{bmatrix} \varphi_1 & \varphi_2 & \varphi_3 \\ \varphi_1' & \varphi_2' & \varphi_3' \\ \varphi_1'' & \varphi_2'' & \varphi_3'' \end{bmatrix} \qquad \Phi_1 = \det\begin{bmatrix} 0 & \varphi_2 & \varphi_3 \\ 0 & \varphi_2' & \varphi_3' \\ f & \varphi_2'' & \varphi_3'' \end{bmatrix}$$

$$\Phi_2 = \det\begin{bmatrix} \varphi_1 & 0 & \varphi_3 \\ \varphi_1' & 0 & \varphi_3' \\ \varphi_1'' & f & \varphi_3'' \end{bmatrix} \qquad \Phi_3 = \det\begin{bmatrix} \varphi_1 & \varphi_2 & 0 \\ \varphi_1' & \varphi_2' & 0 \\ \varphi_1'' & \varphi_2'' & f \end{bmatrix}$$

a Show that

$$y = \varphi_1 \int \frac{\Phi_1}{\Phi} + \varphi_2 \int \frac{\Phi_2}{\Phi} + \varphi_3 \int \frac{\Phi_3}{\Phi}$$

is a solution to $Ly = f$.

b Use this method to find a solution to $y''' - y' = \log x$.

☐ 7 State a result analogous to the previous three exercises for fourth-order operators.

☐ 8 How should the method of Exercises **4** and **6** be modified in order to solve each of the following?

a $2y'' - 5y = x$ b $3y''' - y' = \log x$

Use this modification to find one solution to each equation.

SECTION 6 Applications

Linear differential equations with constant coefficients occur often in applied problems. We give some examples of these problems in this section.

EXAMPLE 1 *Growth and Decay.*

In many problems involving growth or decay (such as radioactive decay) experimentation has shown that the rate of growth or decay can often be taken to be proportional to the amount of material on hand. This principle can be given a mathematical formulation. Thus, if $y = y(t)$ represents the amount of material at time t, then y' is the rate of growth (or decay), and the basic assumption consequently is

1 $$y' = ky$$

where k is constant. Growth problems correspond to $k > 0$, decay problems to $k < 0$.

As shown in Section 1, equation **1** has the general solution

2 $$y = ce^{kt}$$

If we set $t = 0$, then $y(0) = c$; thus c is the *initial* amount of material.

If we know the amount of material at some later time then we can find the value of k.

DECAY: For example, suppose equation **1** represents radioactive decay (so that $k < 0$). If at time t_0 the amount of material present is $c/2$, then

$$\frac{c}{2} = y(t_0) = ce^{kt_0}$$

which gives

3 $$t_0 = -\frac{1}{k}\log 2$$

This is called the *half-life* of the material. Therefore, if we know the half-life, we can solve equation **3** for k.

GROWTH: In certain situations bacteria increase at a rate proportional to the number present; thus, if y represents the number present at time t, then $y' = ky$, with $k > 0$. Therefore, $y = ce^{kt}$, and, consequently, if we know the amount present at two different times, we can determine the precise nature of y. For example, if we begin with $y = 10{,}000$ and find from measurement that 1 million are present one hour later, we have

$$10{,}000 = ce^{0k}$$
$$1{,}000{,}000 = ce^{60k}$$

(assuming that t is given in terms of minutes). These give

$$c = 10{,}000$$
$$k = \tfrac{1}{30} \log 10 \quad \text{which is approximately } .078$$

FRICTION: In some situations friction can be taken to exert a force proportional to the velocity. Thus, if a mass m moves with a velocity $v = v(t)$, subject to such a friction, Newton's second law (force is mass times acceleration) gives

$$m \frac{dv}{dt} = kv$$

If an external force $f(t)$ is applied, we have

4 $$m \frac{dv}{dt} = kv + f(t)$$

Suppose, for example, that f is constant; let us say $f = b$. The general solution to equation **4** is (the student should verify this):

$$v = -\frac{b}{k} + ce^{at} \qquad \text{where} \qquad a = \frac{k}{m}$$

Since k is negative (for friction tends to slow the object) we see that as $t \to \infty$ the term $ce^{at} \to 0$. Thus, after a long time interval, the velocity is nearly constant.

Suppose f is periodic; let us say $f = A \sin \omega t$. The general solution to equation **4** is then seen to be (again, the student should verify this):

$$v = -\frac{A}{a^2 + \omega^2} (\omega \cos \omega t + a \sin \omega t) + ce^{at} \qquad a = \frac{k}{m}$$

In this case, if t is large, the velocity is nearly equal to

$$-\frac{A}{a^2 + \omega^2} (\omega \cos \omega t + a \sin \omega t)$$

EXAMPLE 2 *Mechanical Vibrations.*

Suppose a mass m is attached to a spring and the spring is pulled back a little and released. We should expect the mass to oscillate back and forth, supposing, of course, that the mass of the spring and the friction present are both so small, relative to m, that their effect can be neglected, and that the oscillations are small (so that the spring is not stretched too far).

Suppose we choose coordinates such that the center of gravity of the mass, with no tension in the spring, is at 0 and that $y = y(t)$ represents the distance from the center of gravity of m from 0 at time t (as shown in Figure 1).

Figure 1 *At* 0 *there is no tension in the spring.*

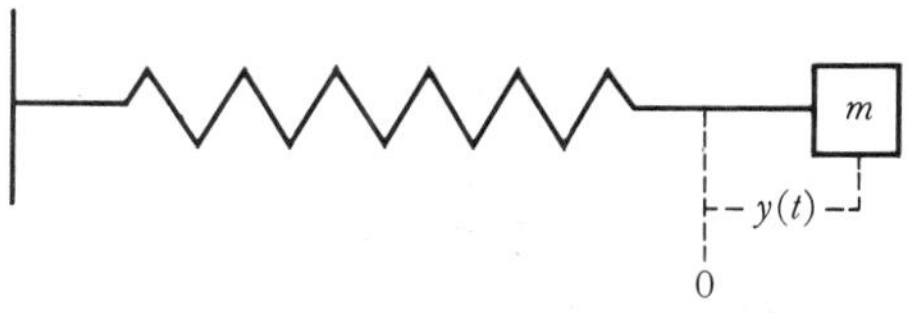

Newton's second law gives

$$my'' = \text{force acting on the mass}$$

Hooke's law (also a result of much experimental observation) asserts that if no external forces are present, this force is proportional to the distance from 0; therefore, we have

5 $$my'' = -ky$$

where k is a positive constant known as the *spring constant* dependent upon the nature of the spring. The minus sign indicates that the force tends to restore the mass to 0.

The general solution to equation **5** is

6 $$y = A \sin \alpha t + B \cos \alpha t \qquad \alpha = \sqrt{k/m}$$

We have

$$y(0) = B \qquad y'(0) = \alpha A$$

which give the initial position and velocity of the mass. In particular, if we assume that the mass is moved to y_0 and released from rest so that

7 $$y(0) = y_0 \qquad y'(0) = 0$$

then the unique solution to equation **5** which satisfies initial conditions **7** is $y = y_0 \cos \alpha t$.

We can use trigonometric identities to rewrite equation **6** as

$$y = C \sin(\alpha t + \beta)$$

where

$$C = \sqrt{A^2 + B^2} \qquad \text{and} \qquad \beta = \tan^{-1} \frac{A}{B}$$

(See Exercise **8**, below.)

The number C is called the *amplitude*, β/α is called the *displacement* or *phase angle*, and $2\pi/\alpha$ is called the *period* of the vibration. Figure 2 indicates the graph of y.

Figure 2 $y = c \sin(\alpha t + \beta)$.

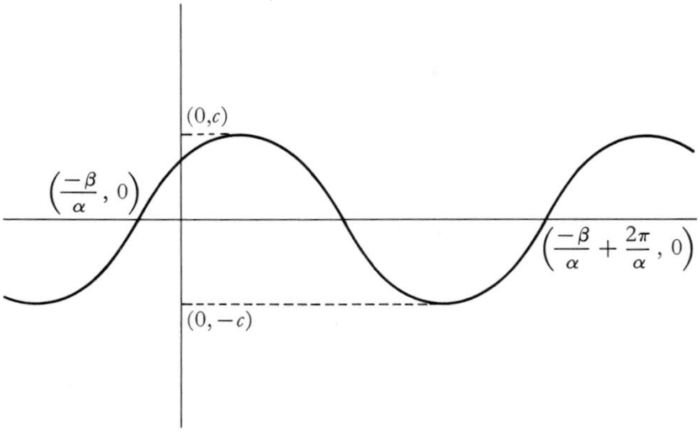

Suppose we attach a damper to the spring of Figure 1 that exerts a damping force proportional to the velocity. We then have the equation $my'' = -ky - k_1 y'$, where k and k_1 are both positive. We can write this as

8 $$y'' + by' + cy = 0 \quad \text{where } b = \frac{k_1}{m}, \; c = \frac{k}{m}$$

The nature of the general solution to equation **8** depends upon whether $b^2 - 4c > 0$, $b^2 - 4c = 0$, or $b^2 - 4c < 0$. We thus have the following possible general solutions to equation **8**:

9

a $y = c_1 e^{\alpha t} + c_2 e^{\beta t}$ if $b^2 - 4c > 0$

b $y = c_1 e^{\alpha t} + c_2 t e^{\alpha t}$ if $b^2 - 4c = 0$

c $y = c_1 e^{\alpha t} \sin kt + c_2 e^{\alpha t} \cos kt$ if $b^2 - 4c < 0$

The behavior of each of these solutions is indicated in Exercise **9**, below.

EXAMPLE 3 *A Simple Pendulum.*

Applied problems often give rise to nonlinear differential equations. In some cases, approximate solutions can be obtained by using approximations to obtain a linear equation. An example of this is the differential equation of motion of a pendulum.

Suppose a mass m is hung on a stiff wire, pulled to one side, and released. Let Θ be the angle the wire makes with the vertical. We shall determine the force which is attempting to restore the wire to a vertical position. Neglecting air resistance and the weight of the wire, we find that the only force acting on the mass is the force of gravity, which is vertical.(See Figure 3.)

Figure 3 *The pendulum.*

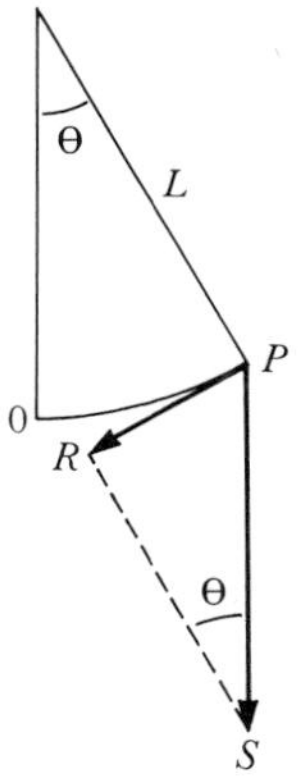

Physical observation has shown that the acceleration due to gravity is (in the English system of measurement) approximately equal to 32 feet per second, so that Newton's second law gives $32m$ as the force due to gravity. We wish to calculate the tangential force PR due to the vertical force $PS = 32m$. We see from Figure 3 that this force is $32m \sin \Theta$. Assuming that the downward direction is negative we therefore find that the tangential, or restoring, force is $-32m \sin \Theta$.

Applying Newton's second law of motion again, we see that if $y = y(t)$ is the distance OP at time t, then $my'' = -32m \sin \Theta$. Since $OP = L\Theta$ we have

$$y'' = L \frac{d^2\Theta}{dt^2}$$

Thus, the equation of motion is

10 $$mL\frac{d^2\Theta}{dt^2} = -32m \sin \Theta$$

If we assume that the mass is released from rest, with the wire at an angle Θ_0 with the vertical, we have

11 $$\Theta(0) = \Theta_0 \qquad \Theta'(0) = 0$$

Thus our function of position, $\Theta = \Theta(t)$, satisfies equation **10** and the initial conditions **11**. We note immediately that the motion does not depend on the amount of mass m, as can be seen from equation **10** (subject, of course, to m being large in relation to the wire weight, so that this equation will hold.)

Equation **10** is nonlinear and very difficult to solve. We recall from calculus that

$$\lim_{\Theta \to 0} \frac{\sin \Theta}{\Theta} = 1$$

so we would expect that a reasonable approximation for small oscillations to the motion $\Theta = \Theta(t)$ can be obtained by replacing equation **10** by

12 $$mL\frac{d^2\Theta}{dt^2} = -32m\Theta$$

Since the general solution to this is

$$\Theta = c_1 \sin\left(\sqrt{\frac{32}{L}}\right) t + c_2 \cos\left(\sqrt{\frac{32}{L}}\right) t$$

the solution that satisfies the initial conditions **11** is

$$\Theta(t) = \Theta_0 \cos\left(\sqrt{\frac{32}{L}}\right) t$$

Experimentation has shown that this solution gives a good approximation to the motion for small oscillations.

EXERCISES

1 Suppose conditions are such that bacterial growth is proportional to the amount present.

a If 1 million bacteria double in one minute, how long will it take for 10 million to be present?

b If 1 million bacteria are present at noon and 10 million are present at 1 P.M., how many are present at 12:30 P.M.?

2 A radioactive material decays at a rate proportional to the amount present. If it takes 1 million years for 10 grams to decay to one gram, what is the half-life of the substance?

3 It is often assumed that human population growth is proportional to the amount present. Use this assumption and the fact that the population of the United States was 106.5 million in 1920 and 132 million in 1940 to determine the population of the United States in 1960. What does this result lead you to conjecture about the truth or falsity of this growth assumption? (The actual population in 1960 was 180 million.)

4 A mass m is dropped in the atmosphere, subject only to air resistance (assumed proportional to velocity) and gravity. Suppose that as $t \to \infty$ the velocity approaches 64 feet per second. How fast is the mass falling after two seconds?

5 Newton's law of cooling states that the rate at which the temperature of a body changes is proportional to the difference between its temperature and that of the surrounding medium. Let y denote the temperature of the body at time t and T the temperature of the surrounding medium.

a Show that $y' = k(y - T)$ where k is constant.

b What happens to the solution as $t \to \infty$?

6 A mass with $m = 10$ is connected to a spring with spring constant $k = 2$. Find the equation of motion if

a The mass is distorted three units to the right and released from rest.

b The mass is distorted four units to the left and released from rest.

c The mass is released from the point of no tension with an initial velocity of five units, directed to the right.

d The mass is distorted three units to the right and released with an initial velocity of two units to the left.

7 Find the period, amplitude, and phase angle for each of the motions of Exercise **6**.

8 Show that

$$a \sin \varphi + b \cos \varphi = c \sin (\varphi + \theta)$$

where $c = \sqrt{a^2 + b^2}$ and $\theta = \tan^{-1} a/b$. (Hint: Apply the sum formula to the right-hand side.)

9 A damper is applied to the mass of Figure 1 so that equation **8** holds. The spring is distorted d units to the right and released from rest.

a Does the mass cross the point 0 if $m = 2$, $k_1 = 6$, and $k = 4$? Does your answer depend upon d (if $d \neq 0$)? What happens as $t \to \infty$?

b Answer the questions of part **a** if $m = 2$, $k_1 = 4$, and $k = 2$.

c Answer the questions of part **a** if $m = 2$, $k_1 = 4$, and $k = 4$. Does the mass cross 0 more than once?

10 Suppose the mass of the previous exercise is also subject to an external force $f(t)$. Find the equation of motion if

a $m = 2$, $k_1 = 6$, $k = 4$, and $f(t) = 1$

b $m = 2$, $k_1 = 6$, $k = 4$, and $f(t) = e^{-t}$
c $m = 2$, $k_1 = 4$, $k = 2$, and $f(t) = \sin t$
d $m = 2$, $k_1 = 4$, $k = 2$, and $f(t) = e^{-t}$
e $m = 2$, $k_1 = 4$, $k = 4$, and $f(t) = \sin t$

11 Experimentation has shown that the current I (measured in amperes) in a simple electric series circuit containing a capacitor of capacity C farads, a coil of inductance of L henrys, a resistance of R ohms, and a voltage source of E volts satisfies the equation

$$L\frac{d^2I}{dt^2} + R\frac{dI}{dt} + \frac{1}{C}I = \frac{dE}{dt}$$

What is the general solution to this if
a $R = 0$ and E is constant?
b $R > 0$ and E is constant?
c $R = 0$ and $E = A \sin \omega t$?
d $R > 0$ and $E = A \sin \omega t$?
In which cases does the current become small for large t?

12 Suppose a mass $m = 10$ hangs from a wire 8 units long, is moved to an angle of $\pi/100$ with the vertical, and released. Find the approximate equation of motion. What is the period of the motion? Is the period affected if the mass is doubled? if the initial position is changed?

13 A circular disk of mass m and radius r is suspended on a wire through its center, twisted through an angle Θ_0, and then released from rest. Experience has shown that the torsion in the wire is directly proportional to the amount $\Theta(t)$ of twist and inversely proportional to the square of the radius. Derive the differential equation of motion satisfied by $\Theta(t)$ and find the solution which satisfies the given initial conditions.

APPENDIX

DIMENSION THEORY

We have purposely remained at a fairly elementary level in this book. Some proofs have been omitted, and others have been given only for the two-dimensional case. A number of our arguments have been dependent upon the method of row reduction, and have tacitly assumed the uniqueness of the reduced form of a matrix. This dependence upon the particular method for solving systems of equations may, in some cases, obscure the ideas involved.

In this Appendix, we develop the theory of dimension directly, that is, without using any particular method for solving systems of equations. This discussion will introduce the student to many of the theoretical techniques and notational devices used in more advanced works on linear algebra.

Summation Notation

For convenience in our discussion of dimension we first present a notation for linear combinations which is commonly used. This notation is more compact than the notation

$$c_1\bar{u}_1 + c_2\bar{u}_2 + \cdots + c_n\bar{u}_n$$

which we used previously, and it will greatly simplify our discussions. The usual abbreviation for this sum is

$$\sum_{i=1}^{n} c_i\bar{u}_i$$

For example,

$$\sum_{i=1}^{3} c_i\bar{u}_i = c_1\bar{u}_1 + c_2\bar{u}_2 + c_3\bar{u}_3$$

and

$$\sum_{i=1}^{7} d_i\bar{v}_i = d_1\bar{v}_1 + d_2\bar{v}_2 + d_3\bar{v}_3 + d_4\bar{v}_4 + d_5\bar{v}_5 + d_6\bar{v}_6 + d_7\bar{v}_7$$

The letter i is a dummy index and can be replaced by any other letter not already in use. For example, we could write

$$\sum_{j=1}^{n} c_j\bar{u}_j \qquad \text{or} \qquad \sum_{k=1}^{n} c_k\bar{u}_k \qquad \text{rather than} \qquad \sum_{i=1}^{n} c_i\bar{u}_i$$

We will sometimes encounter a sum such as

$$c_2\bar{u}_2 + c_3\bar{u}_3 + \cdots + c_n\bar{u}_n$$

in which case we will use the symbol

$$\sum_{i=2}^{n} c_i\bar{u}_i \qquad \text{or} \qquad \sum_{j=2}^{n} c_j\bar{u}_j$$

As further examples of this we have

$$\sum_{i=2}^{4} d_i\bar{v}_i = d_2\bar{v}_2 + d_3\bar{v}_3 + d_4\bar{v}_4$$

$$\sum_{j=6}^{9} a_j\bar{u}_j = a_6\bar{u}_6 + a_7\bar{u}_7 + a_8\bar{u}_8 + a_9\bar{u}_9$$

The following rules for manipulating with summation can be shown to be true:

1

a $\displaystyle\sum_{i=1}^{n} c_i\bar{u}_i + \sum_{i=1}^{n} d_i\bar{u}_i = \sum_{i=1}^{n} (c_i + d_i)\bar{u}_i$

b $\displaystyle a\sum_{i=1}^{n} c_i\bar{u}_i = \sum_{i=1}^{n} ac_i\bar{u}_i$

The proof of part **b**, for example, proceeds as follows:

$$\begin{aligned} a\sum_{i=1}^{n} c_i\bar{u}_i &= a(c_1\bar{u}_1 + c_2\bar{u}_2 + \cdots + c_n\bar{u}_n) \\ &= (ac_1)\bar{u}_1 + (ac_2)\bar{u}_2 + \cdots + (ac_n)\bar{u}_n \\ &= \sum_{i=1}^{n} ac_i\bar{u}_i \end{aligned}$$

The definitions of linear combination, independence, dependence, spanning, and basis can be restated in terms of this summation notation and are summarized as follows:

2

a $\bar{u}$ is a *linear combination* of $\bar{u}_1, \bar{u}_2, \ldots, \bar{u}_n$ if we can find scalars $c_1, c_2, \ldots, c_n$ such that $\bar{u} = \displaystyle\sum_{i=1}^{n} c_i\bar{u}_i$.

b $\bar{u}_1, \bar{u}_2, \ldots, \bar{u}_n$ are *independent* if the relation

$$\sum_{i=1}^{n} c_i\bar{u}_i = \bar{0}$$

necessarily implies that $c_1 = c_2 = \cdots = c_n = 0$.

2 **c** $\bar{u}_1, \bar{u}_2, \ldots, \bar{u}_n$ are *dependent* if we can find scalars $c_1, c_2, \ldots, c_n$, not all zero, such that $\sum_{i=1}^{n} c_i\bar{u}_i = \bar{0}$.

d $\bar{u}_1, \bar{u}_2, \ldots, \bar{u}_n$ *span* the vector space V if every vector in V is a linear combination of $\bar{u}_1, \bar{u}_2, \ldots, \bar{u}_n$.

e $\bar{u}_1, \bar{u}_2, \ldots, \bar{u}_n$ are a *basis* for V if they are independent and span V.

We gave a somewhat different definition of basis in Section 2 of Chapter 2, but then showed, in the case $n = 2$, that the definition was the same as that given here. The student is asked to give a complete proof of this equivalence in Exercise **7**, below.

As an example of the use of this summation notation we give a complete proof of the following statement: (See statement **11**, page 73.)

3 The vectors $\bar{u}_1, \bar{u}_2, \ldots, \bar{u}_n$ are dependent if and only if one of them is a linear combination of the others.

For suppose that $\bar{u}_1, \bar{u}_2, \ldots, \bar{u}_n$ are dependent. It is then possible to find $c_1, c_2, \ldots, c_n$, not all zero, such that

$$\sum_{i=1}^{n} c_i\bar{u}_i = \bar{0}.$$

By relabeling, if necessary, we can suppose that $c_1 \neq 0$. We then have

$$(-c_1)\bar{u}_1 = \sum_{i=2}^{n} c_i\bar{u}_i$$

so that dividing by $-c_1$ and using property **1b** gives

$$\bar{u}_1 = \sum_{i=2}^{n} (-c_i/c_1)\bar{u}_i$$

This expresses $\bar{u}_1$ as a linear combination of $\bar{u}_2, \bar{u}_3, \ldots, \bar{u}_n$.

Conversely, suppose one of the $\bar{u}_i$ is a linear combination of the others. Again we can relabel, if necessary, in order to assume that $\bar{u}_1$ is a linear combination of $\bar{u}_2, \bar{u}_3, \ldots, \bar{u}_n$. Thus we can find $d_2, d_3, \ldots, d_n$ such that

$$\bar{u}_1 = \sum_{i=2}^{n} d_i\bar{u}_i$$

With $c_2 = -d_2$, $c_3 = -d_3, \ldots, c_n = -d_n$, and $c_1 = 1$, we can rewrite this as

$$\sum_{i=1}^{n} c_i\bar{u}_i = \bar{0}$$

Since at least one of $c_1, c_2, \ldots, c_n$ is not zero (for $c_1 = 1$), we can conclude that $\bar{u}_1, \bar{u}_2, \ldots, \bar{u}_n$ are dependent and thus complete the proof of statement **3**.

With the aid of this new notation we now proceed to the central dimension theorems.

The Three Fundamental Lemmas

We first prove three technical results about independence. These will be referred to as *lemmas*, a common name for theorems that are useful tools for obtaining further results. The first lemma gives an upper bound for the number of independent vectors in a space.

LEMMA 1 Suppose $\bar{u}_1, \bar{u}_2, \ldots, \bar{u}_n$ span V, and $\bar{v}_1, \bar{v}_2, \ldots, \bar{v}_k$ are an independent set in V. Then $k \leq n$.

PROOF: This lemma is actually a disguised form of Theorem **1**, part **b**. We shall first deduce it from that theorem, then show how it can be established directly. Exercise **11** below asks for a proof of Theorem **1b**, using this lemma.

By hypothesis, $\bar{u}_1, \bar{u}_2, \ldots, \bar{u}_n$ span V so that each of the vectors $\bar{v}_1$, $\bar{v}_2, \ldots, \bar{v}_k$ is a linear combination of the vectors $\bar{u}_1, \bar{u}_2, \ldots, \bar{u}_n$. Use this fact to find scalars A_{ij} such that

$$\begin{aligned}
\bar{v}_1 &= A_{11}\bar{u}_1 + A_{12}\bar{u}_2 + \cdots + A_{1n}\bar{u}_n \\
\bar{v}_2 &= A_{21}\bar{u}_1 + A_{22}\bar{u}_2 + \cdots + A_{2n}\bar{u}_n \\
&\vdots \\
\bar{v}_k &= A_{k1}\bar{u}_1 + A_{k2}\bar{u}_2 + \cdots + A_{kn}\bar{u}_n.
\end{aligned}$$

An expression $\sum_{i=1}^{k} c_i\bar{v}_i$ can now be rewritten in the form

$$\sum_{i=1}^{k} c_i\bar{v}_i = \sum_{i=1}^{k} c_i \sum_{j=1}^{n} A_{ij}\bar{u}_j = \sum_{j=1}^{n} \left(\sum_{i=1}^{k} A_{ij}c_i\right) \bar{u}_j$$

If k were larger than n, then the system of n equations in k unknowns,

$$\sum_{i=1}^{k} A_{ij}x_i = 0, \qquad j = 1, 2, \ldots, n$$

would have non-trivial solutions (Theorem **1b**); that is, there would be numbers $c_1, c_2, \ldots, c_k$ *not* all zero such that

$$\sum_{i=1}^{k} A_{ij}c_i = 0, \qquad i = 1, 2, \ldots, n$$

Since $\sum_{i=1}^{k} c_i\bar{v}_i = \sum_{j=1}^{n}\left(\sum_{i=1}^{k} A_{ij}c_i\right)\bar{u}_j$, we would then have

$$\sum_{i=1}^{k} c_i\bar{v}_i = \sum_{j=1}^{n} 0 \cdot \bar{u}_j = \bar{0}$$

This would contradict the assumption that $\bar{v}_1, \bar{v}_2, \ldots, \bar{v}_k$ are independent, for not all the c's are zero. Thus the assumption that $k \geq n$ leads to a contradiction, and must therefore be false.

Now we give a proof of Lemma **1** which makes no use of Theorem **1b**. Suppose again that $\bar{u}_1, \bar{u}_2, \ldots, \bar{u}_n$ span V, and $\bar{v}_1, \bar{v}_2, \ldots, \bar{v}_k$ are independent. Since $\bar{u}_1, \bar{u}_2, \ldots, \bar{u}_n$ span V, we can find numbers $a_1, a_2, \ldots, a_n$ such that

$$\bar{v}_1 = a_1\bar{u}_1 + a_2\bar{u}_2 + \cdots + a_n\bar{u}_n$$

The vector $\bar{v}_1$ is not zero, for no independent set can contain the zero vector. (See Exercise **9**.) Thus, at least one of the coefficients a_i is not zero. We can suppose that $a_1 \neq 0$ (otherwise we relabel so that this is so). We can then solve for $\bar{u}_1$ to obtain

4 $$\bar{u}_1 = (1/a_1)\bar{v}_1 + \sum_{i=2}^{n} (-a_i/a_1)\bar{u}_i$$

We also know that we can find $b_1, b_2, \ldots, b_n$ such that

$$\bar{v}_2 = \sum_{i=1}^{n} b_i\bar{u}_i$$

again using the assumption that $\bar{u}_1, \bar{u}_2, \ldots, \bar{u}_n$ span V. Now substitute in the relation **4** to obtain a relation of the form

5 $$\bar{v}_2 = c_1\bar{v}_1 + \sum_{i=2}^{n} c_i\bar{u}_i$$

We have, of course, changed notation by putting

$$c_1 = -b_1/a_1 \qquad \text{and} \qquad c_i = b_i - b_1 a_i/a_1 \qquad \text{for } i > 1$$

Observe that we must have $n \geq 2$, and cannot have

$$\sum_{i=2}^{n} c_i \bar{u}_i = \bar{0}$$

for then relation **5** would give $\bar{v}_2 = c_1 \bar{v}_1$, which from statement **3**, would contradict the assumption that $\bar{v}_1, \bar{v}_2, \ldots, \bar{v}_k$ are independent. Thus, at least one of the numbers $c_2, c_3, \ldots, c_k$ is not zero. We can therefore suppose that $c_2 \neq 0$ (otherwise relabel so that this is so). We can then rewrite relation **5** as

6
$$\bar{u}_2 = (-c_1/c_2)\bar{v}_1 + (1/c_2)\bar{v}_2 + \sum_{i=3}^{n} (-c_i/c_2)\bar{u}_i$$

The vector $\bar{v}_3$ is a linear combination of $\bar{u}_1, \bar{u}_2, \ldots, \bar{u}_n$, since these span V, so we can write

$$\bar{v}_3 = \sum_{i=1}^{n} d_i \bar{u}_i$$

Replacing $\bar{u}_1$ by its expression **4** and then replacing $\bar{u}_2$ by its expression **6**, we obtain (after changing the notation for the coefficients)

7
$$\bar{v}_3 = e_1\bar{v}_1 + e_2\bar{v}_2 + \sum_{i=3}^{n} e_i \bar{u}_i$$

We again conclude that n is at least 3 and that we cannot have

$$\sum_{i=3}^{n} e_i \bar{u}_i = \bar{0}$$

for, otherwise, equation **7** would be

$$\bar{v}_3 = e_1\bar{v}_1 + e_2\bar{v}_2$$

which would contradict the assumption that $\bar{v}_1, \bar{v}_2, \ldots, \bar{v}_k$ are independent.

Proceeding in this manner we can then (if k is at least 4) obtain for $\bar{v}_4$ the expression

8
$$\bar{v}_4 = f_1\bar{v}_1 + f_2\bar{v}_2 + f_3\bar{v}_3 + \sum_{i=4}^{n} f_i \bar{u}_i$$

where again we must have $n \geq 4$ and

$$\sum_{i=4}^{n} f_i\bar{u}_i \neq \bar{0}$$

We can continue this substitution and relabeling process, each time concluding that n is at least as large as k. We are forced to conclude that indeed k cannot exceed n. This completes the second proof of Lemma **1**.

The next lemma provides a means for increasing the number of vectors in an independent set.

LEMMA 2 If $\bar{v}_1, \bar{v}_2, \ldots, \bar{v}_k$ are independent, and $\bar{v}$ is *not* a linear combination of $\bar{v}_1, \bar{v}_2, \ldots, \bar{v}_k$, then $\bar{v}_1, \bar{v}_2, \ldots, \bar{v}_k, \bar{v}$ is an independent set.

PROOF: Suppose this lemma were not true. Then there are numbers $c_1, c_2, \ldots, c_k, c$ *not* all zero such that

9 $$c\bar{v} + \sum_{i=1}^{k} c_i\bar{v}_i = \bar{0}$$

We shall show that the assumption that one of $c_1, c_2, \ldots, c_k, c$ is not zero leads to a contradiction. First, we show that c must be zero. If $c \neq 0$, then we can solve the expression **9** for $\bar{v}$:

$$\bar{v} = \sum_{i=1}^{k} (-c_i/c)\bar{v}_i$$

This contradicts the hypothesis that $\bar{v}$ is *not* a linear combination of $\bar{v}_1, \bar{v}_2, \ldots, \bar{v}_k$. It must therefore be true that $c = 0$. Since $c = 0$, the equation **9** becomes the equation

$$\sum_{i=1}^{k} c_i\bar{v}_i = \bar{0}$$

and this implies (since $\bar{v}_1, \bar{v}_2, \ldots, \bar{v}_k$ are independent) that $c_1 = c_2 = \cdots = c_k = 0$.

In summary, the assumption that one of the numbers $c_1, c_2, \ldots, c_k, c$ is *not* zero has led us to the conclusion that $c = 0$, and that $c_1 = c_2 = \cdots = c_k = 0$. This contradiction forces us to conclude that $\bar{v}_1, \bar{v}_2, \ldots, \bar{v}_k, \bar{v}$ is an independent set, which is precisely what Lemma **2** asserts.

Our third lemma enables us to select from a spanning set a set of vectors which span and are also independent.

LEMMA 3 Suppose $\bar{v}_1, \bar{v}_2, \ldots, \bar{v}_n$ span V, and at least one $\bar{v}_i \neq \bar{0}$. These vectors can be relabeled so that, for some k, $1 \leq k \leq n$, the vectors $\bar{v}_1, \bar{v}_2, \ldots, \bar{v}_k$ are independent *and* span V.

PROOF: Let k be the largest integer such that at least k of the vectors $\bar{v}_1, \bar{v}_2, \ldots, \bar{v}_n$ are independent. If $k = n$, we are finished, for the set $\bar{v}_1, \bar{v}_2, \ldots, \bar{v}_n$ will be independent (since $k = n$) and spans V (by hypothesis). Let us assume therefore that $k < n$.

We can relabel, if necessary, so that we can suppose that

10 $$\bar{v}_1, \bar{v}_2, \ldots, \bar{v}_k \text{ are independent}$$

From the definition of k we then know that the vectors $\bar{v}_1, \bar{v}_2, \ldots, \bar{v}_k, \bar{v}_{k+1}$ are dependent. Therefore, numbers $c_1, c_2, \ldots, c_k, c_{k+1}$, *not* all zero, can be found such that

$$\sum_{i=1}^{k+1} c_i\bar{v}_i = \bar{0}$$

We cannot have $c_{k+1} = 0$, for it would then follow that

$$\sum_{i=1}^{k} c_i\bar{v}_i = \bar{0}$$

and, therefore, assumption **10** would require that $c_1 = c_2 = \cdots = c_k = 0$ also hold. Therefore we must have $c_{k+1} \neq 0$, and hence we can solve for $\bar{v}_{k+1}$ to obtain

$$\bar{v}_{k+1} = \sum_{i=1}^{k} (-c_i/c_{k+1})\bar{v}_i$$

Similar arguments can be used to show that each of the vectors

$$\bar{v}_{k+2}, \bar{v}_{k+3}, \ldots, \bar{v}_n$$

can be expressed as linear combinations of the $\bar{v}_1, \bar{v}_2, \ldots, \bar{v}_k$. Thus in any relation of the form

11 $$\bar{u} = \sum_{i=1}^{n} a_i\bar{v}_i$$

the expressions of $\bar{v}_{k+1}, \bar{v}_{k+2}, \ldots, \bar{v}_n$ as linear combinations of $\bar{v}_1, \bar{v}_2, \ldots, \bar{v}_k$ can be substituted to obtain a relation of the form (after suitable notation change for the coefficients)

12 $$\bar{u} = \sum_{i=1}^{k} b_i\bar{v}_i$$

By assumption, every vector in V can be written in the form **11** and hence also in the form **12**. In other words, the vectors $\bar{v}_1, \bar{v}_2, \ldots, \bar{v}_k$ *must span* V. Therefore, we have completed the proof of Lemma **3**.

Proofs of the Basic Dimension Theorems

The above three lemmas enable us to prove the basic dimension theorems. We first prove Theorem **4**, page 86.

THEOREM 4 If $\bar{u}_1, \bar{u}_2, \ldots, \bar{u}_n$ are a basis for V, then any other basis for V contains exactly n vectors.

PROOF: Suppose $\bar{u}_1, \bar{u}_2, \ldots, \bar{u}_n$ and $\bar{v}_1, \bar{v}_2, \ldots, \bar{v}_k$ are both bases for V. Therefore, in particular, $\bar{u}_1, \bar{u}_2, \ldots, \bar{u}_n$ span V and $\bar{v}_1, \bar{v}_2, \ldots, \bar{v}_k$ are independent. Lemma **1** then implies that $k \leq n$. We also know that $\bar{v}_1, \bar{v}_2, \ldots, \bar{v}_k$ span V and that $\bar{u}_1, \bar{u}_2, \ldots, \bar{u}_n$ are independent (because both sets are assumed to be bases for V). Applying Lemma **1** again shows that $n \leq k$ and completes the proof of Theorem **4**.

If there is a finite set of vectors that is a basis for V we say that V is *finite-dimensional*. The number of vectors in such a basis is called the *dimension* of V. Theorem **4** tells us that the dimension of V does not depend upon the basis chosen.

Now we prove Theorem **5**, page 87.

THEOREM 5 A set with n vectors in a space V of dimension n is independent if and only if it spans V.

PROOF: Assume that V has dimension n and that $\bar{v}_1, \bar{v}_2, \ldots, \bar{v}_n$ are an independent set in V. Let $\bar{v}$ be a vector in V. We assert that $\bar{v}$ must be a linear combination of $\bar{v}_1, \bar{v}_2, \ldots, \bar{v}_n$. If this were *not* true, then Lemma **2** implies that $\bar{v}_1, \bar{v}_2, \ldots, \bar{v}_n, \bar{v}$ is an independent set, and we have found $n + 1$ independent vectors in V. This would contradict Lemma **1**, for V has dimension n and hence there is a set containing n vectors which spans V (namely, any basis for V). We conclude that any vector $\bar{v}$ in V must be a linear combination of $\bar{v}_1, \bar{v}_2, \ldots, \bar{v}_n$, and so this set indeed spans V.

To prove the converse, suppose again that V has dimension n, and now assume that $\bar{v}_1, \bar{v}_2, \ldots, \bar{v}_n$ spans V. After relabeling if necessary, we can assume (from Lemma **3**) that there is a k, $k \leq n$, such that $\bar{v}_1, \bar{v}_2, \ldots, \bar{v}_k$ are independent *and* span V. Theorem **4** tells us that k must be n, so that the entire set $\bar{v}_1, \bar{v}_2, \ldots, \bar{v}_n$ must be independent. This completes the proof of Theorem **5**.

Other Results about Bases

We can use our techniques to prove the following statements (see statement **5**, page 89).

If V is spanned by n vectors, then

13
- **a** V has a basis containing no more than n vectors.
- **b** No independent set in V can contain more than n vectors.
- **c** Any nonzero subspace of V has a basis containing no more than n vectors.

Part **a** is a simple consequence of Lemma **3**, and part **b** is a restatement of Lemma **1**. Thus the only new result is part **c**, the proof of which follows.

Suppose M is a subspace of V which contains nonzero vectors. We let k be the largest integer such that there are k independent vectors in M. We know from part **b** that such a k exists and that we must have $k \leq n$. Since M contains nonzero vectors we must also have that $k \geq 1$ (for any nonzero vector is independent—see Exercise **12**, below). There are (from the definition of k) k independent vectors $\bar{v}_1, \bar{v}_2, \ldots, \bar{v}_k$ in M.

We now show that, in fact, the vectors $\bar{v}_1, \bar{v}_2, \ldots, \bar{v}_k$ must span M and hence are a basis for M. If these vectors did not span M, we could apply Lemma **2** to find $k + 1$ independent vectors in M. This would contradict the defining property of k. We can only conclude that, indeed, $\bar{v}_1, \bar{v}_2, \ldots, \bar{v}_k$ are a basis for M. Since $k \leq n$, the proof of part **c** and, therefore, the proof of statement **13** has been completed.

The following result is also useful.

14 Let $\bar{v}_1, \bar{v}_2, \ldots, \bar{v}_k$ be an independent set in the space V. Assume that the dimension of V is n. Then there are vectors $\bar{v}_{k+1}, \bar{v}_{k+2}, \ldots, \bar{v}_n$ such that $\bar{v}_1, \bar{v}_2, \ldots, \bar{v}_k, \bar{v}_{k+1}, \ldots, \bar{v}_n$ is a basis for V.

The proof of this is left as an exercise. (See Exercise **15** below.)

The Rank of a Linear Transformation

The basic results about rank described in Section 4 of Chapter 2 were established using the reduced form of a matrix. At this point, we would like to give a proof of Theorem **7**, page 93, which does not make use of the method of row reduction. We shall formulate and prove this for linear transformations rather than matrices.

Let T be a linear transformation from V in W. We assume that V and W are finite-dimensional spaces. The *null space* of T is the set of vectors $\bar{v}$ in V such that $T\bar{v} = \bar{0}$. The *nullity* of T is the dimension of its null space, The *range* of T is the set of all vectors $\bar{w}$ in W for which the equation $T\bar{v} = \bar{w}$ has at least one solution $\bar{v}$ in V. The *rank* of T is the dimension of its range.

We now prove the following version of Theorem **7**, page 93.

THEOREM 7 Dimension V = Rank T + Nullity T.

PROOF: Let us suppose that V has dimension n and that nullity $T = k$. We shall prove this theorem by constructing a basis for the range of T, which contains exactly $n - k$ vectors. First, choose a basis $\bar{v}_1, \bar{v}_2, \ldots, \bar{v}_k$ for the null space of T. The result **14** tells us that there are vectors $\bar{v}_{k+1}, \bar{v}_{k+2}, \ldots, \bar{v}_n$ so that

15 $\bar{v}_1, \bar{v}_2, \ldots, \bar{v}_k, \bar{v}_{k+1}, \ldots, \bar{v}_n$ is a basis for V.

We assert that

16 The set $T\bar{v}_{k+1}, T\bar{v}_{k+2}, \ldots, T\bar{v}_n$ is a basis for the range of T.

Since the set in result **16** contains $n - k$ vectors, this will establish Theorem **7**. We now prove result **16**.

Let $\bar{w}$ be a vector in the range of T, and choose $\bar{v}$ so that $\bar{w} = T\bar{v}$. Now use statement **15** to choose numbers $c_1, c_2, \ldots, c_n$ so that

$$\bar{v} = \sum_{i=1}^{n} c_i\bar{v}_i$$

Apply T to this to obtain

$$\bar{w} = T\bar{v} = T\left(\sum_{i=1}^{n} c_i\bar{v}_i\right) = \sum_{i=1}^{n} c_iT\bar{v}_i$$

The sum $\sum_{i=1}^{n} c_iT\bar{v}_i$ can be written as $\sum_{i=1}^{k} c_iT\bar{v}_i + \sum_{i=k+1}^{n} c_iT\bar{v}_i$. Furthermore, $\sum_{i=1}^{k} c_iT\bar{v}_i = \bar{0}$, since each vector $T\bar{v}_1, T\bar{v}_2, \ldots, T\bar{v}_k$ is in the null space of T. We therefore have

$$\bar{w} = \sum_{i=k+1}^{n} c_iT\bar{v}_i$$

that is, $\bar{w}$ is a linear combination of the vectors $T\bar{v}_{k+1}, T\bar{v}_{k+2}, \ldots, T\bar{v}_n$. This set therefore spans the range of T.

To complete the proof of result **16**, we will show that $T\bar{v}_{k+1}, T\bar{v}_{k+2}, \ldots, T\bar{v}_n$ is an independent set. Suppose

$$\sum_{i=k+1}^{n} c_i T\bar{v}_i = \bar{0}$$

and use the linearity of T to rewrite this as

$$T\left(\sum_{i=k+1}^{n} c_i\bar{v}_i\right) = \bar{0}$$

This shows that the vector $\bar{v} = \sum_{i=k+1}^{n} c_i\bar{v}_i$, which is a linear combination of $\bar{v}_{k+1}, \bar{v}_{k+2}, \ldots, \bar{v}_n$, is in the null space of T and is therefore a linear combination of $\bar{v}_1, \bar{v}_2, \ldots, \bar{v}_k$. In order not to contradict statement **15**, this means that $\bar{v}$ must be $\bar{0}$, and that $c_{k+1} = c_{k+2} = \cdots = c_n = 0$ (see Exercise **13** below). Thus the set $T\bar{v}_{k+1}, T\bar{v}_{k+2}, \ldots, T\bar{v}_n$ is independent, and Theorem **7** is proved.

Invertible Linear Transformations

Let us recall that the linear transformation T from V into V is *invertible* if and only if there is a linear transformation S from V into V such that $ST = I$. We wish to give here a proof of the following version of Theorem **11**, page 205.

THEOREM 11 If V is finite-dimensional, then T is invertible if and only if nullity $T = 0$.

PROOF: If $T\bar{u} = \bar{0}$ and $ST = I$, then certainly $\bar{u}$ must be $\bar{0}$, for

$$\bar{u} = I\bar{u} = (ST)\bar{u} = S(T\bar{u}) = S\bar{0} = \bar{0}$$

Thus if T is invertible, then nullity $T = 0$.

The converse is more difficult to prove, for we must construct an S satisfying $ST = I$, knowing only that nullity $T = 0$. This will be done by making use of the rank plus nullity theorem (Theorem **7**). If nullity $T = 0$, then dimension $V =$ rank T; that is, the range of T *must be equal to* V. Let $\bar{u}_1, \bar{u}_2, \ldots, \bar{u}_n$ be a basis for V. Since the range of T is V, we can find vectors $\bar{v}_i$ so that

$$T\bar{v}_i = \bar{u}_i \qquad i = 1, 2, \ldots, n$$

It is now easy to define S. If $\bar{u}$ is an arbitrary vector in V, first express $\bar{u}$ as a linear combination of the basis vectors $\bar{u}_1, \bar{u}_2, \ldots, \bar{u}_n$:

$$\bar{u} = \sum_{i=1}^{n} c_i \bar{u}_i$$

Then define $S\bar{u}$ to be the vector $\sum_{i=1}^{n} c_i \bar{v}_i$.

The proof that S is linear is left to the exercises. It is obvious that $TS = I$ for, if $\bar{u} = \sum_{i=1}^{n} c_i \bar{u}_i$, then

$$(TS)\bar{u} = T(S\bar{u}) = T\left(\sum_{i=1}^{n} c_i \bar{v}_i\right) = \sum_{i=1}^{n} c_i T\bar{v}_i = \sum_{i=1}^{n} c_i \bar{u}_i = \bar{u}$$

We have thus proved

17 If nullity $T = 0$ and V is finite-dimensional, then there is an S such that $TS = I$.

The fact that $TS = I$ does not immediately tell us that T is invertible, for we need to know that $ST = I$. Thus Theorem **11** is established once we have proved

THEOREM 12 If S and T are linear transformations from V into V, if V is finite-dimensional, and $TS = I$, then $ST = I$.

PROOF: Suppose $TS = I$. This tells us that S is invertible, and hence nullity $S = 0$. Therefore, the result **17** can be applied to S to construct a T' such that $ST' = I$. The following trick shows that $T' = T$:

$$T' = IT' = (TS)T' = T(ST') = TI = T$$

Thus $ST' = I$ and $T' = T$ so that $ST = I$. This proves Theorem **12**.

The Rank of a Product

A number of results can be established using the fact that

18
a Rank $ST \leq$ rank S
b Rank $ST \leq$ rank T.

Here we assume that S and T are both linear transformations from V into V, *and that* V *is finite-dimensional.*

Let us prove the result **18**. To show that rank $ST \leq$ rank S is quite easy, for we have

19 The range of ST is contained in the range of S.

This is merely a restatement of the fact that, if the equation $(ST)\bar{v} = \bar{w}$ has a solution $\bar{v}$, then the equation $S\bar{u} = \bar{w}$ has a solution, namely, $\bar{u} = T\bar{v}$.

The proof of the inequality rank $ST \leq$ rank T makes use of the rank plus nullity theorem (Theorem **7**) and a result analogous to result **19**, relating the null space of ST with that of T, namely

The null space of T is contained in the null space of ST.

This is a restatement of the fact that, if $T\bar{u} = \bar{0}$, then $ST\bar{u}$ is also $\bar{0}$ [for $(ST)\bar{u} = S(T\bar{u}) = S(\bar{0}) = \bar{0}$]. In particular, we must have

20 Nullity $T \leq$ Nullity ST.

To prove statement **18b**, we use Theorem **7** first to obtain

$$\text{rank } T = \text{dimension } V - \text{nullity } T$$

then use inequality **20** to obtain

$$\text{dimension } V - \text{nullity } T \geq \text{dimension } V - \text{nullity } ST$$

so that

$$\text{rank } T \geq \text{dimension } V - \text{nullity } ST = \text{rank } ST$$

This proves statement **18b**.

One can use Theorem **7** and result **18** to show that nullity $S \leq$ nullity ST (see Exercise **26**).

We now prove

21
a If T is invertible, then rank $ST =$ rank S.
b If S is invertible, then rank $ST =$ rank T.

We shall prove part **a**, the proof of part **b** being quite similar. The proof of part **a** is quite easy. First, use inequality **18a** to obtain

$$\text{rank } ST \leq \text{rank } S$$

Then write $S = (ST)T^{-1}$, and apply inequality **18a** with S replaced by ST and T by T^{-1} to obtain

$$\text{rank } (ST)T^{-1} \leq \text{rank } ST$$

so that

$$\text{rank } S = \text{rank } (ST)T^{-1} \leq \text{rank } ST \leq \text{rank } S$$

and hence rank S and rank ST must be the same.

None of these results requires that S and T be defined on the same space V. All that is required is that the product ST be defined. For example, if T is a linear transformation from V into W, and S is a transformation from W into U, then ST will be a transformation from V into U, and all the above results hold (assuming that U, V and W are all finite-dimensional spaces).

One consequence of these observations is that, if T is a linear transformation from R^n to R^m with matrix A, and A has reduced form R, then

22 Rank $T =$ Number of nonzero rows of R.

In other words, rank as defined in this Appendix gives the same value as rank as defined in Chapter 1. To prove this, let B be an invertible matrix such that $BA = R$ (see page 176ff), and let S be the linear transformation from R^m to R^m defined by $S\bar{u} = B\bar{u}$, $\bar{u}$ in R^m. The transformation S is invertible and R is the matrix of ST. Result **21** tells us that rank $ST =$ rank T. It is easy to see that rank ST is the dimension of the column space of R (see result **2**, page 93), and that this is the same as the number of columns of R which contain leading nonzero entries of rows of R. This proves the result **22**.

A Proof of Theorem 16

We now prove

THEOREM 16 Suppose A has n rows and n columns and its characteristic polynomial has n distinct real roots $a_1, a_2, \ldots, a_n$. If $\bar{v}_1, \bar{v}_2, \ldots, \bar{v}_n$ are corresponding characteristic vectors, they are a basis for R^n.

PROOF: The hypothesis tells us that

23 $a_i - a_j \neq 0$, if $i \neq j$, and there are vectors $\bar{v}_1, \bar{v}_2, \ldots,$ $\bar{v}_n$ such that $A\bar{v}_i = a_i\bar{v}_i$, and $\bar{v}_i \neq \bar{0}$ for $1 \leq i \leq n$.

We shall show that these conditions imply that $\bar{v}_1, \bar{v}_2, \ldots, \bar{v}_n$ are independent. Theorem **5** can then be applied to establish that they are a basis for the n-dimensional space R^n.

Suppose that it is possible for $\bar{v}_1, \bar{v}_2, \ldots, \bar{v}_n$ to be dependent. We can then apply Lemma **3** and relabel so that $\bar{v}_1, \bar{v}_2, \ldots, \bar{v}_k$ are independent and $\bar{v}_{k+1}$ is a linear combination of $\bar{v}_1, \bar{v}_2, \ldots, \bar{v}_k$. Therefore, we can find scalars $c_1, c_2, \ldots, c_k$ such that

24 $$\bar{v}_{k+1} = \sum_{i=1}^{k} c_i \bar{v}_i$$

Multiply this by A and use the fact that such multiplication is a linear transformation to obtain

$$A\bar{v}_{k+1} = \sum_{i=1}^{k} c_i A\bar{v}_i$$

Use assumption **23** to rewrite this as

$$a_{k+1}\bar{v}_{k+1} = \sum_{i=1}^{k} c_i a_i \bar{v}_i$$

We then multiply equation **24** by a_{k+1} and subtract from this to obtain (using the rules **1**)

$$\bar{0} = \sum_{i=1}^{k} c_i(a_{k+1} - a_i)\bar{v}_i$$

Since the vectors $\bar{v}_1, \bar{v}_2, \ldots, \bar{v}_k$ are independent it must follow that

$$c_1(a_{k+1} - a_1) = 0, \quad c_2(a_{k+1} - a_2) = 0, \ldots, c_k(a_{k+1} - a_k) = 0$$

Each of the numbers

$$a_{k+1} - a_1, \quad a_{k+1} - a_2, \ldots, a_{k+1} - a_k$$

is not zero (from hypothesis **23**); therefore, we must have

$$c_1 = c_2 = \cdots = c_k = 0$$

Substituting these into equation **24** gives $\bar{v}_{k+1} = \bar{0}$, which contradicts the assumption that each $\bar{v}_i$ is a characteristic vector and hence nonzero.

Thus the assumption that $\bar{v}_1, \bar{v}_2, \ldots, \bar{v}_n$ are dependent has led to a contradiction of the hypothesis. We must therefore conclude that these vectors are independent. This completes the proof of Theorem **16**.

A Proof of Theorem 19

We now prove

THEOREM 19 Suppose S and T are linear transformations from V into V, each having a finite-dimensional null space. Then the null space of ST is finite-dimensional, and the dimension of the null space of ST cannot exceed the sum of the dimensions of the null space of S and the null space of T, that is

$$\text{Nullity } ST \leq \text{Nullity } S + \text{Nullity } T$$

PROOF: This result is fairly easy to prove, using Theorem 7, if one assumes that V is finite-dimensional. In our differential equations applications in Chapter 6, the space V will not be finite-dimensional hence we give here a direct proof.

Suppose $\bar{u}_1, \bar{u}_2, \ldots, \bar{u}_n$ are a basis for the null space of S and that $\bar{v}_1, \bar{v}_2, \ldots, \bar{v}_k$ are a basis for the null space of T. We have

$$(ST)\bar{v}_i = S(T\bar{v}_i) = S\bar{0} = \bar{0} \qquad \text{for } i = 1, 2, \ldots, k$$

Therefore, $\bar{v}_1, \bar{v}_2, \ldots, \bar{v}_k$ must also be in the null space of ST.

If $\bar{v}_1, \bar{v}_2, \ldots, \bar{v}_k$ span the null space of ST, then this space has dimension k, which does not exceed $k + n$, and we are finished. If $\bar{v}_1, \bar{v}_2, \ldots, \bar{v}_k$ do not span the null space of ST, we can apply Lemma 2 to find a vector $\bar{w}_1$ in the null space of ST such that $\bar{v}_1, \bar{v}_2, \ldots, \bar{v}_k$, and $\bar{w}_1$ are independent. If these do not span the null space of ST, we apply Lemma 2 again to find $\bar{w}_2$ in the null space of ST such that $\bar{v}_1, \bar{v}_2, \ldots, \bar{v}_k, \bar{w}_1$, and $\bar{w}_2$ are independent. We can proceed in this manner to obtain vectors $\bar{w}_1, \bar{w}_2, \ldots, \bar{w}_m$ in the null space of ST such that

25 $\bar{v}_1, \bar{v}_2, \ldots, \bar{v}_k, \bar{w}_1, \bar{w}_2, \ldots, \bar{w}_m$ are independent

as long as we do not reach a set which spans this null space. Hence, if the null space of ST were not finite-dimensional or if its dimension exceeded $k + n$, we could then find m vectors $\bar{w}_1, \bar{w}_2, \ldots, \bar{w}_m$ in the null space of ST such that property **25** holds, *and* $m > n$. Suppose we have found such vectors; namely, suppose

26 There are vectors $\bar{w}_1, \bar{w}_2, \ldots, \bar{w}_m$ in the null space of ST such that $m > n$ and $\bar{v}_1, \bar{v}_2, \ldots, \bar{v}_k, \bar{w}_1, \bar{w}_2, \ldots, \bar{w}_m$ are independent.

It will be shown that this statement leads to a contradiction. First, note that

If $\overline{w}$ is in the null space of ST, then $T\overline{w}$ is in the null space of S.

Since the hypothesis tells us that $(ST)\overline{w} = \overline{0}$, it follows that $S(T\overline{w}) = (ST)\overline{w} = \overline{0}$, and thus $T\overline{w}$ must be in the null space of S. We conclude from this observation that

$$T\overline{w}_1,\ T\overline{w}_2,\ \ldots,\ T\overline{w}_m \text{ are in the null space of } S$$

We now have m vectors in the null space of S, which is, by assumption, n-dimensional. Furthermore, we have assumed that $m > n$. These vectors must therefore be dependent (from Lemma **1**). Therefore we can find $c_1, c_2, \ldots, c_m$ *not* all zero such that

$$\sum_{i=1}^{m} c_i T\overline{w}_i = \overline{0}$$

Since T is linear this can be rewritten as

$$T\left(\sum_{i=1}^{m} c_i\overline{w}_i\right) = \overline{0}$$

This tells us that the vector

$$\overline{v} = \sum_{i=1}^{m} c_i\overline{w}_i$$

must be in the null space of T. Since $\overline{v}_1, \overline{v}_2, \ldots, \overline{v}_k$ were assumed to be a basis for this null space it is possible to find $d_1, d_2, \ldots, d_k$ such that

$$\overline{v} = \sum_{i=1}^{k} d_i\overline{v}_i$$

We therefore have

$$\sum_{i=1}^{k} d_i\overline{v}_i = \sum_{i=1}^{m} c_i\overline{w}_i$$

which can be rewritten as

$$\sum_{i=1}^{k} d_i\overline{v}_i + \sum_{i=1}^{m} (-c_i)\overline{w}_i = \overline{0}$$

The assumption **26** then tells us that

$$d_1 = d_2 = \cdots = d_k = -c_1 = -c_2 = \cdots = -c_m = 0$$

This contradicts the fact that at least one of the c_i is *not* zero. Since we have finally reached a contradiction from assumption **26**, we conclude that this assumption must be false. Recall that assumption **26** was a consequence of denying the validity of Theorem **19**; we can only conclude that this theorem must be true.

A Related Operator Result

Arguments similar to those above can be used to prove the following statement:

27 Suppose S and T are linear transformations from V into V such that each has a finite-dimensional null space. Suppose also that the dimension of the null space of ST equals the sum of the dimensions of the null space of S and the null space of T. If $\bar{u}$ is in the null space of S, then there is a vector $\bar{v}$ in the null space of ST such that $T\bar{v} = \bar{u}$.

[Suppose K and L are constant coefficient linear differential operators. We know that the dimension of the null space of such an operator is the order of the operator (Theorem **20**). Furthermore, the order of KL is the sum of the orders of K and L. (See Exercise **4**, page 315.) Thus result **27** tells us that if f is in the null space of K, there is a function y in the null space of KL such that $Ly = f$. This is result **4**, page 317, which was basic to the method of undetermined coefficients.]

Proceeding to prove statement **27**, we assume that $\bar{u}$ is a vector in the null space of S. Choose a basis, $\bar{u}_1, \bar{u}_2, \ldots, \bar{u}_n$ for the null space of T. Since, for each i,

$$(ST)\bar{u}_i = S(T\bar{u}_i) = S\bar{0} = \bar{0}$$

it follows that each $\bar{u}_i$ is also in the null space of ST. If the dimension of the null space of S is k, the hypothesis tells us that the dimension of the null space of ST is $n + k$.

We know that $\bar{u}_1, \bar{u}_2, \ldots, \bar{u}_n$ are independent and belong to the null space of ST. If $k = 0$, we can apply Theorem **5** to conclude that these vectors are a basis for the null space of ST. If $k \neq 0$, Theorem **4** shows that they cannot span this null space. Therefore, Lemma **2** can be applied to find a vector $\bar{v}_1$ in the null space of ST such that

$$\bar{u}_1, \bar{u}_2, \ldots, \bar{u}_n \text{ and } \bar{v}_1 \text{ are independent}$$

If these do not span the null space of ST, Lemma 2 can be applied again, and further vectors, $\bar{v}_2, \bar{v}_3, \ldots$, in the null space of ST can be obtained so that

$$\bar{u}_1, \bar{u}_2, \ldots, \bar{u}_n, \bar{v}_1, \bar{v}_2, \bar{v}_3, \ldots, \text{ are independent}$$

We have assumed that the null space of ST has dimension $n + k$, where k is the dimension of the null space of S. Therefore, this process will stop after k such vectors are chosen. For we cannot have more than $n + k$ independent vectors in the null space of ST (from Lemma **1**), and no independent set with less than $n + k$ vectors can span this null space (from Theorem **4**).

We have shown that with the hypothesis of statement **27** we can find vectors

28 $\bar{v}_1, \bar{v}_2, \ldots, \bar{v}_k$ in the null space of ST such that $\bar{u}_1, \bar{u}_2, \ldots, \bar{u}_n, \bar{v}_1, \bar{v}_2, \ldots, \bar{v}_k$ are independent.

We will show that

29 $T\bar{v}_1, T\bar{v}_2, \ldots, T\bar{v}_k$ are a basis for the null space of S,

a fact that will enable us to prove statement **27**. Suppose $\bar{u}$ is in the null space of S. Use property **29** to find $c_1, c_2, \ldots, c_k$ such that

$$\bar{u} = \sum_{i=1}^{k} c_i T\bar{v}_i$$

Thus if we put

$$\bar{v} = \sum_{i=1}^{k} c_i \bar{v}_i$$

then $T\bar{v} = \bar{u}$, and $\bar{v}$ belongs to the null space of ST (from statement **28**). Therefore we need only prove that property **29** is true. Since we have assumed that the null space of S has dimension k, it is enough (because of Theorem **5**) to show that

30 $T\bar{v}_1, T\bar{v}_2, \ldots, T\bar{v}_k$ are independent and belong to the null space of S.

Establishing the fact that they belong to the null space of S is easy, for we know that

$$S(T\bar{v}_i) = (ST)\bar{v}_i = \bar{0} \qquad i = 1, 2, \ldots, k$$

because the $\bar{v}_i$'s belong to the null space of ST. The proof of independence is somewhat more difficult, however. Suppose

$$\sum_{i=1}^{k} c_i T\bar{v}_i = \bar{0}$$

Rewrite this as

$$T\left(\sum_{i=1}^{k} c_i\bar{v}_i\right) = \bar{0}$$

so that the vector

$$\bar{w} = \sum_{i=1}^{k} c_i\bar{v}_i$$

belongs to the null space of T. Since the vectors $\bar{u}_1, \bar{u}_2, \ldots, \bar{u}_n$ were chosen to be a basis for this null space, we can write

$$\bar{w} = \sum_{i=1}^{n} d_i\bar{u}_i$$

Equating these two expressions for $\bar{w}$ gives

$$\sum_{i=1}^{k} c_i\bar{v}_i = \sum_{i=1}^{n} d_i\bar{u}_i$$

and rewriting gives

$$\sum_{i=1}^{k} c_i\bar{v}_i + \sum_{i=1}^{n} (-d_i)\bar{u}_i = \bar{0}$$

Property **28** then tells us that

$$c_1 = c_2 = \cdots = c_k = -d_1 = -d_2 = \cdots = -d_n = 0$$

We have therefore shown that if

$$\sum_{i=1}^{k} c_i T\bar{u}_i = \bar{0}$$

then $c_1 = c_2 = \cdots = c_k = 0$. This concludes the proof of statement **30**.

EXERCISES The first six exercises are concerned with summation notation.

1 Establish rule **1a**.

2 Show that $$\sum_{i=1}^{n} c_i\bar{u}_i - \sum_{i=1}^{n} d_i\bar{u}_i = \sum_{i=1}^{n} (c_i - d_i)\bar{u}_i$$

3 Show that $$\sum_{j=1}^{k} \sum_{i=1}^{n} a_{ij} = \sum_{i=1}^{n} \sum_{j=1}^{k} a_{ij}$$

4 Show that if A_{ij} denotes the entry in the ith row and jth column of A and $C = AB$, then

$$C_{ij} = \sum_{k=1}^{n} A_{ik}B_{kj}$$

5 Use Exercises **3** and **4** to show directly that $(AB)C = A(BC)$.

6 Suppose

$$\bar{v}_j = \sum_{i=1}^{m} b_{ij}\bar{u}_i \qquad \text{for } j = 1, 2, \ldots, n$$

Let A denote the matrix whose columns are $\bar{u}_1, \bar{u}_2, \ldots, \bar{u}_n$, B the matrix whose entry in the ith row and jth column is b_{ij}, and C the matrix whose columns are $\bar{v}_1, \bar{v}_2, \ldots, \bar{v}_n$. Show that $AB = C$.

The next nine exercises are concerned with independence, dependence, and bases. The student may find it helpful to first prove two-dimensional versions of each result.

7 Show that, if $\bar{u}_1, \bar{u}_2, \ldots, \bar{u}_n$ is a basis for V and $\bar{u}$ is in V, then there is a unique n-tuple $c_1, c_2, \ldots, c_n$ such that $\bar{u} = \sum_{i=1}^{n} c_i\bar{u}_i$. Prove the converse; that is, if each $\bar{u}$ in V has a unique expression $\bar{u} = \sum_{i=1}^{n} c_i\bar{u}_i$, then $\bar{u}_1, \bar{u}_2, \ldots, \bar{u}_n$ is independent and spans V.

8 Show that the set $\bar{u}_1, \bar{u}_2, \ldots, \bar{u}_n$ of nonzero vectors is dependent if and only if there is a k, $1 < k \leq n$, such that $\bar{u}_k$ is a linear combination of $\bar{u}_1, \bar{u}_2, \ldots, \bar{u}_{k-1}$.

9 **a** Show that, if the set $\bar{u}_1, \bar{u}_2, \ldots, \bar{u}_n$ is independent, then any non-empty subset of this set and any rearrangement of this set is independent.
b Show that, if $\bar{u}_1, \bar{u}_2, \ldots, \bar{u}_n$ is a dependent set, then any set which contains this set is also dependent.

10 Let $\bar{v}_1, \bar{v}_2, \ldots, \bar{v}_n$ be an independent set, and suppose that, for any $\bar{v}$ in V, the set $\bar{v}_1, \bar{v}_2, \ldots, \bar{v}_n, \bar{v}$ is *not* independent. Show that $\bar{v}_1, \bar{v}_2, \ldots, \bar{v}_n$ is a basis for V.

11 Show that Lemma **1** implies Theorem **1** part **b**. (Hint: Denote the columns of A by $\bar{v}_1, \bar{v}_2, \ldots, \bar{v}_k$. If A has n rows and $k > n$, apply Lemma **1** to obtain $c_1, c_2, \ldots, c_k$ not all zero such that $\sum_{i=1}^{k} c_i\bar{v}_i = \bar{0}$. Rewrite this as a matrix equation.)

12 **a** Suppose $\bar{v}_1, \bar{v}_2, \ldots, \bar{v}_k$ are independent. Show that each $\bar{v}_i \neq \bar{0}$.
b Suppose $\bar{v} \neq \bar{0}$. Show that $\bar{v}$ is independent.

13 Suppose $\bar{v}_1, \bar{v}_2, \ldots, \bar{v}_k, \bar{v}_{k+1}, \ldots, \bar{v}_n$ is an independent set and $\bar{v}$ is a linear combination of $\bar{v}_1, \bar{v}_2, \ldots, \bar{v}_n$. Show that if $\bar{v}$ is also a linear combination of $\bar{v}_{k+1}, \bar{v}_{k+2}, \ldots, \bar{v}_k$ then $\bar{v} = 0$. (Hint: Use Exercise **7**.)

14 **a** Show that if $\bar{v}_1, \bar{v}_2, \ldots, \bar{v}_n$ span V and $a \neq 0$, then $\bar{v}_1, \bar{v}_2, \ldots, \bar{v}_{i-1}, a\bar{v}_i, \bar{v}_{i+1}, \ldots, \bar{v}_n$ also span V.

b Show that if $\bar{v}_1, \bar{v}_2, \ldots, \bar{v}_n$ span V and $i \neq j$, then $\bar{v}_1, \bar{v}_2, \ldots, \bar{v}_{i-1}, \bar{v}_i + a\bar{v}_j, \bar{v}_{i+1}, \ldots, \bar{v}_n$ span V.

c The *row space* of a matrix A is the subspace spanned by the rows of A. Show that if B is obtained from A by a row operation of the type used in Section 1, Chapter 1, then A and B have the same row space. (Hint: Use parts **a** and **b**.)

d Show that if A is reduced to the reduced matrix R, then the nonzero rows of R are a basis for the row space of A. (Hint: Show that these rows are independent and use part **c**.)

15 Suppose the subspace M of V has the basis $\bar{v}_1, \bar{v}_2, \ldots, \bar{v}_k$ and that $\bar{v}$ is not in M. Show that $\bar{v}_1, \bar{v}_2, \ldots, \bar{v}_k$ and $\bar{v}$ are independent. Use this to show that if V is finite-dimensional, then there are vectors $\bar{u}_1, \bar{u}_2, \ldots, \bar{u}_n$ such that $\bar{v}_1, \bar{v}_2, \ldots, \bar{v}_k, \bar{u}_1, \bar{u}_2, \ldots, \bar{u}_n$ is a basis for V. This proves statement **14**.

The remaining exercises are concerned with linear transformations and matrices.

16 Let T be a linear transformation from V into V, and let $\bar{u}_1, \bar{u}_2, \ldots, \bar{u}_n$ be a basis for V. Assume rank $T = k$, and show that the set $T\bar{u}_1, T\bar{u}_2, \ldots, T\bar{u}_n$ contains a subset with k vectors which is a basis for the range of T.

17 Show that, if T is linear and $T\bar{u}_1, T\bar{u}_2, \ldots, T\bar{u}_n$ are independent, then $\bar{u}_1, \bar{u}_2, \ldots, \bar{u}_n$ are independent.

18 Show that, if $\bar{v}_1, \bar{v}_2, \ldots, \bar{v}_k$ are independent vectors in the null space of T and $T\bar{v}_{k+1}, T\bar{v}_{k+2}, \ldots, T\bar{v}_n$ is an independent set, then $\bar{v}_1, \bar{v}_2, \ldots, \bar{v}_k, \bar{v}_{k+1}, \ldots, \bar{v}_n$ is an independent set.

19 Show that, if T is linear and $T\bar{u}_1, T\bar{u}_2, \ldots, T\bar{u}_n$ is an independent set when $\bar{u}_1, \bar{u}_2, \ldots, \bar{u}_n$ is independent, then nullity $T = 0$.

20 Suppose $\bar{u}_1, \bar{u}_2, \ldots, \bar{u}_n$ is a basis for V, and $\bar{v}_1, \bar{v}_2, \ldots, \bar{v}_n$ are n given vectors in W. Show that there is exactly one linear transformation S from V into W such that $S\bar{u}_1 = \bar{v}_2, S\bar{u}_2 = \bar{v}_2, \ldots, S\bar{u}_n = \bar{v}_n$. (Hint: For $\bar{u} = \sum_{i=1}^{n} c_i\bar{u}_i$, define $S\bar{u}$ to be $\sum_{i=1}^{n} c_i\bar{v}_i$, and show that this S is indeed linear.)

21 Let $\bar{u}_1, \bar{u}_2, \ldots, \bar{u}_n$ be a basis for V, and let T have nullity zero. Define S by

$$S \sum_{i=1}^{n} (c_i\bar{u}_i) = \sum_{i=1}^{n} c_i\bar{v}_i,$$

where $T\bar{v}_i = \bar{u}_i$, $1 \leq i \leq n$. Note that $ST\bar{v}_i = \bar{v}_i$, $1 \leq i \leq n$, and use Exercises **17** and **20** to prove that $ST = I$.

22 A linear transformation T from V into V is *one-to-one* if $T\bar{u}$ can be equal to $T\bar{v}$ only when $\bar{u} = \bar{v}$. T is said to be *onto* V if the range of T is V. Assume V is finite-dimensional. Show that

a if T is one-to-one, then T is onto;

b if T is onto, then T is one-to-one.

23 **a** Show that, if S or T is invertible, then rank ST = rank TS.

b Find S and T so that rank $ST <$ rank TS.

24 If S and T are linear transformations from V into V and rank ST = dimension V, then S and T are invertible.

25 Show that, if T has the property that, for all S, rank ST = rank S, then T must be invertible.

26 Show that, if S and T are linear transformations from V into V, and V is finite-dimensional, then nullity $S \leq$ nullity ST.

□ 27 Let $Tf(x) = \int_0^x f(t)\,dt$, $0 \leq x \leq 1$, and $Sf = f'$, where f is a polynomial. Show that nullity $S = 1$ and nullity $ST = 0$.

28 Suppose the characteristic polynomial f of A has the n distinct real roots $a_1, a_2, \ldots, a_n$, where A has n rows and n columns. Show that $f(A) = 0$. (Hint: Let $\bar{v}_1, \bar{v}_2, \ldots, \bar{v}_n$ be a basis consisting of characteristic vectors, and show that $f(A)\bar{v}_i = 0$, $i = 1, 2, \ldots, n$.)

29 Let T be a linear transformation from V into W, and suppose M is a subspace of V. Let T_M be the transformation from M into W, defined for $\bar{u}$ in M by $T_M\bar{u} = T\bar{u}$. T_M is called the *restriction* of T to M.

a Show that nullity $T_M \leq$ nullity T.

b Show that rank $T_M \leq$ rank T.

30 Suppose S and T are linear transformations from V into V, where V is finite-dimensional. Let M be the range of T, and let S_M denote the restriction of S to M. (See Exercise **29**.)

a Show that rank T = nullity S_M + rank S_M.

b Show that rank ST = rank S_M.

c Use **a**, **b**, to show that nullity ST = nullity T + nullity S_M.

d Show that nullity $ST \leq$ nullity S + nullity T.

e Suppose nullity ST = nullity S + nullity T. Show that the null space of S is contained in the range of T. (Hint: The null space of S includes the null space of S_M and nullity S = nullity S_M.)

ANSWERS TO EXERCISES

CHAPTER 1

SECTION 1

1 (A) ↔ (d) (B) ↔ (f) (C) ↔ (e)
(D) ↔ (g) (E) ↔ (b) (F) ↔ (a) and (c)

2 **a** $x = 0$, $y = 0$ **b** $x = 0$, $y = 0$ **c** $x = \frac{1}{4}y$, $y = y$ **d** $x = 0$, $y = \frac{7}{2}z$, $z = z$ **e** $x = -\frac{1}{6}y$, $y = y$, $z = 0$ **f** $x = 0$, $y = 0$

g $x = 0$, $y = 0$, $z = 0$ **h** $x = -\frac{1}{2}y$, $y = y$, $z = 0$ **i** $x = -\frac{65}{106}z$, $y = \frac{63}{106}z$, $z = z$ **j** $x = 0$, $y = 0$, $z = 0$ **k** $x = \frac{2}{3}y - \frac{1}{3}z$, $y = y$, $z = z$

3 $x = \frac{7}{3}z$, $y = \frac{5}{3}z$, $z = z$ $x = \frac{7}{5}y$, $y = y$, $z = \frac{3}{5}y$ $x = x$, $y = \frac{5}{7}x$, $z = \frac{3}{7}x$

4 **a** $x = -\frac{3}{7}z$, $y = \frac{8}{7}z$, $z = z$ **b** $x = 0$, $y = 0$ **c** $x_1 = -\frac{3}{7}x_3$, $x_2 = \frac{8}{7}x_3$, $x_3 = x_3$ **d** $x = 0$, $y = 0$

e $x = -2y - 3z - w$, $y = y$, $z = z$, $w = w$ **f** $x_1 = -2x_5$, $x_2 = 2x_5$, $x_3 = 0$, $x_4 = 0$, $x_5 = x_5$ **g** $x = 0$, $y = 0$

5 (See Exercise 2.)

6 **a** $x = 0$, $y = 0$, $z = 0$, $w = 0$

b $x_1 = \frac{7}{3}x_4 - \frac{5}{6}x_6$, $x_2 = -\frac{2}{3}x_4 - \frac{1}{3}x_6$, $x_3 = -\frac{4}{3}x_4 + \frac{5}{6}x_6$, $x_4 = x_4$, $x_5 = -\frac{1}{2}x_6$, $x_6 = x_6$

c $x_1 = \frac{7}{5}x_3 - \frac{16}{5}x_4 + \frac{1}{5}x_5 + \frac{6}{5}x_6$, $x_2 = \frac{7}{5}x_3 + \frac{19}{5}x_4 - \frac{4}{5}x_5 - \frac{9}{5}x_6$, $x_3 = x_3$, $x_4 = x_4$, $x_5 = x_5$, $x_6 = x_6$

7 Yes, the solution in which each variable is zero.

8 **a**, **b**

10 $x = 0$, $y = 0$

16 System B is system A with two variables renamed.

21 **a** $x = \frac{1}{3}iz$ **b** $x = \left(\frac{4 - 6i}{13}\right)z$

$y = \frac{5}{6}(1 - i)z$ $y = \left(\frac{16 - 2i}{13}\right)z$

$z = z$ $z = z$

SECTION 2

1 **a** $x = \frac{11}{5}$, $y = \frac{2}{5}$ **b** No solution.   **c** No solution.   **d** $x = z + \frac{2}{3}w + 1$, $y = -2z - \frac{7}{3}w$, $z = z$, $w = w$

e No solution.

f $x_1 = \frac{33}{12}$, $x_2 = -\frac{47}{6}$, $x_3 = \frac{25}{6}$, $x_4 = \frac{5}{6}$, $x_5 = \frac{1}{2}$, $x_6 = -2$

g

$$\begin{aligned} x_1 &= -\tfrac{3}{2}x_4 \qquad\qquad - \tfrac{3}{2}x_6 + 1 \\ x_2 &= -4x_4 - 3x_5 + x_6 - 1 \\ x_3 &= 2x_4 + x_5 - x_6 \\ x_4 &= x_4 \\ x_5 &= x_5 \\ x_6 &= x_6 \end{aligned}$$

4 **a** Unique solution.   **b** More than one solution.
c More than one solution.   **d** Unique solution.

5 **a** No solution.   **b** Unique solution.
c No solution.   **d** No solution.

6 **a** Unique solution for any choice of the constants.
b Cannot always be solved; if one solution exists, then more than one solution exists.
c More than one solution for any choice of the constants.
d Cannot always be solved; if one solution exists, then the solution is unique.

9 **a** $x = \dfrac{5 + 7i}{37}$, $y = \dfrac{22 + 16i}{37}$ **b** No solution.

c $x = \dfrac{-2i}{3}z + \dfrac{4 + i}{3}$, $y = \dfrac{-4 - 2i}{3}z + \dfrac{3 - 7i}{3}$, $z = z$

SECTION 3

1 **a** 2   **b** 1   **c** 2   **d** 2   **e** 4

2 **a** Rank 1   **b** Rank 1   **c** Rank 3   **d** Rank 2   **e** Rank 4

3 **a** $x = y/2$, $y = y, y \neq 0$ **b** $x = \frac{1}{2}y$, $y = y, y \neq 0$ **c** $x = -\frac{10}{41}w$, $y = \frac{23}{41}w$, $z = \frac{35}{41}w$, $w = w, w \neq 0$

d $x = -z$
$y = 0$
$z = z,\ z \neq 0$

e $x_1 = 0$
$x_2 = -x_5$
$x_3 = -x_5$
$x_4 = 0$
$x_5 = x_5,\ x_5 \neq 0$

4 The trivial solution in which all the variables are zero.

5 Unique: **a** and **c**

6 **a** 6 **b** minimum = 5, maximum = 5

7 Yes

8 **a** Yes, No **b** smallest = 0, largest = 5

9 **b, d, f, h**

10 No—all solutions are of the form: $x = -2y, \quad y = y, \quad z = 0$

12 **a** Line (plane) passes through origin. If unique solution, intersection is exactly the origin. Otherwise, intersection is a line or plane through the origin.
c Intersection is empty (if no solution), consists of exactly one point (if exactly one solution), is a line or plane (if more than one solution).
d Change of the constant term gives a parallel line or plane.

15 **a** Unique solution: $x = y = 0$. **b** No.

SECTION 4

1 **a** $(0,\frac{5}{4},\frac{10}{3})$ **b** $(-\frac{3}{2},\frac{3}{4},8)$ **c** $(-1,-\frac{3}{4},2)$

2 **a** $(-5,-3,-3,-1,1)$ **b** $(3,-3,9,9,-3)$ **c** $(-2,2,-2,2,-2)$

5 **a** $x(1,2,1,0) + y(-1,1,-1,2)$
b $x_3(-\frac{5}{3},-\frac{1}{5},1,0,0) + x_4(0,0,0,1,0) + x_5(\frac{1}{5},-\frac{3}{5},0,0,1)$

6 **a** $a(1,-\frac{1}{2})$, a any real number. **b** $y = \frac{1}{3}x$ **c** $y\begin{bmatrix}2\\1\\0\end{bmatrix} + z\begin{bmatrix}-1\\0\\1\end{bmatrix}$ **d** No.

7 $(2x_1 - 3y_1 + z_1,\ 2x_2 - 3y_2,\ 2x_3 - 3y_3 + 1,\ 2x_4 - 3y_4 + z_4)$

10 **a** 5*b* **b** 5*c* **c** 5*a*,*b*,*c*

11 **a** $-2 - 2x + 2x^2 + x^3$ **b** $18x - 6x^2$ **c** $-2 + 7x - x^2 - x^3$

14 $(-1 + 4i, 4 + i, 5 + i) = \bar{u} - 2i\bar{v} + (1 + i)\bar{w}$

$$\bar{x} = \left(0, \frac{-3 + i}{2}, \frac{1 - i}{2}\right)$$

SECTION 5

1 **a** $\begin{bmatrix} -5 \\ -3 \\ 4 \end{bmatrix}$ **b** $\begin{bmatrix} 9 \\ 4 \\ -5 \end{bmatrix}$ **c** $\bar{x} = \begin{bmatrix} -1 \\ 2 \\ 1 \end{bmatrix}$

2 **a** $(-5, -3, 4)$ **b** $(9, 4, -5)$ **c** $\bar{x} = (-1, 2, 1)$

3 **a** $\begin{bmatrix} 7 \\ 2 \end{bmatrix}$ **b** $\begin{bmatrix} 4 \\ 6 \end{bmatrix}$ **c** $\begin{bmatrix} 2 \\ 2 \\ 1 \end{bmatrix}$ **d** $\begin{bmatrix} 0 \\ 0 \\ 0 \end{bmatrix}$ **e** $\begin{bmatrix} 1 \\ 3 \\ 1 \end{bmatrix}$ **f** $\begin{bmatrix} 0 \\ 17 \\ 34 \end{bmatrix}$ **g** $\begin{bmatrix} 37 \\ 24 \\ 3 \\ 60 \\ 0 \end{bmatrix}$

4 **a** $\begin{bmatrix} 2 & -1 & 1 \\ 1 & -3 & 1 \\ 1 & 1 & -1 \\ 4 & -1 & 2 \end{bmatrix} \begin{bmatrix} x \\ y \\ z \end{bmatrix} = \begin{bmatrix} 1 \\ 2 \\ 0 \\ 1 \end{bmatrix}$ **b** $[1 \quad -1] \begin{bmatrix} x \\ y \end{bmatrix} = 0$

c $\begin{bmatrix} 0 & 0 & 0 \\ 0 & 0 & 0 \\ 0 & 0 & 0 \end{bmatrix} \begin{bmatrix} x \\ y \\ z \end{bmatrix} = \begin{bmatrix} 0 \\ 0 \\ 0 \end{bmatrix}$

5 **a** No. **b** Yes. **c** Yes. **d** No.

6 No.

8 $A(\bar{u} + \bar{v}) = \begin{bmatrix} 6 \\ 3 \end{bmatrix}$ $A(3\bar{u}) = \begin{bmatrix} 3 \\ -6 \end{bmatrix}$

10 Yes.

11 No.

12 $\begin{bmatrix} 1 & -1 \\ 2 & 3 \end{bmatrix} \begin{bmatrix} c_1 \\ c_2 \end{bmatrix} = \begin{bmatrix} 4 \\ 2 \end{bmatrix}$

13 0

15 Yes.

16 **a** $\begin{bmatrix} -1 + i \\ 2(1 + i) \end{bmatrix}$ **b** $\begin{bmatrix} 5i \\ -1 + 3i \end{bmatrix}$ **c** $\begin{bmatrix} i & -2i \\ 3 - i & i - 1 \end{bmatrix} \begin{bmatrix} x \\ y \end{bmatrix} = \begin{bmatrix} i \\ i/2 \end{bmatrix}$

SECTION 6

1 **a** No. **b** No. **c** Yes. **d** No. **e** Yes.

2 **a** Yes. **b** No. **c** Yes. **d** Yes. **e** No.

3 **a** $a\begin{bmatrix} -3 \\ 5 \\ 1 \end{bmatrix}$ **b** $a\begin{bmatrix} -\frac{1}{3} \\ 1 \\ 0 \\ 0 \end{bmatrix} + b\begin{bmatrix} -\frac{1}{3} \\ 0 \\ 1 \\ 0 \end{bmatrix}$ **c** $\begin{bmatrix} 0 \\ 0 \end{bmatrix}$

d $a_1\begin{bmatrix} 0 \\ 0 \\ 1 \\ 0 \\ 0 \\ 0 \\ 0 \end{bmatrix} + a_2\begin{bmatrix} 0 \\ -6 \\ 0 \\ 1 \\ 0 \\ 0 \\ 0 \end{bmatrix} + a_3\begin{bmatrix} -\frac{1}{2} \\ 0 \\ 0 \\ 0 \\ 1 \\ 0 \\ 0 \end{bmatrix} + a_4\begin{bmatrix} -\frac{1}{2} \\ 0 \\ 0 \\ 0 \\ 0 \\ 1 \\ 0 \end{bmatrix} + a_5\begin{bmatrix} \frac{1}{2} \\ -3 \\ 0 \\ 0 \\ 0 \\ 0 \\ 1 \end{bmatrix}$

4 Other choices of a solution to $A\bar{x} = \bar{c}$ will give different first vectors in each case.

a $\begin{bmatrix} 1 \\ -1 \\ 0 \end{bmatrix} + a\begin{bmatrix} -3 \\ 5 \\ 1 \end{bmatrix}$ **b** $\begin{bmatrix} \frac{1}{3} \\ 0 \\ 0 \\ -1 \end{bmatrix} + a\begin{bmatrix} -\frac{1}{3} \\ 1 \\ 0 \\ 0 \end{bmatrix} + b\begin{bmatrix} -\frac{1}{3} \\ 0 \\ 1 \\ 0 \end{bmatrix}$ **c** $\begin{bmatrix} 1 \\ 0 \end{bmatrix}$

d $\begin{bmatrix}0\\1\\0\\0\\0\\0\\0\end{bmatrix} + a_1\begin{bmatrix}0\\0\\1\\0\\0\\0\\0\end{bmatrix} + a_2\begin{bmatrix}0\\-6\\0\\1\\0\\0\\0\end{bmatrix} + a_3\begin{bmatrix}-\frac{1}{2}\\0\\0\\0\\1\\0\\0\end{bmatrix} + a_4\begin{bmatrix}-\frac{1}{2}\\0\\0\\0\\0\\1\\0\end{bmatrix} + a_5\begin{bmatrix}\frac{1}{2}\\-3\\0\\0\\0\\0\\1\end{bmatrix}$

7 Any $\bar{c}$ of the form $\begin{bmatrix}c_1\\c_2\\c_3\end{bmatrix}$ with $c_3 \neq 0$.

9 Does not have a unique solution.

10 **a** Plane through the origin. **b** Plane parallel to the plane of part **a**.

11 **a** A plane or a line through the origin.
b The planes corresponding to the system of equations are parallel. If the system has a solution, then these planes meet in a line or are coincident.

12 **a** If $A = 0$, all vectors are solutions. If $A \neq 0$, solutions form a plane or a line through the origin or the origin only.
b The planes of the system have no common intersection.
c The origin only. **d** The line through the terminal point of $\bar{u}_1$, parallel to $\bar{u}_2$.

13 **a** One solution. **b** One solution.

14 Other solutions to $A\bar{x} = \begin{bmatrix}1\\i\end{bmatrix}$ will give different first vectors. One form is:

$$\bar{x} = \begin{bmatrix}\frac{1-i}{2}\\0\\2\end{bmatrix} + a\begin{bmatrix}-\frac{1+i}{2}\\1\\0\end{bmatrix}$$

SUPPLEMENT

1 $\bar{w} = (2,1) + t(1,-3) = (2 + t, 1 - 3t)$

2 $\bar{w} = (\frac{7}{2},0) + t(\frac{7}{2},-\frac{7}{5}) = (\frac{7}{2} + \frac{7}{2}t, -\frac{7}{5}t)$

3 $\overline{w} = (1,-5) + t(1,2) = (1 + t, -5 + 2t)$

4 $\dfrac{x}{1} = \dfrac{y-1}{-2} = \dfrac{z}{1}$

5 $x = t, y = 1 - 4t, z = 2 + t$

6 $\overline{w} = t(5,7,7) + (1 - t)(3,-1,2),\ 0 \leq t \leq 1$

7 A ray, namely, the points on the line $y = -\frac{3}{2}x + \frac{7}{2}$ which are to the right of $(-7,14)$.

8 The line through the terminal points of $\overline{u}$ and $\overline{v}$.

9 $(\frac{11}{5},\frac{6}{5},-\frac{3}{5}), (\frac{7}{5},\frac{7}{5},-\frac{1}{5}), (\frac{3}{5},\frac{8}{5},\frac{1}{5}), (\frac{1}{5},\frac{9}{5},\frac{3}{5})$

10 $s(-1,1,0) + t(-1,0,1) \quad 0 \leq s \leq 1 \quad 0 \leq t \leq 1$

11 Triangle with vertices $\overline{u}$, $\overline{v}$, $(0,0,0)$

12 $s(0,-3,-1) + t(1,-3,0),\ 0 \leq s \leq 1,\ 0 \leq t \leq 1,\ s + t \leq 1$

CHAPTER 2

SECTION 1

1 Yes.

2 **a** Dependent. **b** Independent. **c** Independent. **d** Dependent. **e** Independent.

3 For example:
a $(-4,3,-5) = 2(1,3,-1) - 3(2,1,1)$ **b** $(1,-1) = -\frac{5}{11}(1,3) + \frac{4}{11}(4,1)$
c $(1,0,5,3,2) = 3(1,2,1,1,0) - 2(1,3,-1,0,-1)$

4 No.

5 No.

6 $(1,0,3) = \frac{1}{3}(2,1,1) + \frac{1}{3}(1,-1,1) + \frac{7}{3}(0,0,1)$

7 $(x - 1)^2 = x^2 - 2x + 1$

13 **a** No. **b** Yes. **c** Yes.

15 **c** $(3 + i, i) = \dfrac{5 - 2i}{4}(1 + i, 2i) + \dfrac{1 - 5i}{4}(i, 1 - i)$

SECTION 2

4 Given in row form. Of course, other bases are possible.

a $(-3,5,1)$ **b** $(-\frac{1}{3},1,0,0)$ **c** $(0,0,1,0,0,0,0)$

$(-\frac{1}{3},0,1,0)$ $(0,-6,0,1,0,0,0)$

$(-\frac{1}{2},0,0,0,1,0,0)$

$(-\frac{1}{2},0,0,0,0,1,0)$

$(\frac{1}{2},-3,0,0,0,0,1)$

5 a $\dfrac{-x+3y-z}{14}, \dfrac{3x-2y+3z}{7}, \dfrac{-5x+y+9z}{14}$

b $x, -x+y, -y+z$ **c** $\dfrac{-x+2y-z}{11}, \dfrac{7x-3y+7z}{22}, \dfrac{14x-6y+3z}{55}$

6 a $\dfrac{x+y+z-w}{4}, \dfrac{x+y-z+w}{4}, \dfrac{x-y+z+w}{4}, \dfrac{-x+y+z+w}{4}$

b $x, -2x+y, x-2y+z, y-2z+w$

7 For example, $(0,1,0)$, $(0,0,1)$

10 $\dfrac{a+b}{2}, \dfrac{a-b}{2}$

SECTION 3

1 a Yes. **b** No. **c** No. **d** Yes.

3 2

4 5

5 Yes.

7 2

9 **a, b, e**

10 a 4

b $\dfrac{(x-1)(x-2)(x-3)}{-6}, \dfrac{x(x-2)(x-3)}{2}, \dfrac{x(x-1)(x-3)}{-2}$

$\dfrac{x(x-1)(x-2)}{6}$

c $\dfrac{(x+1)(x-1)(x-2)}{-12}, \dfrac{(x+2)(x-1)(x-2)}{6},$

$\dfrac{(x+2)(x+1)(x-2)}{-6}, \dfrac{(x+2)(x+1)(x-1)}{12}$

14 a No. **b** No. **c** Yes. **d** Yes.

SECTION 4

3 **a** $\begin{bmatrix} 1 & 0 & 0 & 0 \\ 0 & 1 & 0 & 0 \\ 0 & 0 & 1 & 0 \\ 0 & 0 & 0 & 1 \\ 0 & 0 & 0 & 0 \end{bmatrix}$ **b** $\begin{bmatrix} 1 & 0 & 1 & 0 \\ 0 & 1 & 1 & 0 \\ 0 & 0 & 0 & 1 \\ 0 & 0 & 0 & 0 \\ 0 & 0 & 0 & 0 \end{bmatrix}$

c $\begin{bmatrix} 1 & 0 & 0 & 0 & 2 \\ 0 & 1 & 0 & 0 & -2 \\ 0 & 0 & 1 & 0 & -1 \\ 0 & 0 & 0 & 1 & 0 \\ 0 & 0 & 0 & 0 & 0 \end{bmatrix}$

4 **b** If column two is a multiple of column one and column four is a linear combination of columns one and two. Other answers are possible.

5 **a** 2 **b** 2 **c** 2

6 **a** $\bar{u}_1, \bar{u}_2, \bar{u}_3$ **b** $\bar{v}_1, \bar{v}_2, \bar{v}_4$

7 R^m

11 $\bar{u}_1 = (1,0,0,0), \bar{u}_2 = (0,1,-1,0), \bar{u}_3 = (0,0,0,1)$

12 **a** (1,0,2)
(0,1,−6) **b** $\begin{bmatrix} 4 \\ 8 \\ 3 \end{bmatrix}, \begin{bmatrix} 1 \\ 2 \\ 1 \end{bmatrix}$

Other answers are possible; these are the ones found using result **5** for part **a** and the method of Example **3** for part **b**.

14 rank $A' \leq$ rank A

15 **a** No. **b** $\bar{u}$ and $\bar{v}$ are dependent **c** No. **d** 6

16 $A = 0$

17 **a** The first and second columns are a basis (found by using the method of Example **3**).
b Using result **5** gives the basis $(1, i, 1 - i, -i)$

SECTION 5

1 **a** 11 **b** 1 **c** $3\overline{w}$

2 **a** 3 **b** 5 **c** $2\overline{w}$

3 **a** 11 **b** 4 **c** $2\overline{w}$

4 **a** $\sqrt{2}$ **b** 3 **c** 6

6 **a** $\frac{1}{\sqrt{5}}\overline{u}$ **b** $\frac{1}{\sqrt{3}}\overline{u}$ **c** $\frac{1}{\sqrt{11}}\overline{u}$ **d** $\frac{1}{3}\overline{u}$

7 **a** $\frac{3}{5}\overline{v}$ **b** $\overline{0}$ **c** $\frac{1}{5}\overline{v}$ **d** $\frac{5}{12}\overline{v}$

8 **a** $(\frac{2}{5},-\frac{1}{5})$ **b** $\overline{u}$ **c** $(\frac{3}{5},2,\frac{6}{5})$ **d** $(\frac{19}{12},-\frac{3}{12},\frac{5}{12},-\frac{5}{12},6)$

9 **a** $\overline{w}_1 = (\frac{3}{5},\frac{6}{5})$ $\overline{w}_2 = (\frac{2}{5},-\frac{1}{5})$ **b** $\overline{w}_1 = \overline{0}$ $\overline{w}_2 = \overline{u}$
c $\overline{w}_1 = (\frac{2}{5},0,-\frac{1}{5})$ $\overline{w}_2 = (\frac{3}{5},2,\frac{6}{5})$
d $\overline{w}_1 = (\frac{5}{12},\frac{15}{12},-\frac{5}{12},\frac{5}{12},0)$ $\overline{w}_2 = (\frac{19}{12},-\frac{3}{12},\frac{5}{12},-\frac{5}{12},6)$

10 Projection of $\overline{u}$ onto $\overline{v}$ is $\overline{0}$; projection of $\overline{u}$ orthogonal to $\overline{v}$ is $\overline{u}$.

12 **a** $\sqrt{2}/2$
c The projections onto and orthogonal to 1 are $\frac{1}{2}$ and $x - \frac{1}{2}$, onto and orthogonal to $\cos \pi x$ are $\frac{-4}{\pi^2}\cos \pi x$ and $x + \frac{4}{\pi^2}\cos \pi x$.

14 $(-1,3,5)$

16 **d** Not in general.

17 **b** $1 - i,\ -1 - i,\ 6 - 8i$ **c** $\sqrt{2}, \sqrt{2}$
d $(\frac{1}{2},i/2),\ (-\frac{1}{2} + i,\ 1 + i/2)$ **e** $\overline{u} \cdot \overline{u} > 0$ if $\overline{u} \neq \overline{0}$

SECTION 6

2 **a** $\frac{1}{\sqrt{5}}(1,2);\ \frac{1}{\sqrt{5}}(-2,1)$ **b** $\frac{1}{\sqrt{2}}(1,0,1);\ \frac{1}{\sqrt{3}}(1,1,-1);\ \frac{1}{\sqrt{6}}(-1,2,1)$

c $\frac{1}{\sqrt{2}}(1,0,1,0);\ \frac{1}{\sqrt{3}}(1,1,-1,0);\ \frac{1}{\sqrt{7}}(1,-2,-1,1)$

3 $(1,-1)$

4 $(2,-1,2)$

5 $(3,2,1) = 2(1,0,1) + \frac{4}{3}(1,1,-1) + \frac{1}{3}(-1,2,1)$

6 $\frac{4}{3}(-1,2,1)$

9 Try $\bar{u} = (0,0,1,0)$.

11 **c** $\bar{u}_1 = (1,1,1)$, $\bar{u}_2 = (-\frac{1}{3},\frac{2}{3},-\frac{1}{3})$
$\bar{u}_3 = (-\frac{1}{2},0,\frac{1}{2})$

d $\bar{u}_1 = 1$
$\bar{u}_2 = x - \frac{1}{2}$
$\bar{u}_3 = x^2 - x + \frac{1}{6}$

12 **b** $\frac{1}{\sqrt{3}}(1,i,1);\ \frac{1}{\sqrt{2}}(i,0,-i);\ \frac{1}{\sqrt{6}}(i,2,i)$

c $(1,1,i) = \frac{1}{3}(1,i,1) - \frac{i+1}{2}(i,0,-i) + \frac{3-i}{6}(i,2,i)$

SUPPLEMENT

1 $2(x - 1) - 1(y - 2) - 1(z - 1) = 0$

2 $\bar{n} = 2\bar{i} - 3\bar{j} + 4\bar{k}$

3 $\bar{w} = (0,0,\frac{1}{4}) + s(\frac{3}{2},1,0) + t(-2,0,1)$

4 $y - 1 = 0$

5 $2(x + 1) + 7(y - 1) + 6(z - 1) = 0$

6 $1(x - 6) - 3(y - 1) + 2(z - 2) = 0$

8 $(0,0,2)$

9 $4x + 5y - 8(z - 2) = 0$

10 $\frac{1}{13}\sqrt{104}$

11 $2\sqrt{3}$

12 $\frac{1}{14}\sqrt{27^2 + 30^2 + 11^2}$

13 $\sqrt{14}$

14 $\frac{1}{2}\sqrt{34}$

15 $\frac{2}{7}\sqrt{14}$

16 $\frac{1}{29}\sqrt{35^2 + 79^2 + 41^2}$

17 $\frac{1}{3}\sqrt{6}$

18 $\frac{2}{3}\sqrt{6}$

19 $\frac{1}{14}\sqrt{1 + 23^2 + 10^2}$

CHAPTER 3

SECTION 1

1 **a** $(\frac{5}{9},\frac{10}{9},\frac{10}{9})$ **b** $(\frac{2}{9},\frac{4}{9},\frac{4}{9})$ **c** $(\frac{7}{9},\frac{14}{9},\frac{14}{9})$ **d** $(\frac{20}{9},\frac{40}{9},\frac{40}{9})$ **e** $(\frac{20}{9},\frac{40}{9},\frac{40}{9})$

2 **a** $(-3,2)$ **b** $(1,-4)$ **c** $(-2,-2)$ **d** $(-2,-2)$

4 **a** $\frac{10}{9}(1,-2,2)$ **b** $\frac{2}{9}(1,-2,2)$ **c** $(\frac{2}{9},-\frac{13}{9},\frac{31}{9})$ **d** $(-\frac{14}{9},\frac{1}{9},\frac{17}{9})$

5 $T(1,0) = (1/\sqrt{2},1/\sqrt{2})$ $T(0,1) = (-1/\sqrt{2},1/\sqrt{2})$ $T(1,1) = (0,\sqrt{2})$

6 **a** Reflection in the line $y = x$. **b** Projection onto the y, z plane.
c Reflection in the x-axis. **d** Projection onto the x-axis.

7 $P\bar{u} = \bar{u} - (\bar{u} \cdot \bar{w})\bar{w}$

9 $T(1,2) = (1,2)$
$T(-1,2) = (1,2)$
No; No.

11 **a** $\cos x$ **b** e^x **c** $e^x + \cos x$ **d** $ae^x + b \cos x$

12 **a** $\sin x$ **b** $x^3/3$ **c** $e^x - 1$ **d** $x^3/3 + 3 \sin x + 2(e^x - 1)$

13 **a** $1, x, x^2$ is one such basis. **b** $c_1 + c_2x + c_3x^2 + \cos x$

14 The solutions to $Lf = x$ are the functions of the form $c_1e^x + c_2e^{-x} - x$.

15 The solutions to $Lf = 2 + x^2$ are the functions of the form $c_1 \sin x + c_2 \cos x + x^2$.

16 The null space of T consists of all functions f in $C[0,1]$ such that $f(x) = 0$ if $g(x) \neq 0$.

SECTION 2

1 **a** $\begin{bmatrix} 1 & 0 & 0 & 0 \\ 1 & 1 & 0 & 0 \\ 1 & 1 & 1 & 0 \\ 1 & 1 & 1 & 1 \end{bmatrix}$ **b** $\begin{bmatrix} 1 & -1 & 0 \\ 0 & 0 & 1 \end{bmatrix}$ **c** $\begin{bmatrix} 1 & 0 & 0 & 0 \\ 1 & 0 & 0 & 0 \\ 1 & 0 & 0 & 0 \\ 1 & 0 & 0 & 0 \\ 1 & 0 & 0 & 0 \end{bmatrix}$

2 **a** $\begin{bmatrix} 1/\sqrt{2} & -1/\sqrt{2} \\ 1/\sqrt{2} & 1/\sqrt{2} \end{bmatrix}$ **b** $\begin{bmatrix} 0 & -1 \\ 1 & 0 \end{bmatrix}$ **c** $\begin{bmatrix} -1 & 0 \\ 0 & -1 \end{bmatrix}$

d $\begin{bmatrix} -1 & 0 \\ 0 & -1 \end{bmatrix}$ **e** $\begin{bmatrix} 1 & 0 \\ 0 & 1 \end{bmatrix}$

3 **a** $T(1,1) = (0,\sqrt{2})$
$T(2,-3) = (5/\sqrt{2}, -1/\sqrt{2})$

b $T(1,1) = (-1,1)$
$T(2,-3) = (3,2)$

c $T(1,1) = (-1,-1)$
$T(2,-3) = (-2,3)$

d $T(1,1) = (-1,-1)$
$T(2,-3) = (-2,3)$

e $T(1,1) = (1,1)$
$T(2,-3) = (2,-3)$

4 $$P = \begin{bmatrix} \frac{1}{2} & \frac{1}{2} \\ \frac{1}{2} & \frac{1}{2} \end{bmatrix}$$

5 $$T = \begin{bmatrix} 0 & 1 \\ 1 & 0 \end{bmatrix}$$

6 $$P = \begin{bmatrix} \frac{8}{9} & -\frac{2}{9} & -\frac{2}{9} \\ -\frac{2}{9} & \frac{5}{9} & -\frac{4}{9} \\ -\frac{2}{9} & -\frac{4}{9} & \frac{5}{9} \end{bmatrix}$$

7 $$\begin{bmatrix} 0 & 0 \\ 0 & 0 \end{bmatrix}$$

8 $$\begin{bmatrix} 1 & 0 \\ 0 & 1 \end{bmatrix}$$

9 $T(x,y,z) = (2x + y + 3z, -y + z)$

10 **a** Reflection in the line $y = x$. **b** Projection onto the x-axis.
c Reflection in the line $y = -x$. **d** Rotation through π.

11 **a** $\begin{bmatrix} 1 \\ 1 \\ 0 \end{bmatrix}$ $\begin{bmatrix} 1 \\ 1 \end{bmatrix}$ **b** The third column of A. **c** $\bar{u} = \begin{bmatrix} 0 \\ 0 \\ 0 \\ 0 \\ 1 \end{bmatrix}$

12 **a** R^n
b R^m
c Null space A = null space T, range T = column space A, rank T = rank A.
d No.
e The null space of T contains nonzero vectors.

13 $$\begin{bmatrix} -\frac{8}{13} & -\frac{2}{13} \\ \frac{19}{13} & \frac{21}{13} \end{bmatrix}$$

15 **a** $\begin{bmatrix} i & -1 \\ 1+i & 0 \end{bmatrix}$ **b** $\begin{bmatrix} 0 & 0 \\ 0 & 0 \\ i & 0 \\ i-1 & 2 \end{bmatrix}$ **c** $\begin{bmatrix} \frac{1}{2} & i/2 \\ -i/2 & \frac{1}{2} \end{bmatrix}$

SECTION 3

1 **a** $\begin{bmatrix} -1 & 3 \\ 1 & 4 \end{bmatrix}$ **b** $\begin{bmatrix} -5 & -2 & -1 & 4 \\ 3 & 7 & -1 & 0 \end{bmatrix}$ **c** $\begin{bmatrix} 0 & 0 \\ 0 & 0 \\ 0 & 0 \end{bmatrix}$

3 **a** [15,6] **b** [−8,4] **c** $\begin{bmatrix} 2 & 1 \\ 0 & 0 \end{bmatrix}$ $\begin{bmatrix} 3 & -1 \\ 2 & 0 \end{bmatrix}$ $\begin{bmatrix} 5 & 0 \\ 2 & 0 \end{bmatrix}$

4 $\begin{bmatrix} 5 & 0 & 0 \\ 1 & 0 & 0 \\ 2 & 1 & 1 \\ 1 & 0 & 1 \end{bmatrix}$ $\begin{bmatrix} -3 & 0 & 0 \\ 1 & 0 & 0 \\ 0 & -1 & -1 \\ 1 & 0 & -1 \end{bmatrix}$ $\begin{bmatrix} 8 & 0 & 0 \\ 0 & 0 & 0 \\ 2 & 2 & 2 \\ 0 & 0 & 2 \end{bmatrix}$ $\begin{bmatrix} -5 & 0 & 0 \\ 3 & 0 & 0 \\ 1 & -2 & -2 \\ 3 & 0 & -2 \end{bmatrix}$

5 $m = p \quad n = q$

6 If P is projection onto $\bar{w}$, then $I - P$ is projection onto M, and $2(I - P) - I = I - 2P$ is reflection through M.

7 **a** 0 **b** $6x$ **c** $1 + 7e^{2x}$ $\cos x$ belongs to the null space of $T + D$

8 $T = \frac{3}{2}R - S - \frac{1}{2}I$

9 $\begin{bmatrix} \lambda - a_{11} & -a_{12} \\ -a_{21} & \lambda - a_{22} \end{bmatrix}$ $A\bar{u} = \lambda\bar{u}$, if $\bar{u}$ is in the null space of $\lambda I - A$.

11 $m \cdot n$

12 No.

13 **a** $((3i + 1)x + (2i + 2)y, -x - iy)$ **c** $\begin{bmatrix} -3+i & -1 \\ 0 & 0 \end{bmatrix}$

SECTION 4

1 **a** Counterclockwise rotation through $\pi/2$.
b Counterclockwise rotation through 2π.
c Counterclockwise rotation through π.

d Counterclockwise rotation through $5\pi/4$.
e Counterclockwise rotation through $5\pi/2$.
f Counterclockwise rotation through 2π.

2 Do not commute.

3 **a** Do not commute. **b** S_1S is reflection through the origin $S_1S = SS_1$.
c $T_1T = TT_1$

6 **a** 1 **b** $x^3/3$ **c** $-\cos x$ **d** $\frac{1}{2}x - (\sin 2x)/4$ **e** $\cos x - 1$ **f** $\cos x$
g $6 + \frac{2}{3}x^3$ D and T do not commute.

7 **a** $\begin{bmatrix} -2 & 2 \\ -1 & 2 \end{bmatrix}$ **b** $\begin{bmatrix} 0 & 5 \\ 0 & -1 \end{bmatrix}$ **c** $\begin{bmatrix} 19 & 22 \\ 43 & 50 \end{bmatrix}$ **d** $\begin{bmatrix} 1 & 0 \\ 0 & 1 \end{bmatrix}$

8 **a** $\begin{bmatrix} 5 & -7 & -3 \\ 7 & -3 & -1 \\ 7 & 1 & 1 \end{bmatrix}$ **b** $\begin{bmatrix} -3 & 1 & 1 \\ 1 & -3 & 1 \\ 1 & 1 & -3 \end{bmatrix}$ **c** $\begin{bmatrix} 14 & -16 & 18 \\ 26 & -31 & 36 \\ 38 & -46 & 54 \end{bmatrix}$ **d** $\begin{bmatrix} 1 & 1 & 2 \\ 2 & 1 & 1 \\ 1 & 2 & 1 \end{bmatrix}$

9 **a** $\begin{bmatrix} 3 & 0 & 7 \\ 1 & 3 & 4 \end{bmatrix}$ **b** 1 **c** $\begin{bmatrix} 0 & 4 & -6 & -1 & -5 \\ 6 & 7 & 7 & 4 & -9 \\ -5 & -12 & -1 & -5 & -6 \\ 0 & 7 & 3 & 7 & -4 \\ 3 & 2 & 1 & 0 & 2 \end{bmatrix}$ **d** $\begin{bmatrix} -7 & 7 \\ -4 & 4 \\ -6 & 6 \end{bmatrix}$

10 **a** $\begin{bmatrix} 1 & 2 \\ 0 & 1 \end{bmatrix} \begin{bmatrix} 1 & 3 \\ 0 & 1 \end{bmatrix} \begin{bmatrix} 1 & 4 \\ 0 & 1 \end{bmatrix} \begin{bmatrix} 1 & 4 \\ 0 & 1 \end{bmatrix}$ **b** $\begin{bmatrix} 1 & 0 \\ 0 & 1 \end{bmatrix} \begin{bmatrix} 0 & 1 \\ 1 & 0 \end{bmatrix} \begin{bmatrix} 1 & 0 \\ 0 & 1 \end{bmatrix} \begin{bmatrix} 1 & 0 \\ 0 & 1 \end{bmatrix}$

c $\begin{bmatrix} 7 & 10 \\ 15 & 22 \end{bmatrix} \begin{bmatrix} 37 & 54 \\ 81 & 118 \end{bmatrix} \begin{bmatrix} 199 & 290 \\ 435 & 634 \end{bmatrix} \begin{bmatrix} 199 & 290 \\ 435 & 634 \end{bmatrix}$

d $\begin{bmatrix} 5 & 1 & 1 \\ 2 & 4 & 0 \\ 5 & 1 & 3 \end{bmatrix} \begin{bmatrix} 12 & 6 & 2 \\ 8 & -2 & 4 \\ 14 & 10 & 4 \end{bmatrix} \begin{bmatrix} 32 & 10 & 8 \\ 18 & 18 & 2 \\ 42 & 12 & 14 \end{bmatrix} \begin{bmatrix} 32 & 10 & 8 \\ 18 & 18 & 2 \\ 42 & 12 & 14 \end{bmatrix}$

11 **a** $\begin{bmatrix} 8 & 1 \\ -1 & 0 \end{bmatrix}$ **b** $\begin{bmatrix} 15 & 1 \\ -2 & 0 \end{bmatrix}$ **c** $\begin{bmatrix} 15 & 1 \\ -2 & 0 \end{bmatrix}$ **d** $\begin{bmatrix} 16 & 2 \\ -9 & -1 \end{bmatrix}$

e $\begin{bmatrix} 5 & 4 & 2 \\ 5 & 1 & -2 \\ -1 & 1 & 2 \end{bmatrix}$ **f** Not defined. **g** Not defined. **h** Not defined. **i** $\begin{bmatrix} 4 & 1 \\ 2 & -1 \end{bmatrix}$ **j** Not defined.

12 Number of columns of A = number of rows of B. AB has the same number of rows as A and the same number of columns as B.

13 Third row of A and fourth column of B.

15 $TS(x,y) = (6x + 3y, 4x + 2y)$, $ST(x,y) = (8x - 2y, 0)$

$$\begin{bmatrix} 2 & 1 \\ 0 & 0 \end{bmatrix} \begin{bmatrix} 3 & -1 \\ 2 & 0 \end{bmatrix} \begin{bmatrix} 8 & -2 \\ 0 & 0 \end{bmatrix} \begin{bmatrix} 6 & 3 \\ 4 & 2 \end{bmatrix}$$

16 **a**

$$P_0 = \begin{bmatrix} \frac{1}{9} & \frac{2}{9} & \frac{2}{9} \\ \frac{2}{9} & \frac{4}{9} & \frac{4}{9} \\ \frac{2}{9} & \frac{4}{9} & \frac{4}{9} \end{bmatrix} \qquad T_0 = \begin{bmatrix} -\frac{7}{9} & \frac{4}{9} & \frac{4}{9} \\ \frac{4}{9} & -\frac{1}{9} & \frac{4}{9} \\ \frac{4}{9} & \frac{4}{9} & -\frac{1}{9} \end{bmatrix}$$

17 **a**

$$P_0 = \begin{bmatrix} \frac{1}{5} & \frac{2}{5} \\ \frac{2}{5} & \frac{4}{5} \end{bmatrix} \qquad Q_0 = \begin{bmatrix} \frac{4}{5} & -\frac{2}{5} \\ -\frac{2}{5} & \frac{1}{5} \end{bmatrix}$$

18 **a** $\begin{bmatrix} 2 & 2i - 1 \\ 2 - 3i & 2i + 1 \end{bmatrix}$ **b** $\begin{bmatrix} 3 + 2i & 2 \\ 2 - 2i & 1 + 2i \end{bmatrix}$

c $\begin{bmatrix} i & 1 + i \\ i & 0 \end{bmatrix} \begin{bmatrix} 0 & -1 + i \\ -1 & 0 \end{bmatrix} \begin{bmatrix} 0 & -1 + i \\ 0 & 0 \end{bmatrix}$

SECTION 5

2 $B = \frac{2}{3}A - \frac{1}{3}AC$

4 No.

5 $R(ST)$ and $(RS)T$ are counterclockwise rotations through $\theta + \varphi + \psi$. $R(S^2T^3)$ and $(RS^2)T^3$ are counterclockwise rotations through $3\varphi + 2\psi + \theta$.

6 **a** $\begin{bmatrix} 0 & 0 \\ 0 & 0 \end{bmatrix}$ **b** $\begin{bmatrix} 0 & 0 & 0 \\ 0 & 0 & 0 \end{bmatrix}$ **c** $\begin{bmatrix} 1 & 2 \\ 1 & 1 \end{bmatrix}$ **d** $\begin{bmatrix} 1 & 2 & 1 \\ 1 & 1 & 1 \end{bmatrix}$ **e** $\begin{bmatrix} 0 & 0 \\ 0 & 0 \end{bmatrix}$

f $\begin{bmatrix} 3 & 1 \\ 2 & -1 \\ 0 & 1 \\ 1 & 1 \end{bmatrix}$ **g** Not defined.

7 **a** $\begin{bmatrix} 7 & 5 \\ 0 & 0 \end{bmatrix}$ **b** $\begin{bmatrix} -1 & 0 \\ 0 & 0 \end{bmatrix}$ **c** $\begin{bmatrix} 0 & 13 & 6 & 25 \\ 0 & 0 & 0 & 0 \\ 0 & 0 & 0 & 0 \end{bmatrix}$ **d** $\begin{bmatrix} 9 & 3 & 0 \\ 0 & 0 & 0 \\ 3 & 1 & 0 \end{bmatrix}$

8 **a** $\begin{bmatrix} 12 & 6 & 18 & \frac{3}{2} \\ 2 & -1 & 0 & \frac{1}{2} \end{bmatrix}$ **b** $\begin{bmatrix} -24 & -24 & -4 & -24 \\ -\frac{1}{2} & 0 & \frac{1}{2} & \frac{1}{2} \\ 0 & 3 & 9 & 0 \\ 0 & -1 & \frac{1}{2} & -\frac{1}{2} \end{bmatrix}$

c $\begin{bmatrix} -24 & 3 & 3 & -3 \\ 4 & 0 & 3 & -\frac{1}{2} \\ 0 & \frac{1}{2} & 9 & 0 \\ 0 & 1 & -3 & -\frac{1}{2} \end{bmatrix}$ **d** $\begin{bmatrix} 1 & \frac{3}{2} \\ \frac{1}{2} & 0 \\ \frac{1}{2} & \frac{3}{2} \\ \frac{1}{2} & -\frac{3}{2} \end{bmatrix}$

9 **a** $\begin{bmatrix} 1 & 0 & 0 \\ 0 & 8 & 0 \\ 0 & 0 & 27 \end{bmatrix}$ **b** $\begin{bmatrix} 9 & 0 & 0 \\ 0 & 0 & 0 \\ 0 & 0 & -8 \end{bmatrix}$ **c** $\begin{bmatrix} 1 & 0 & 0 \\ 0 & -1 & 0 \\ 0 & 0 & 1 \end{bmatrix}$

d $\begin{bmatrix} 0 & 0 & 0 & 0 \\ 0 & 1 & 0 & 0 \\ 0 & 0 & 27 & 0 \\ 0 & 0 & 0 & -1 \end{bmatrix}$

10 **a** $p(A) = \begin{bmatrix} 1 & 4 \\ 0 & -3 \end{bmatrix}$ $q(A) = \begin{bmatrix} -2 & 2 \\ 0 & -4 \end{bmatrix}$ $r(A) = \begin{bmatrix} -2 & 0 \\ 0 & -2 \end{bmatrix}$

b $p(A) = \begin{bmatrix} 30 & 9 & 3 \\ 21 & -3 & 6 \\ 24 & 9 & 9 \end{bmatrix}$ $q(A) = \begin{bmatrix} 0 & 1 & 0 \\ 2 & -4 & 1 \\ 1 & 1 & -1 \end{bmatrix}$ $r(A) = \begin{bmatrix} 18 & 4 & 2 \\ 10 & 4 & 2 \\ 14 & 4 & 6 \end{bmatrix}$

c $p(A) = \begin{bmatrix} 52 & 0 & 0 \\ 0 & -3 & 0 \\ 0 & 0 & 6 \end{bmatrix}$ $q(A) = \begin{bmatrix} 1 & 0 & 0 \\ 0 & -4 & 0 \\ 0 & 0 & -1 \end{bmatrix}$ $r(A) = \begin{bmatrix} 28 & 0 & 0 \\ 0 & -2 & 0 \\ 0 & 0 & 4 \end{bmatrix}$

11 $p(A) = \begin{bmatrix} p(a) & 0 & 0 \\ 0 & p(b) & 0 \\ 0 & 0 & p(c) \end{bmatrix}$ In particular, $p(A) = \begin{bmatrix} 0 & 0 & 0 \\ 0 & 0 & 0 \\ 0 & 0 & 0 \end{bmatrix}$

when $p(\lambda) = (\lambda - a)(\lambda - b)(\lambda - c)$.

12 $B = \begin{bmatrix} 2 & 0 & 0 \\ 0 & 3 & 0 \\ 0 & 0 & -2 \end{bmatrix}$ is one such matrix.

13 $\begin{bmatrix} -3 & 0 & 0 \\ 0 & -4 & 0 \\ 0 & 0 & 1 \end{bmatrix}$ and $\begin{bmatrix} 0 & 6 \\ 6 & 6 \end{bmatrix}$

20 **a** $\begin{bmatrix} 1 & -1 & 1 \\ 3 & 2 & 1 \\ 2 & 1 & 1 \end{bmatrix}$ **b** $\begin{bmatrix} 3 & 2 & 1 \\ \frac{1}{2} & -\frac{1}{2} & \frac{1}{2} \\ 2 & 1 & 1 \end{bmatrix}$ **c** $\begin{bmatrix} 3 & 2 & 1 \\ 10 & 5 & 4 \\ 2 & 1 & 1 \end{bmatrix}$ **d** $\begin{bmatrix} 3 & 2 & 1 \\ -8 & -7 & -2 \\ 2 & 1 & 1 \end{bmatrix}$

21 **a** $\begin{bmatrix} 0 & 1 & 0 \\ 1 & 0 & 0 \\ 0 & 0 & 1 \end{bmatrix}$ **b** $\begin{bmatrix} 1 & 0 & 0 \\ 0 & 2 & 0 \\ 0 & 0 & 1 \end{bmatrix}$ **c** $\begin{bmatrix} 1 & 0 & 0 \\ -3 & 1 & 0 \\ 0 & 0 & 1 \end{bmatrix}$ **d** $\begin{bmatrix} 1 & 0 & 0 \\ 3 & 1 & 0 \\ 0 & 0 & 1 \end{bmatrix}$

22 $EDCBA = \begin{bmatrix} 1 & 0 \\ 0 & 1 \end{bmatrix}$. Put $A' = EDCB$.

24 **a** $\begin{bmatrix} 0 & 0 \\ 0 & 0 \end{bmatrix}$ **b** $\begin{bmatrix} i & 2i \\ 1+i & 3-i \end{bmatrix}$ **c** $\begin{bmatrix} 1+3i & 3+3i \\ 0 & 0 \end{bmatrix}$ **d** $\begin{bmatrix} 0 & -1+2i \\ 0 & 2 \end{bmatrix}$

e $\begin{bmatrix} -1 & -2 \\ -2+2i & 2+6i \end{bmatrix}$ **f** $\begin{bmatrix} -1 & -4 \\ -1+i & 2+6i \end{bmatrix}$ **g** $\begin{bmatrix} 1 & 0 \\ 0 & 1 \end{bmatrix}$

SUPPLEMENT

1 It is mapped into the parallelogram

$$t\begin{bmatrix} 0 \\ -1 \end{bmatrix} + s\begin{bmatrix} 5 \\ 2 \end{bmatrix}, \quad 0 \leq t \leq 1, \quad 0 \leq s \leq 1$$

3 It is mapped onto the line segment from (0,0) to (2,4).

4 T maps the set onto the line segment from $(\frac{3}{5},\frac{6}{5})$ to $(\frac{6}{5},\frac{12}{5})$.

6 T maps this cube into the parallelopiped $s\bar{u} + t\bar{v} + r\bar{w}$, $0 \leq s \leq 1$, $0 \leq t \leq 1$, $0 \leq r \leq 1$ where

$$\bar{u} = \begin{bmatrix} 2 \\ 2 \\ 1 \end{bmatrix} \quad \bar{v} = \begin{bmatrix} 1 \\ 1 \\ 0 \end{bmatrix} \quad \bar{w} = \begin{bmatrix} -1 \\ 0 \\ 1 \end{bmatrix}$$

7 A parallelogram, a line or a single point.

8 Yes.

CHAPTER 4

SECTION 1

2 **a** $\begin{bmatrix} \frac{2}{11} & -\frac{1}{22} \\ -\frac{1}{11} & \frac{3}{11} \end{bmatrix}$ **b** $\begin{bmatrix} -\frac{1}{2} & \frac{1}{2} \\ -\frac{1}{2} & -\frac{1}{2} \end{bmatrix}$ **c** $\begin{bmatrix} -2 & 1 \\ \frac{3}{2} & -\frac{1}{2} \end{bmatrix}$ **d** $\begin{bmatrix} 1 & 0 \\ -2 & 1 \end{bmatrix}$

e $\begin{bmatrix} 2 & -2 & 1 \\ \frac{1}{2} & -\frac{1}{2} & \frac{1}{2} \\ -\frac{7}{2} & \frac{9}{2} & -\frac{5}{2} \end{bmatrix}$ **f** $\begin{bmatrix} 1 & -1 & 0 \\ 0 & 1 & -1 \\ 0 & 0 & 1 \end{bmatrix}$ **g** $\begin{bmatrix} -\frac{1}{9} & \frac{1}{9} & \frac{8}{9} \\ \frac{10}{9} & -\frac{1}{9} & -\frac{26}{9} \\ \frac{1}{9} & -\frac{1}{9} & \frac{1}{9} \end{bmatrix}$

h $\begin{bmatrix} 1 & -1 & 2 & 2 \\ 0 & \frac{1}{2} & 0 & \frac{1}{2} \\ 0 & 0 & -2 & -2 \\ 0 & 0 & 0 & -1 \end{bmatrix}$ **i** $\begin{bmatrix} -\frac{1}{2} & \frac{1}{2} & \frac{1}{2} & \frac{1}{2} \\ \frac{1}{2} & 0 & 0 & -\frac{1}{2} \\ \frac{1}{2} & 0 & -\frac{1}{2} & 0 \\ \frac{1}{2} & -\frac{1}{2} & 0 & 0 \end{bmatrix}$

3 **a** $\begin{bmatrix} 1 & -1 \\ 0 & 1 \end{bmatrix}$ **b** $\begin{bmatrix} 1 & -1 & 0 \\ 0 & 1 & -1 \\ 0 & 0 & 1 \end{bmatrix}$ **c** $\begin{bmatrix} 1 & -1 & 0 & 0 & 0 \\ 0 & 1 & -1 & 0 & 0 \\ 0 & 0 & 1 & -1 & 0 \\ 0 & 0 & 0 & 1 & -1 \\ 0 & 0 & 0 & 0 & 1 \end{bmatrix}$

4 **a** $\begin{bmatrix} \frac{1}{2} & 0 \\ 0 & -\frac{1}{2} \end{bmatrix}$ **b** $\begin{bmatrix} 4 & 0 & 0 \\ 0 & 2 & 0 \\ 0 & 0 & 3 \end{bmatrix}$ **c** $\begin{bmatrix} 1 & 0 & 0 & 0 & 0 \\ 0 & \frac{1}{4} & 0 & 0 & 0 \\ 0 & 0 & \frac{1}{2} & 0 & 0 \\ 0 & 0 & 0 & -1 & 0 \\ 0 & 0 & 0 & 0 & \frac{1}{3} \end{bmatrix}$

5 **a** Not invertible. **b** $\begin{bmatrix} \frac{1}{3} & -\frac{1}{3} & -\frac{1}{3} \\ 0 & \frac{1}{2} & 1 \\ 0 & 0 & -1 \end{bmatrix}$ **c** $\begin{bmatrix} \frac{1}{3} & -\frac{1}{3} & 0 & -\frac{1}{3} \\ 0 & 1 & -\frac{1}{2} & \frac{5}{2} \\ 0 & 0 & \frac{1}{2} & \frac{1}{2} \\ 0 & 0 & 0 & -1 \end{bmatrix}$

d Not invertible. **e** $\begin{bmatrix} \frac{1}{2} & 0 & 0 & 0 \\ -\frac{1}{6} & \frac{1}{3} & 0 & 0 \\ \frac{1}{3} & -\frac{1}{6} & \frac{1}{2} & 0 \\ -\frac{1}{2} & -\frac{1}{2} & -\frac{1}{2} & 1 \end{bmatrix}$

9 **a** $\begin{bmatrix} i & 1 \\ 2 & -i \end{bmatrix}$ **b** $\begin{bmatrix} (1-i)/2 & 0 \\ 0 & (1+i)/2 \end{bmatrix}$

SECTION 2

1 **a** $B = CA^{-1}$ **c** $A = PBP^{-1}$

2 **a** $A + aI$ invertible. $B = (A + aI)^{-1}D$
b $A - D$ invertible. $B = 2(D - A)^{-1}C$
c $A - 3I$ invertible. $B = (A - 3I)^{-1}C$
d $A + C$ invertible. $B = C(A + C)^{-1}$
e A invertible. $B = C$

3 **a** $D^{-1}C^{-1}B^{-1}A^{-1}$
b $BAB^{-1}A^{-1}$

4 **a** $(A^{-1})^3$ **b** $\frac{1}{3}A^{-1}$ **c** $-(A^{-1})^4$ **d** A^2

5 **a** $\frac{1}{3}(A - 2I)$ **b** $-\frac{1}{2}(A^5 - 5A^3 + 3A)$

6 **a** No **b** No **c** No

7 **a** A **b** $A^t + B^t$ **c** $(A^t)^2$

8 **a** $A^{-1} = \begin{bmatrix} 0 & \frac{1}{3} \\ 1 & -\frac{2}{3} \end{bmatrix}$ $B^{-1} = \begin{bmatrix} -\frac{1}{3} & \frac{2}{3} \\ \frac{1}{3} & \frac{1}{3} \end{bmatrix}$

14 $\begin{bmatrix} \frac{3}{2} & 2 & \frac{5}{2} \\ \frac{1}{2} & -1 & \frac{3}{2} \end{bmatrix}$

SECTION 3

1 **a** Counterclockwise rotation through $\theta = -\pi/4, 3\pi/2, -\pi$.
b Counterclockwise rotation through -2θ.
c Multiples of π. **d** Multiples of $\pi/2$.

2 T^{-1} is reflection followed by the opposite rotation.

3 $\begin{bmatrix} 2 & 1 & -5 \\ -1 & -1 & 4 \\ 1 & 1 & -3 \end{bmatrix}$

4 $T^{-1}(x,y) = (\frac{2}{7}x + \frac{1}{7}y,\ -\frac{1}{7}x + \frac{3}{7}y)$

6 **a** $T^{-1} = T$ **b** $A = \begin{bmatrix} 0 & 1 \\ 0 & 0 \end{bmatrix} = A^{-1}$

7 $\begin{bmatrix} \frac{1}{2} & \frac{1}{2} \\ \frac{1}{2} & \frac{1}{2} \end{bmatrix}$

10 **a, b, d**

13 $T^{-1}f = \dfrac{1}{g}f$

14 **a** $\begin{bmatrix} \dfrac{-1-7i}{2} & 2+i \\ -\dfrac{i}{2} & 1 \end{bmatrix}$ **b** $\dfrac{1-2i}{5}y,\ \dfrac{1+i}{2}x - \dfrac{1+3i}{10}y$

SECTION 4

1 The columns of P are independent

2 **a** $c_1 = -\frac{1}{3}x - \frac{2}{3}y$ **b** $c_1 = -2$
$c_2 = \frac{1}{3}x - \frac{1}{3}y$ $c_2 = -1$

3 **b** $c_1 = \frac{3}{8}x + \frac{1}{8}y - \frac{2}{8}z$ **c** $-\frac{7}{8}, \frac{21}{8}, \frac{18}{8}$
$c_2 = -\frac{1}{8}x - \frac{3}{8}y + \frac{6}{8}z$
$c_3 = -\frac{2}{8}x + \frac{2}{8}x + \frac{4}{8}z$

4 $c_1 = -\frac{5}{3}d_1 - \frac{6}{3}d_2$ $d_1 = \frac{3}{2}c_2$
$c_2 = \frac{2}{3}d_1$ $d_2 = -\frac{2}{4}c_1 - \frac{5}{4}c_2$

5 **b** $P^{-1} = \begin{bmatrix} \dfrac{1}{\sqrt{3}} & \dfrac{1}{\sqrt{3}} & \dfrac{1}{\sqrt{3}} \\ \dfrac{1}{\sqrt{2}} & -\dfrac{1}{\sqrt{2}} & 0 \\ \dfrac{1}{\sqrt{6}} & \dfrac{1}{\sqrt{6}} & -\dfrac{2}{\sqrt{6}} \end{bmatrix}$ **c** $c_1 = \dfrac{1}{\sqrt{3}}(x+y+z)$

$c_2 = \dfrac{1}{\sqrt{2}}(x-y)$

$c_3 = \dfrac{1}{\sqrt{6}}(x+y-2z)$

6 **c** $f(0), \dfrac{f(1)-f(-1)}{2}, \dfrac{f(1)+f(-1)}{2} - f(0)$

d $f(-1) = f(2) - 3f(1) + 3f(0)$
$f(0) = f(0)$
$f(1) = f(1)$

9 Let (c_1,c_2), (d_1,d_2), (e_1,e_2) be coordinates with respect to $\{\bar{u}_1,\bar{u}_2\}$, $\{\bar{v}_1,\bar{v}_2\}$, $\{\bar{w}_1,\bar{w}_2\}$, respectively. Then

$$\begin{bmatrix} c_1 \\ c_2 \end{bmatrix} = P^{-1}Q\begin{bmatrix} d_1 \\ d_2 \end{bmatrix} = P^{-1}R\begin{bmatrix} e_1 \\ e_2 \end{bmatrix}$$

$$\begin{bmatrix} d_1 \\ d_2 \end{bmatrix} = Q^{-1}P\begin{bmatrix} c_1 \\ c_2 \end{bmatrix} = Q^{-1}R\begin{bmatrix} e_1 \\ e_2 \end{bmatrix}$$

$$\begin{bmatrix} e_1 \\ e_2 \end{bmatrix} = R^{-1}Q\begin{bmatrix} d_1 \\ d_2 \end{bmatrix} = R^{-1}P\begin{bmatrix} c_1 \\ c_2 \end{bmatrix}$$

10 The columns of PA^{-1} are $\bar{v}_1$ and $\bar{v}_2$.

11 $$\begin{bmatrix} c_1 \\ c_2 \end{bmatrix} = \begin{bmatrix} -\frac{i}{2} & \frac{1}{2} \\ \frac{1}{2} & -\frac{i}{2} \end{bmatrix}\begin{bmatrix} x \\ y \end{bmatrix}$$

SECTION 5

1 28

3 **a** 3 **b** 0 **c** 0 **d** −23 **e** 0 **f** 0 **g** 6 **h** 0 **i** 0

4 **a**, **d** and **g** are invertible.

9 **a** $\lambda^2 - 5\lambda + 7$ **b** $\lambda^3 - 2\lambda^2 - 4\lambda + 6$

12 $-1 - i$; Yes.

13 2

14 **a** 0; Dependent. **b** 0; Independent.

17 $$\begin{bmatrix} 0 & 0 & 0 \\ 14 & -4 & 2 \\ -7 & 2 & -1 \end{bmatrix}$$

18 **a** $x = \frac{5}{3}$, $y = -\frac{7}{3}$ **b** $x = -2$, $y = \frac{1}{2}$ **c** $x = 5$, $y = 0$, $z = -3$ **d** $x = -\frac{1}{15}$, $y = -\frac{13}{15}$, $z = \frac{1}{3}$

SUPPLEMENT

3 **a** 1 **b** Columns are interchanged. **f** Counterclockwise rotation through $\pi/3$ $(5\pi/6)$ followed by reflection in $y = x$ (in the y-axis).

8 Yes, No **9** $\sqrt{3}$, $\sqrt{14}$

CHAPTER 5

SECTION 1

1 a $B = \begin{bmatrix} 3 & 0 \\ 0 & -2 \end{bmatrix}$ b $\begin{bmatrix} 1 & 1 \\ 0 & 1 \end{bmatrix}$

2 a $\begin{bmatrix} -1 & 0 & 0 \\ 0 & -1 & 0 \\ 0 & 0 & 2 \end{bmatrix}$ b $\begin{bmatrix} 1 & 0 & 0 \\ 0 & 1 & 0 \\ 0 & 0 & 64 \end{bmatrix}$

3 a $\begin{bmatrix} 3 & 5 \\ 1 & 2 \end{bmatrix}$ b $\begin{bmatrix} -1 & -1 & -3 \\ 1 & 0 & -1 \\ 0 & 1 & 3 \end{bmatrix}$

4 For 5**a** a $\begin{bmatrix} 27 & 0 \\ 0 & -8 \end{bmatrix}$ b $\begin{bmatrix} 4 & 0 \\ 0 & -1 \end{bmatrix}$ c $\begin{bmatrix} 657 & 0 \\ 0 & 52 \end{bmatrix}$ d $\begin{bmatrix} \frac{1}{3} & 0 \\ 0 & -\frac{1}{2} \end{bmatrix}$

For 5**b** a $\begin{bmatrix} 1 & 3 \\ 0 & 1 \end{bmatrix}$ b $\begin{bmatrix} 2 & 1 \\ 0 & 2 \end{bmatrix}$ c $\begin{bmatrix} 1 & 4 \\ 0 & 1 \end{bmatrix}$ d $\begin{bmatrix} 1 & -1 \\ 0 & 1 \end{bmatrix}$

5 Columns are interchanged.

6 The matrix I; the matrix 0; the matrix aI.

7 $$D = \begin{bmatrix} 0 & 1 & 0 & 0 \\ 0 & 0 & 2 & 0 \\ 0 & 0 & 0 & 3 \\ 0 & 0 & 0 & 0 \end{bmatrix}$$

8 $$D^2 = \begin{bmatrix} 0 & 0 & 2 & 0 \\ 0 & 0 & 0 & 6 \\ 0 & 0 & 0 & 0 \\ 0 & 0 & 0 & 0 \end{bmatrix} \quad D^3 = \begin{bmatrix} 0 & 0 & 0 & 6 \\ 0 & 0 & 0 & 0 \\ 0 & 0 & 0 & 0 \\ 0 & 0 & 0 & 0 \end{bmatrix} \quad D^4 = \begin{bmatrix} 0 & 0 & 0 & 0 \\ 0 & 0 & 0 & 0 \\ 0 & 0 & 0 & 0 \\ 0 & 0 & 0 & 0 \end{bmatrix}$$

9 $$\begin{bmatrix} -\frac{11}{6} & 3 & -\frac{3}{2} & \frac{1}{3} \\ -\frac{1}{3} & -\frac{1}{2} & 1 & -\frac{1}{6} \\ \frac{1}{6} & -1 & \frac{1}{2} & \frac{1}{3} \\ -\frac{1}{3} & \frac{3}{2} & -3 & \frac{11}{6} \end{bmatrix}$$

10 **a** $x^2/3 + 1$ **b** $1 - x/2$ **c** $A = \begin{bmatrix} 1 & 0 & 0 \\ 0 & \frac{1}{2} & 0 \\ 0 & 0 & \frac{1}{3} \end{bmatrix}$ **d** $A^{-1} = \begin{bmatrix} 1 & 0 & 0 \\ 0 & 2 & 0 \\ 0 & 0 & 3 \end{bmatrix}$

$T^{-1}(1) = 1 \qquad T^{-1}(x) = 2x \qquad T^{-1}(x^2) = 3x^2$

11 $\begin{bmatrix} 1 & 0 & 0 \\ \frac{5}{12} & \frac{2}{3} & -\frac{1}{12} \\ \frac{1}{6} & \frac{2}{3} & \frac{1}{6} \end{bmatrix}$

13 $P = \begin{bmatrix} 1 & 0 & 0 \\ -\frac{3}{2} & 2 & -\frac{1}{2} \\ \frac{1}{2} & -1 & \frac{1}{2} \end{bmatrix} \qquad Q = \begin{bmatrix} 1 & 0 & 0 \\ 1 & 1 & 1 \\ 1 & 2 & 4 \end{bmatrix}$

14 **a** $\begin{bmatrix} 1+i & 0 \\ 0 & -1 \end{bmatrix}$ **b** $\begin{bmatrix} 4i & -3 \\ -1 & -i \end{bmatrix}$

SECTION 2

1 Because P and A may not commute.

2 **a** $\begin{bmatrix} 3 & -1 & 0 \\ 1 & 0 & -1 \\ -3 & 1 & 1 \end{bmatrix}$ **b** $\begin{bmatrix} 6 & 9 & 8 \\ 2 & 4 & 3 \\ -5 & -9 & -7 \end{bmatrix}$

3 **a** $\begin{bmatrix} -1 & 0 & -1 \\ 3 & 3 & 4 \\ -1 & -1 & -1 \end{bmatrix}$ **b** $\begin{bmatrix} -1 & 0 & 0 \\ 0 & -1 & 0 \\ 0 & 0 & 2 \end{bmatrix}$ **e** $\begin{bmatrix} 190 & 189 & 189 \\ 63 & 64 & 63 \\ -189 & -189 & -188 \end{bmatrix}$

f $q(A) = \begin{bmatrix} 0 & 0 & 0 \\ 0 & 0 & 0 \\ 0 & 0 & 48 \end{bmatrix} \qquad q(B) = \begin{bmatrix} 144 & 144 & 144 \\ 48 & 48 & 48 \\ -144 & -144 & -144 \end{bmatrix}$

4 **b** $\begin{bmatrix} 9\sqrt{3} - 8 & -6\sqrt{3} + 6 \\ 12\sqrt{3} - 12 & -8\sqrt{3} + 9 \end{bmatrix}$ **c** $\begin{bmatrix} 9\sqrt[3]{3} - 8 & -6\sqrt[3]{3} + 6 \\ 12\sqrt[3]{3} - 12 & -8\sqrt[3]{3} + 9 \end{bmatrix}$

5 abc

6 I

7 0

8 **a** 0

9 **b** $\begin{bmatrix} 0 & 0 & 0 \\ 0 & 0 & 0 \\ 0 & 0 & 0 \end{bmatrix}$ **c** $\begin{bmatrix} 0 & 0 & 0 \\ 0 & 0 & 0 \\ 0 & 0 & 0 \end{bmatrix}$

16 **a** $P^{-1} = \begin{bmatrix} 4i & -3 \\ -1 & -i \end{bmatrix}$ $B = \begin{bmatrix} 7 + 4i & -12 + 24i \\ -1 + 2i & -7 - 3i \end{bmatrix}$

b $B^8 = \begin{bmatrix} 61 & 180i \\ 15i & -44 \end{bmatrix}$

SECTION 3

1 **a** The nonzero vectors parallel to $\overline{w}$ are characteristic vectors belonging to $\lambda = 1$; those perpendicular to $\overline{w}$ belong to $\lambda = -1$.

b The nonzero vectors parallel to $\overline{w}$ are characteristic vectors belonging to $\lambda = 0$; those perpendicular to $\overline{w}$ belong to $\lambda = 1$.

c See answer to part **a**.

d The nonzero vectors parallel to $\overline{w}$ are characteristic vectors belonging to $\lambda = 1$; those perpendicular to $\overline{w}$ belong to $\lambda = 0$.

e See answer to part **b**.

f Nonzero multiples of $(\cos \pi/8, -\sin \pi/8)$ belong to $\lambda = 1$; while nonzero multiples of $(\cos 3\pi/8, \sin 3\pi/8)$ belong to $\lambda = -1$.

3 3 and 1, respectively.

4 **a** $\lambda^2 - \lambda$, $\lambda = 0, 1$ **b** $\lambda^2 - 6\lambda + 8$, $\lambda = 2, 4$ **c** $\lambda^3 - 8\lambda^2 + 20\lambda - 16$, $\lambda = 2, 4$

d $\lambda^3 - 8\lambda^2 + 16\lambda$, $\lambda = 0, 4$ **e** $(\lambda - 1)(\lambda - 2)(\lambda - 3)$, $\lambda = 1, 2, 3$ **f** λ^4, $\lambda = 0$

5 These answers are not unique.

a For $\lambda = 0$, $\begin{bmatrix} -1 \\ 1 \end{bmatrix}$ For $\lambda = 1$, $\begin{bmatrix} 1 \\ 1 \end{bmatrix}$

b For $\lambda = 2$, $\begin{bmatrix} -1 \\ 1 \end{bmatrix}$ For $\lambda = 4$, $\begin{bmatrix} 1 \\ 1 \end{bmatrix}$

c For $\lambda = 2$, $\begin{bmatrix} -1 \\ 1 \\ 0 \end{bmatrix}$ or $\begin{bmatrix} 0 \\ 0 \\ 1 \end{bmatrix}$ For $\lambda = 4$, $\begin{bmatrix} 1 \\ 1 \\ 0 \end{bmatrix}$

d For $\lambda = 0$, $\begin{bmatrix} -1 \\ 3 \\ 2 \end{bmatrix}$ For $\lambda = 4$, $\begin{bmatrix} -1 \\ 1 \\ 0 \end{bmatrix}$ or $\begin{bmatrix} -1 \\ 0 \\ 1 \end{bmatrix}$

e For $\lambda = 1$, $\begin{bmatrix} 1 \\ 0 \\ 0 \end{bmatrix}$ For $\lambda = 2$, $\begin{bmatrix} 2 \\ 1 \\ 0 \end{bmatrix}$ For $\lambda = 3$, $\begin{bmatrix} 3 \\ 2 \\ 2 \end{bmatrix}$

f For $\lambda = 0$, $\begin{bmatrix} 1 \\ 0 \\ 0 \\ 0 \end{bmatrix}$

7 **b** $\lambda_1^2, \lambda_2^2, \ldots, \lambda_k^2$ **c** of A^2, $\lambda = 1, 9$ of A^3, $\lambda = 1, -1, 27$

8 **b** $q(\lambda_1), q(\lambda_2), \ldots, q(\lambda_k)$

10 $\lambda = 2, -1, 3$

11 i and $-i$, respectively.

12 **a** $\lambda^2 - (1 + 2i)\lambda - 1 + i$; $\lambda = 1 + i, i$; $\begin{bmatrix} 2 \\ -1 + 2i \end{bmatrix}$ $\begin{bmatrix} 1 \\ -1 + i \end{bmatrix}$ respectively.

b $\lambda^2 + 1$; $\lambda = i, -i$; $\begin{bmatrix} 1 \\ -i \end{bmatrix}$ $\begin{bmatrix} -i \\ 1 \end{bmatrix}$ respectively.

SECTION 4

1 Other answers are possible.

a $\begin{bmatrix} -1 & 1 \\ 1 & 1 \end{bmatrix}$ **b** $\begin{bmatrix} -1 & 1 \\ 1 & 1 \end{bmatrix}$ **c** $\begin{bmatrix} -1 & 1 & 0 \\ 1 & 1 & 0 \\ 0 & 0 & 1 \end{bmatrix}$ **d** $\begin{bmatrix} 1 & 2 & 3 \\ 0 & 1 & 2 \\ 0 & 0 & 2 \end{bmatrix}$

e $\begin{bmatrix} -1 & -1 & -1 \\ 3 & 1 & 0 \\ 2 & 0 & 1 \end{bmatrix}$ **f** $\begin{bmatrix} 7 & 9 & 2 \\ -3 & -3 & -1 \\ 4 & 5 & 1 \end{bmatrix}$

2 For $A = \begin{bmatrix} 2 & -6 \\ -2 & 1 \end{bmatrix}$ $P = \begin{bmatrix} -2 & 3 \\ 1 & 2 \end{bmatrix}$

For $A = \begin{bmatrix} 3 & -1 \\ 2 & 0 \end{bmatrix}$ $P = \begin{bmatrix} 1 & 1 \\ 1 & 2 \end{bmatrix}$

5 **a** $\frac{1}{2}\begin{bmatrix} 3^{10}+1 & 3^{10}-1 \\ 3^{10}-1 & 3^{10}+1 \end{bmatrix}$ **b** $\begin{bmatrix} 4 & -1 & -1 \\ 9 & -2 & -3 \\ 3 & -1 & 0 \end{bmatrix}$

6 $\begin{bmatrix} -2 & -2 & 0 \\ 1 & 0 & -1 \\ 5 & 10 & 5 \end{bmatrix}$ is one possibility.

7 Other answers are possible (by permutation of the diagonal entries).

a $\begin{bmatrix} 1 & 0 \\ 0 & -1 \end{bmatrix}$ **b** $\begin{bmatrix} 0 & 0 \\ 0 & 1 \end{bmatrix}$ **c** $\begin{bmatrix} 1 & 0 & 0 \\ 0 & -1 & 0 \\ 0 & 0 & -1 \end{bmatrix}$ **d** $\begin{bmatrix} 1 & 0 & 0 \\ 0 & 0 & 0 \\ 0 & 0 & 0 \end{bmatrix}$

e $\begin{bmatrix} 0 & 0 & 0 \\ 0 & 1 & 0 \\ 0 & 0 & 1 \end{bmatrix}$ **f** $\begin{bmatrix} 1 & 0 \\ 0 & -1 \end{bmatrix}$

8 $\begin{bmatrix} 1 & 0 \\ 0 & -1 \end{bmatrix}$

9 **b** $\begin{bmatrix} 1 & -1 & 1 \\ 0 & 2 & -4 \\ 0 & 0 & 4 \end{bmatrix}$ **d** For example, $1, x-1, (x-1)^2$

11 For example, $P = \begin{bmatrix} 2 & 2 \\ 1-i & 1+i \end{bmatrix}$

SECTION 5

1 Possible answers are

a $\begin{bmatrix} -1/\sqrt{2} & 1/\sqrt{2} \\ 1/\sqrt{2} & 1/\sqrt{2} \end{bmatrix}$ **b** $\begin{bmatrix} 1/\sqrt{2} & -1/\sqrt{2} \\ 1/\sqrt{2} & 1/\sqrt{2} \end{bmatrix}$

c $\begin{bmatrix} -1/\sqrt{2} & -1/\sqrt{6} & 1/\sqrt{3} \\ 1/\sqrt{2} & -1/\sqrt{6} & 1/\sqrt{3} \\ 0 & 2/\sqrt{6} & 1/\sqrt{3} \end{bmatrix}$ **d** $\begin{bmatrix} 1/\sqrt{66} & 1/\sqrt{6} & -3/\sqrt{11} \\ -4/\sqrt{66} & 2/\sqrt{6} & 1/\sqrt{11} \\ 7/\sqrt{66} & 1/\sqrt{6} & 1/\sqrt{11} \end{bmatrix}$

e $\begin{bmatrix} 0 & -1/\sqrt{3} & 21\sqrt{6} \\ 1/\sqrt{2} & -1/\sqrt{3} & -1/\sqrt{6} \\ 1/\sqrt{2} & 1/\sqrt{3} & 1/\sqrt{6} \end{bmatrix}$

2 $\begin{bmatrix} 1/\sqrt{3} & 1/\sqrt{3} & 1/\sqrt{3} \\ 1/\sqrt{3} & 1/\sqrt{3} & 1/\sqrt{3} \\ 1/\sqrt{3} & 1/\sqrt{3} & 1/\sqrt{3} \end{bmatrix}$

4 $\dfrac{1}{\sqrt{2}}(1,-1),\ \dfrac{1}{\sqrt{2}}(1,1)$

5 $A = \begin{bmatrix} -\frac{5}{6} & \frac{2}{3} & \frac{13}{6} \\ \frac{2}{3} & \frac{2}{3} & \frac{2}{3} \\ \frac{13}{6} & \frac{2}{3} & -\frac{5}{6} \end{bmatrix}$

7 $\dfrac{1}{2}\begin{bmatrix} 1+3^{10} & 1-3^{10} \\ 1-3^{10} & 1+3^{10} \end{bmatrix}$

11 **f** $\begin{bmatrix} i/\sqrt{2} & i/\sqrt{2} \\ 1/\sqrt{2} & -1/\sqrt{2} \end{bmatrix}$ **g** Same as **f**.

SECTION 6

1 Possible answers are **a** $3x_1^2 + y_1^2 = 1$ **b** $\dfrac{5}{\sqrt{5}}x_1^2 - \dfrac{5}{\sqrt{5}}y_1^2 = 1$

c $\dfrac{x_1^2}{2} - \dfrac{y_1^2}{2} = 1$ **d** $\dfrac{x_1^2}{6} + x_2^2 = 1$

2 **a** hyperbola **b** hyperbola **c** ellipse **d** ellipse

3 **b** The single point (0,0) or one or two lines through (0,0), or the entire plane.
c Empty (if $a + c \leq 0$), otherwise two parallel lines.

4 Possible answers are **a** $x_1^2 + 3y_1^2 - 3z_1^2 = 1$ **b** $x_1^2 - y_1^2 - 2z_1^2 = 1$

7 **a** furthest $\left(\frac{1}{\sqrt{2}}, -\frac{1}{\sqrt{2}}\right), \left(-\frac{1}{\sqrt{2}}, \frac{1}{\sqrt{2}}\right)$

closest $\left(\frac{1}{\sqrt{6}}, \frac{1}{\sqrt{6}}\right), \left(-\frac{1}{\sqrt{6}}, -\frac{1}{\sqrt{6}}\right)$

b $(\sqrt{\tfrac{2}{5}}, 2\sqrt{\tfrac{2}{5}}), (-\sqrt{\tfrac{2}{5}}, -2\sqrt{\tfrac{2}{5}})$

8 **a** $\dfrac{a + c \pm \sqrt{(a - c)^2 + b^2}}{2}$

b $\dfrac{a + c + \sqrt{(a - c)^2 + b^2}}{2}$

c $1/\sqrt{\lambda}$, where $\lambda = \dfrac{a + c + \sqrt{(a - c)^2 + b^2}}{2}$

9 $\dfrac{a + c + \sqrt{(a - c)^2 + b^2}}{2}$

CHAPTER 6

SECTION 1

1 **a** $y = ce^{x}$ **b** $y = ce^{-\sqrt{5}x}$ **c** $y = ce^{-(2/3)x}$
d $y = ce^{(1/2)x}$ **e** $y = ce^{-ex}$

2 **a** $y = e^{2x}$ **b** $y = e^{-2x}$ **c** $y = 2e^{(1/2)x-1/2}$ **d** $y = 3e^{-(5/4)x+5/2}$

3 One, namely, $y = 0$.

4 $y \to 0$; $|y| \to +\infty$; $|y| \to +\infty$

5 **a** $y = ce^{2x} + x/2 + \frac{1}{4}$ **b** $y = ce^{-3x} + \frac{1}{4}e^{x}$

c $y = ce^{-x} + \dfrac{\sin x}{2} - \dfrac{\cos x}{2}$ **d** $y = ce^{x} - x^2 - 2x - 2$

6 **a** $y(0) = 1: y = \frac{3}{4}e^{2x} + x/2 + \frac{1}{4}$
$y(1) = 2: y = \frac{5}{4}e^{2x-2} + x/2 + \frac{1}{4}$
b $y(0) = 1: y = \frac{3}{4}e^{-3x} + \frac{1}{4}e^{x}$
$y(1) = 2: y = 2e^{-3x+3} - \frac{1}{4}e^{-3x+4} + \frac{1}{4}e^{x}$

c $y(0) = 1: y = \frac{3}{2}e^{-x} + \dfrac{\sin x}{2} - \dfrac{\cos x}{2}$

$y(1) = 2: y = \left(2 + \dfrac{\cos 1}{2} - \dfrac{\sin 1}{2}\right)e^{-x+1} + \dfrac{\sin x}{2} - \dfrac{\cos x}{2}$

d $y(0) = 1: y = 3e^x - x^2 - 2x - 2$
$y(1) = 2: y = 7e^{x-1} - x^2 - 2x - 2$

7 **a** $y = ce^{-x^3/3}$ **b** $y = c/x + x^3/4$ **c** $y = x^{-x}(-1 + ce^x)$ **d** $y = ce^{\cos x}$

8 **a** $y = e^{-(1/3)(x^3-1)}$ **b** $y = 3/4x + x^3/4$
c $y = x^{-x}(-1 + 2e^{x-1})$ **d** $y = e^{\cos x - \cos 1}$

SECTION 2

1 **a** $y = c_1e^{6x} + c_2e^{-x}$ **b** $y = c_1e^{3x} + c_2e^{2x}$
c $y = c_1e^{-2x} + c_2xe^{-2x}$ **d** $y = c_1e^{-(3/2+\sqrt{5}/2)x} + c_2e^{-(3/2-\sqrt{5}/2)x}$
e $y = c_1 + c_2x$ **f** $y = c_1e^{\sqrt{2}x} + c_2e^{-\sqrt{2}x}$
g $y = c_1e^{3x} + c_2xe^{3x}$ **h** $y = c_1e^{(1+\sqrt{3})x} + c_2e^{(1-\sqrt{3})x}$

2 **a** $y = -xe^{2x}$ **b** $y = -7e^{-2x} + 9e^{-x}$
c $y = -13e^{2x-6} + 4xe^{2x-6}$ **d** $y = -e^{2x-20} + 2e^{x-10}$

3 **a** 0; No. **b** $y = ce^{\beta x}$ $y = ce^{\alpha x}$ **c** No; Yes.

4 **a** $y = c_1e^{5x} + c_2e^{(2/3)x}$ **b** $y = c_1e^{-(1/2)x} + c_2xe^{-(1/2)x}$

5 **a** No solution. **b** unique: $y = 1 + 2x$
c Not unique: $y = e^x + c_2xe^x$ for any c_2

6 **a** $\begin{bmatrix} 0 & 1 \\ -c & -b \end{bmatrix}$

7 **a** Unique: $y = 2x$ **b** No solution. **c** Not unique: $y = c_2xe^{-x}$ for any c_2

8 **f** $y = c_1e^{-x}\sin x + c_2e^{-x}\cos x$

SECTION 3

1 **a** $y = c_1 \sin 2x + c_2 \cos 2x$
b $y = c_1 \sin \sqrt{b}x + c_2 \cos \sqrt{b}x$ $b > 0$
$y = c_1e^{\sqrt{-b}x} + c_2e^{-\sqrt{-b}x}$ $b < 0$
$y = c_1 + c_2x$ $b = 0$
c $y = c_1 \sin \sqrt{7/3}x + c_2 \cos \sqrt{7/3}x$ **d** $y = c_1 \sin \sqrt{5/2}x + c_2 \cos \sqrt{5/2}x$

2 **a** $y = c_1e^{-x}\sin x + c_2e^{-x}\cos x$ **b** $y = c_1e^{-2x}\sin x + c_2e^{-2x}\cos x$
c $y = c_1e^{(1/2)x}\sin x + c_2e^{(1/2)x}\cos x$ **d** $y = c_1e^{4x}\sin \sqrt{2}x + c_2e^{4x}\cos \sqrt{2}x$

3 **a** $y = c_1e^{-3x} + c_2xe^{-3x}$ **b** $y = c_1e^{-3x}\sin x + c_2e^{-3x}\cos x$
c $y = c_1e^{-4x} + c_2e^{-2x}$ **d** $y = c_1e^{(-3+\sqrt{76}/2)x} + c_2e^{(-3-\sqrt{76}/2)x}$

4 **a** $y = \frac{1}{2}\sin 2x$ **b** $y = -\frac{3}{2}\sin x$

c $y = e^{-x+\pi/2} \sin x$ d $y = \frac{1}{\sqrt{2}} e^{2x} \sin \sqrt{2}x$

5 a $e^{3x}, xe^{3x}, x^2e^{3x}, x^3e^{3x}$ b $e^{-2x}, xe^{-2x}, x^2e^{-2x}, x^3e^{-2x}, x^4e^{-2x}, x^5e^{-2x}$

6 a $e^{3x}D^4 (\log x)$ b $e^{-2x}D^3(x^5)$
c $e^xD^5(e^{3x})$ d $e^{\sqrt{2}x}D^{12} (\sin x)$

SECTION 4

1 a $y = c_1 \sin \sqrt{2}x + c_2 \cos \sqrt{2}x + c_3x \sin \sqrt{2}x + c_4x \cos \sqrt{2}x$
b $y = c_1e^x + c_2xe^x + c_3x^2e^x$
c $y = c_1 + c_2x + c_3x^2 + c_4x^3 + c_5e^{4x} + c_6e^{-x}$
d $y = c_1 + c_2x + c_3e^{-x} \sin x + c_4e^{-x} \cos x$
e $y = c_1e^{-x/\sqrt{2}} \sin (x/\sqrt{2}) + c_2e^{-x/\sqrt{2}} \cos (x/\sqrt{2}) + c_3e^{x/\sqrt{2}} \sin (x/\sqrt{2}) + c_4e^{x/\sqrt{2}} \cos (x/\sqrt{2})$
f $y = c_1e^x + c_2e^{-(1/2)x} \sin (\sqrt{3}/2)x + c_3e^{-(1/2)x} \cos (\sqrt{3}/2)x$

2 a $e^{-3x}, xe^{-3x}, e^x, xe^x$ $e^{-x}, xe^{-x}, \sin x, \cos x$
b $1, x, x^2, x^3, x^4, e^{-2x} \sin x, e^{-2x} \cos x, xe^{-2x} \sin x, xe^{-2x} \cos x, x^2e^{-2x} \sin x, x^2e^{-2x} \cos x$
c $e^{4x}, xe^{4x}, e^{-x}, xe^{-x}, e^{-(1/2)x} \sin (\sqrt{3}/2)x, e^{-(1/2)x} \cos (\sqrt{3}/2)x, xe^{-(1/2)x} \sin (\sqrt{3}/2)x, xe^{-(1/2)x} \cos (\sqrt{3}/2)x$
d $1, x, x^2, x^3, e^{3x} \sin 2x, e^{3x} \cos 2x, xe^{3x} \sin 2x, xe^{3x} \cos 2x$

3 a $y = (3/\sqrt{2}) \sin \sqrt{2}x + \cos \sqrt{2}x + (1/\sqrt{2})x \sin \sqrt{2}x - x \cos \sqrt{2}x$
b $y = \frac{1}{2} \cos x + \frac{1}{4}e^x - \frac{3}{4}e^{-x}$ c $y = (\frac{1}{2} + x - \frac{1}{2}x^2)e^{-x+1}$
d $y = -\frac{3}{10} \sin x + \frac{1}{10} \cos x - \frac{1}{5}e^{-x} \cos x + \frac{1}{10}e^x$

5 a 1, 1, 0

SECTION 5

1 a $y = \frac{1}{5}xe^{-3x} + \frac{6}{25}e^{-3x}$ b $y = \frac{1}{4}xe^{2x}$ c $y = \frac{1}{8}x^2 + \frac{3}{16}x + \frac{15}{64}$
d $y = 2$ e $y = \frac{1}{6}x^3e^{-x}$ f $y = \frac{1}{5}(x - 1) + \frac{1}{8}(\sin x + \cos x)$
g $y = x^2 - \frac{2}{9} - \frac{1}{12}x^2 \cos 3x + \frac{1}{36}x \sin 3x$
h $y = \frac{1}{4}xe^x - \frac{1}{4}xe^{-x} - \frac{1}{4}x \cos x$
i $y = -\frac{1}{12}x^2 - \frac{5}{36}x + \frac{1}{4}e^{-x}$ j $y = 1 - 2e^{2x} + xe^{2x}$

2 a $y = \frac{1}{5}e^x + (\frac{1}{2}e^{-2\pi} - \frac{3}{20}e^{-\pi})e^{2x} + (-\frac{1}{2}e^{2\pi} + \frac{1}{20}e^{3\pi})e^{-2x}$
b $y = -\frac{1}{2}xe^{-x} + \frac{3}{4}e^x + \frac{1}{4}e^{-x}$ c $y = 2x^2 + 6x + 7 + e^{2x} - 8e^x$
d $y = \frac{1}{2}x^2e^{-x}$

3 a $K = (D^2 + I)^3(D - I)$ b $K = (D^2 - 4D + 13I)(D - I)^3$
c $K = (D^2 + 4I)D$ d $K = D^4$

5 **a** $y = x$

b With $z_1 = \int e^{-x} \log x\, dx$, $z_2 = \int e^{x} \log x\, dx$, then

$$y = \tfrac{1}{2}e^{x} z_1 - \tfrac{1}{2}e^{-x} z_2$$

c With $z_1 = \int \sqrt{xe^{-x}}\, dx$, $z_2 = \int \sqrt{xe^{-2x}}\, dx$, then

$$y = -e^{x} z_1 + e^{2x} z_2$$

d With $z_1 = \int xe^{-x}\sqrt{1 - x^2}\, dx$, $z_2 = \int e^{-x}\sqrt{1 - x^2}\, dx$,

$$y = -e^{x} z_1 + xe^{x} z_2$$

6 **b** With $z_1 = \int \frac{\log x}{x}\, dx$, $z_2 = \int \frac{e^{-x}(x + 1) \log x}{2x}\, dx$, and

$$z_3 = \int \frac{e^{x}(x - 1) \log x}{2x}\, dx,\ y = -xz_1 + e^{x} z_2 + e^{-x} z_3$$

8 **a** $y = -\frac{1}{5}x$

b With $z_1 = \int \frac{\log x}{x}\, dx$, $z_2 = \frac{1}{2}\int \frac{e^{-\sqrt{1/3}x}(x + \sqrt{3}) \log x}{x}\, dx$

$$z_3 = \frac{1}{2}\int \frac{e^{\sqrt{1/3}x}(x - \sqrt{3}) \log x}{x}\, dx$$

$$y = -xz_1 + e^{\sqrt{1/3}x} z_2 + e^{-\sqrt{1/3}x} z_3$$

SECTION 6

1 **a** $t = \frac{\log 10}{\log 2}$ minutes **b** $10^6 \cdot \sqrt{10}$

2 $10^6 \frac{\ln 2}{\ln 10}$ years

3 163.7 million

4 $64(1 - e^{-1})$

5 **b** $t \to T$

6 **a** $y = 3 \cos \sqrt{\tfrac{1}{5}}t$ **b** $y = -4 \cos \sqrt{\tfrac{1}{5}}t$ **c** $y = 5\sqrt{5} \sin \sqrt{\tfrac{1}{5}}t$
d $y = -2\sqrt{5} \sin \sqrt{\tfrac{1}{5}}t + 3 \cos \sqrt{\tfrac{1}{5}}t$

BIBLIOGRAPHY

Listed below are a few books that pursue further the topics discussed in this book and related results.

Hoffman, K., and R. Kunze: *Linear Algebra*, Prentice-Hall, Inc., Englewood Cliffs, N.J., 1961.
Includes a proof that the reduced form of a matrix is unique. There are also complete discussions of determinants, polynomial theory, and spectral theory. The authors prove the Cayley-Hamilton theorem and the spectral theorem and, in addition, discuss in detail many deeper results about similar matrices, including a treatment of the Jordan canonical form.

Shields, P.: *Linear Algebra*, Addison-Wesley Publishing Co., Inc., Reading, Mass., 1964.
More elementary and less complete than the above-mentioned book, it does contain a complete discussion of third-order orthogonal matrices and a proof of the spectral theorem that is a direct generalization of the proof given in this text, in Example 3, Section 5, Chapter 5.

Jacobson, N.: *Lectures in Abstract Algebra*, vol. 2, D. Van Nostrand Co., Inc., Princeton, N.J., 1953.
A more general treatment of linear algebra than that of the above books. Included also, is a discussion of infinite-dimensional theory.

Faddeeva, V. N.: *Computational Methods of Linear Algebra*, C. D. Benster, translator, Dover Publications, Inc., New York, 1959.
A discussion of numerical methods for solving systems of equations and for finding characteristic values of a matrix.

Kreider, D. L., R. G. Kuller, D. R. Ostberg, and F. W. Perkins: *Introduction to Linear Analysis*, Addison-Wesley Publishing Co., Inc., Reading Mass., 1966.
A thorough discussion, using the linear algebra framework, of linear differential equations. This book also includes a treatment of inner products and linear partial differential equations.

Robinson, G.: *Vector Geometry*, Allyn and Bacon, Inc., Boston, 1962.
An elementary, yet quite thorough discussion of vector geometry.

Hummel, J.: *Introduction to Vector Functions*, Addison-Wesley Publishing Co., Inc., Reading, Mass., 1967.
An introduction to the calculus of several variables that uses the linear algebra framework. This is the first of a number of books that attempt to present this material at a level suitable for students with a background of one year of calculus.

Ficken, F. A.: *The Simplex Method of Linear Programming*, Holt, Rinehart and Winston, Inc., New York, 1961.
Discusses an important new set of techniques in linear algebra with wide application in the physical and social sciences. Linear programming provides a systematic method for finding maximum and minimum values of linear functions subject to constraints.

LIST OF THEOREMS

THEOREM 1, page 24: A homogeneous linear system has a unique solution if and only if the rank of its coefficient matrix equals the number of variables.

THEOREM 1, Matrix form, page 52: The null space of A contains only the zero vector if and only if the rank of A equals the number of columns of A.

THEOREM 1A, page 31: A nonhomogeneous linear system has a solution if and only if the rank of the augmented matrix equals the rank of the coefficient matrix. If these ranks are equal, then the system has a unique solution if and only if this common rank is equal to the number of unknowns.

THEOREM 2, page 46: If A has m rows and n columns, and $\bar{u}$, $\bar{v}$ are in R^n, and a is any scalar, then

$$A(\bar{u} + \bar{v}) = A\bar{u} + A\bar{v}, \; A(a\bar{u}) = aA\bar{u}.$$

THEOREM 3, page 71: $A\bar{u} = \bar{0}$ has a unique solution if and only if the columns of A are independent.

THEOREM 4, page 86: If V has a basis with n vectors, then any other basis for V has n vectors.

THEOREM 5, page 87: A set with n vectors in a space V of dimension n is independent if and only if it spans V.

THEOREM 6, page 88: If A has n rows and n columns, then the columns of A are independent if and only if they span R^n. In particular, if $A\bar{x} = \bar{0}$ has a unique solution, then for any $\bar{c}$ in R^n, $A\bar{x} = \bar{c}$ has a solution.

THEOREM 7, page 93: Rank A + nullity A = number of columns of A.

THEOREM 8, page 95: Rank A = dimension of column space of A.

THEOREM 9, page 97: Rank A = dimension of row space of A.

THEOREM 10, page 141: If T is a linear transformation from R^n into R^m, then there is a *unique* matrix A, with m rows and n columns, such that $T\bar{u} = A\bar{u}$ for all $\bar{u}$ in R^n.

THEOREM 11, pages 188, 205: A square matrix A is invertible if and only if its rank is the same as the number of its columns; that is, $A\bar{u} = \bar{0}$ has a unique solution.

THEOREM 12, page 189: If A is square and $BA = I$, then $AB = I$.

THEOREM 13, page 205: If T is an invertible linear transformation and $\bar{u}_1, \bar{u}_2, \ldots, \bar{u}_n$ are independent, then $T\bar{u}_1, T\bar{u}_2, \ldots, T\bar{u}_n$ are independent.

THEOREM 14, page 220: A has an inverse if and only if $\det A \neq 0$.

THEOREM 15, page 260: The characteristic values of A are the roots of the characteristic polynomial $f(\lambda) = \det(\lambda I - A)$. If λ is a root of this polynomial, then any nonzero vector in the null space of $\lambda I - A$ is a characteristic vector belonging to λ.

THEOREM 16, page 266: If the characteristic values of A are real and distinct then A is similar to a diagonal matrix.

THEOREM 17, page 274: If A is symmetric then A is similar to a diagonal matrix.

THEOREM 18, page 309: Suppose $p(\lambda)$ is a polynomial and that D is the differentiation operator. If $L = p(D - aI)$ and $L_1 = p(D)$, then $L(e^{ax}y) = e^{ax}L_1y$. In particular the general solution to $Ly = 0$ is

$$e^{ax}(c_1\varphi_1 + c_2\varphi_2 + \cdots + c_n\varphi_n)$$

where $c_1\varphi_1 + c_2\varphi_2 + \cdots + c_n\varphi_n$ is the general solution to $L_1y = 0$.

THEOREM 19, page 312: Suppose S and T are linear transformations defined on V, each having a finite-dimensional null space. Then the dimension of the null space of ST cannot exceed the sum of the dimensions of the null space of S and the null space of T.

THEOREM 20, page 313: The dimension of the null space of a constant coefficient linear differential operator is the order of the operator.

INDEX